AF721414

Geodynamics of the Alps 1

SCIENCES

Geoscience, Field Director – Yves Lagabrielle

Dynamics of the Continental Lithosphere,
Subject Head – Sylvie Leroy

Geodynamics of the Alps 1

Present-Structure and Regional Alpine Studies from Extension to Collision

Coordinated by
Claudio L. Rosenberg
Nicolas Bellahsen

iSTE

WILEY

First published 2023 in Great Britain and the United States by ISTE Ltd and John Wiley & Sons, Inc.

Apart from any fair dealing for the purposes of research or private study, or criticism or review, as permitted under the Copyright, Designs and Patents Act 1988, this publication may only be reproduced, stored or transmitted, in any form or by any means, with the prior permission in writing of the publishers, or in the case of reprographic reproduction in accordance with the terms and licenses issued by the CLA. Enquiries concerning reproduction outside these terms should be sent to the publishers at the undermentioned address:

ISTE Ltd
27-37 St George's Road
London SW19 4EU
UK

www.iste.co.uk

John Wiley & Sons, Inc.
111 River Street
Hoboken, NJ 07030
USA

www.wiley.com

© ISTE Ltd 2023
The rights of Claudio L. Rosenberg and Nicolas Bellahsen to be identified as the authors of this work have been asserted by them in accordance with the Copyright, Designs and Patents Act 1988.

Any opinions, findings, and conclusions or recommendations expressed in this material are those of the author(s), contributor(s) or editor(s) and do not necessarily reflect the views of ISTE Group.

Library of Congress Control Number: 2022941484

British Library Cataloguing-in-Publication Data
A CIP record for this book is available from the British Library
ISBN 978-1-78945-116-0

ERC code:
PE10 Earth System Science
PE10_4 Terrestrial ecology, land cover change
PE10_5 Geology, tectonics, volcanology
PE10_13 Physical geography

Contents

Introduction

Nicolas BELLAHSEN and Claudio L. ROSENBERG
ISTeP, Sorbonne Université,
UMR 7193, CNRS, Paris, France

The European Alps represent an archetypal mountain belt, mainly because its orogenic wedge is the most studied one. It results from a period of Mesozoic divergence forming the Tethyan Ligurian oceanic realm, linking the Neo-Tethys in the east with the Central Atlantic in the west. This followed the Variscan orogenic event affecting the European lithosphere and preceded a period of convergence starting during upper Cretaceous times with the northward movement of the Adria/Africa plates. This convergence was first accommodated by oceanic subduction, and later followed by continental subduction and eventually collision *sensu stricto*. The transition between oceanic and continental subduction is characterized by exhumation of high-pressure (HP) nappes of oceanic origin during early Eocene times, followed by HP metamorphism and exhumation of continental units during middle to late Eocene (Briançonnais and European distal continental margin). Collision started at the Eocene–Oligocene transition and is characterized by barrovian metamorphism, crustal shortening, high topographic relief and molasse-type peri-orogenic basins. During the whole convergence history, the amount of magmatism was very small. Today, convergence has slowed down, varying from almost zero in the Western Alps to approximately 2 mm/yr in the Eastern Alps, although vertical movements, extensive and compressive earthquakes are recorded, even in the Western Alps.

In spite of the latter observations, the European Alps may be considered as simpler than many other mountain belts that also result from the closure of oceanic domains. Indeed, in many other mountain belts, subduction and collision do not

Geodynamics of the Alps 1,
coordinated by Claudio L. ROSENBERG and Nicolas BELLAHSEN. © ISTE Ltd 2023.

follow a simple sequence. For example, in the middle Paleozoic, during the Caledonian cycle and the Iapetus closure and more recently, during Cenozoic times and the Tethys closure between India and Asia, HP metamorphism events occurred long after the onset of collision. This is possibly due to the subduction of larger oceanic domains, and to longer and more vigorous convergence dynamics, which probably did not affect the Alpine realm.

The book *Geodynamics of the Alps* consists of three volumes. The first includes the recent and present-day structure and tectonic setting of the Alpine chain, from lithospheric mantle to brittle crust and surface topography, but also a historical overview of Alpine research and includes two chapters covering specific Alpine regions (Corsica and Eastern Alps), through all phases of Alpine history.

The second volume presents the pre-collisional history of the Alps. It starts with the Variscan orogeny in the Alpine realm, and continues with the inferred paleogeography, structure and extensional processes affecting continental margins, and their mantle structure in the pre-orogenic Alpine realm. This volume finishes with two chapters describing oceanic and continental subduction processes.

The third volume is entirely dedicated to Alpine collision. It describes and interprets the following elements of the Alpine chain: (1) Alpine magmatism, which is mainly contemporaneous to the very first stages of collision; (2) the formation of the external massifs, which represent a peculiarity in the Alpine chain, rarely occurring in other collisional orogens; (3) the foreland (Molasse) basin, in relation to the lithospheric processes that controlled its evolution; (4) exhumation of the internal part of the chain as constrained from the thermo-chronological record of the foreland basin sediments, and finally (5) the northern deformation front of the Alps, the youngest out of the large-scale collisional structures of the orogen.

This book will not provide a single, consistent and unique view on Alpine geodynamics, but rather create a space for experts on Alpine research to present their state of the art of specific subjects, not withdrawing from proposing their own interpretations, whether or not they are entirely consistent with those suggested in other chapters.

Volume 1

The first volume starts with a chapter on the history of Alpine cross-sections. This chapter does not review the historical evolution of tectonic and geodynamic processes in the Alps, but it shows how the style and format of geological cross-sections have evolved through time in relationship and as a consequence of major tectonic discoveries and debates. It also underlines changes through time in the scientific approach of scientists, showing how the very close link to observations in the mid-19th century evolved toward more conceptual interpretations, eventually lacking an observational base at the beginning of Plate Tectonics, when entirely new concepts of Alpine orogeny were formulated. Only in the last decade of the 20th century, have these conceptual models regained a close link to structural and geophysical observable data.

Chapter 2 of Volume 1 describes the present-day deep structure of the Alps, in which several slabs and slab segments can be identified. The dip-direction and the presence or absence of break-offs in the European slab are discussed in this chapter, strongly suggesting continuous, vertical slabs. In addition, the surface of the Alpine Mohos and the first-order structures of the deep crust are reassessed. Finally, based on the integration of receiver function, Vp/Vs seismic tomography and local earthquake tomography, the geometry and orientation of the west-Alpine subduction channel is imaged.

Chapter 3 of Volume 1 provides a quantitative description of the present-day surface topography of the Alps and its inferred uplift history. Several processes affecting the Alps since the Pliocene, such as sedimentation in the foreland basins, cooling ages and geochemically inferred erosion rates are integrated in the discussion of the present-day topographic structure, suggesting that changes in both climatic and tectonic processes must be invoked to explain the uplift history of the Alps.

Chapter 4 of Volume 1 reviews stress patterns inferred by the analysis of the youngest Alpine brittle fault systems, in addition to inferred deformation fields based on seismic and GPS data. These analyses show that the topographically highest portions of the Alpine chain in the Western and Central Alps were affected by orogen-perpendicular extension during Pliocene time, and this state of deformation is still recorded by seismic activity. The areas affected by such orogen-perpendicular extension coincide with high uplift rates in the northern and central parts of the Alpine arc. The possible collisional and post-collisional large-scale geodynamic causes of this deformation and uplift patterns are discussed.

Volume 1 concludes with two chapters which are both regionally focused. Because of their broad approach, providing the geologic al history of the Alpine realm throughout its entire history and up to the present, they are included within the first volume. Chapter 5 of Volume 1 presents the geology of the Eastern Alps discussing its underlying geodynamic processes. It shows that the stacking of nappes within the upper plate of the Alpine orogen can be inferred to represent a syn-collisional Cretaceous event, hence leading to the conclusion that two distinct orogenies, each of them consisting of a subduction/collision process, affected the Eastern Alps. The younger (Cenozoic) history is also described, with its peculiar style of collision, very different from that in the Western and Central Alps. Indeed, syn-collisional extrusion-related faults accommodate a significant part of collisional convergence in the Eastern Alps.

Chapter 6 of Volume 1 gives an overview of the geology of Corsica. Because the Sardo–Corsican block rifted away from the rest of the Alpine chain at the beginning of collision, it is not affected by the latter process, and it is ideally suited to study subduction and Cenozoic extension, and in particular, to relate the Alpine to the Appeninic chain. This chapter describes and illustrates how the Eocene-aged Alpine/Corsican east-directed subduction is followed by slab break-off and the initiation of a new, west-directed subduction system, which eventually generated the Appenninic chain.

Volume 2

Chapter 1 of Volume 2 deals with the Paleozoic evolution of basement rocks in the Alps. The oldest events recorded correspond to the formation of a Neoproterozoic peri-Gondwanian arc before its accretion to Gondwana, presumably during an Ordovician collisional event. The Cambrian–Ordovician period was marked by tectonic subsidence and widespread magmatic activity, prior to the opening of the Rheic Ocean. Convergence between Laurussia and Gondwana started in the Devonian, which marked the closure of Rheic Ocean, followed by the Variscan collision during the Carboniferous. Evidence of this convergence is provided by Devonian HP-LT eclogites preserved in the Austroalpine units, and Devonian–Tournaisian arc-related magmatism in the External Crystalline Massifs (Western Alps). The main period of nappe stacking and crustal thickening took place during the Mississipian and it was followed by strike-slip tectonics and extension from Pennsylvanian to Permian times. The orogen-scale dextral strike-slip east-Variscan shear zone was active from the Late Carboniferous to the Permian. Because of the intense reorganization during the Alpine orogeny, replacing the Alpine basement in the European Variscan puzzle remains a challenge to be solved.

Chapter 2 of Volume 2 provides an overview of the continental margins involved in Alpine orogeny. The reconstructed stratigraphic sequence and structure of both the European and Adriatic margin are presented based on a literature review showing numerous case studies from all over the Alpine chain. The questions on how many ocean basins and continental blocks existed before the onset of Alpine subduction, and their inferred paleo-geographic positions are reviewed, taking into account the nature of the oceanic substratum, that is, exhumed mantle versus oceanic crust. In addition to describing numerous interpretations of this complex paleo-geographic realm, a new Jurassic paleo-geographic map and a Jurassic-to-Cretaceous evolution are proposed. It suggests the existence of a main oceanic axis, which propagated into different parts of adjacent, pre-fractured continental blocks. A similar pattern is inferred to characterize the North Atlantic ocean in the Early Eocene, where numerous oceanization attempts, leading to hyper-extended crust and exhumed mantle, initiated from one major, stable central axis of accretion.

Chapter 3 of Volume 2 summarizes the petrology, the geochemistry and the age data of mantle and oceanic crustal remnants in the Alps and Appennines, and discusses these data in the frame of rifting and seafloor spreading preceding Alpine subduction. The comparison of these data with those of peridotites worldwide, including mantle samples dredged offshore, points to their origin from passive margins and ultra-slow spreading ridges. Their association with lower crustal granulites and mafic intrusions, in addition to their sedimentary cover typical of continental margins, suggest that they originate from the mantle lithosphere beneath a continental crust. The small volumes of gabbroic dikes recording melt extraction and magmatism during formation of the Piemonte–Liguria Ocean, and the similarity between the REE signature of their basalts, and that of MORB from ultra-slow spreading ridges, point to a likely origin from an ultra-slow spreading ridge.

Chapter 4 of Volume 2 is the first of two chapters on Alpine subduction. Based on a large synthesis of P-T conditions and PTt paths determined from HP rocks, this chapter reconstructs the 3D architecture of oceanic subduction in the Alpine realm and draws several conclusions on the dynamics of Alpine subduction. The consistent alignment of compiled P/T data along a P-T gradient corresponding to that of mature subduction zones suggests that Alpine HP rocks record the P-T conditions of subduction without being significantly affected by overpressure. The magmatic age of subducted and exhumed oceanic HP rocks points to the fact that only oceanic domains next to the continental margin and/or OCT were exhumed, whereas the remaining oceanic crust disappeared during the first stages of subduction. Integrating observations on syn-subduction sedimentary deposits all along the Alpine arc to petro-chronological data on HP rocks shows that an along-strike segmentation of the subduction zone both in terms of crustal-lithospheric structure and processes existed, with subduction accretion dominating in the Western Alps and subduction erosion dominating in the Central Alps. These differences in the nature of the subduction processes are discussed in relationship to the inherited structure of the continental margins and the oceanic crust itself. Finally, the Alpine case study is compared with larger scale subduction zones of the Mediterranean and Pacific.

Chapter 5 of Volume 2 reviews the available data about the subduction of continental crust in the European Alps. The type of crustal material that was involved in the subduction process is characterized (lower vs. upper crust, tilted blocks vs. extensional allochthons) and the data on pressure, temperature and age of metamorphic units are summarized. Finally, a discussion of the present geometry of the subducted continental units allows the setting of some constraints on the kinematic evolution of the mountain belt, and on the exhumation of HP-UHP rocks in particular.

Volume 3

Chapter 1 of Volume 3 focuses on Alpine magmatism, reviewing petrological, geochemical and geochronological data of Alpine magmatic bodies in relationship to their distribution all along the Alpine chain. Based on their spatial occurrence, age and geochemical signature, Alpine magmatic bodies are shown to be generated by different geodynamic processes, which are inferred to be (1) subduction of both continental and oceanic crust, whose melting produced magmas that were emplaced in the upper plate in the Middle and Late Eocene; (2) slab break-offs at the transition from subduction to collision, allowing for the rise of asthenosphere and its direct contact to the lower Alpine crust, whose melting and mixing with asthenospheric melts generated the largest volumes of Alpine magmas, forming the Periadriatic Plutons; (3) extensional processes related to mantle dynamics, limited to the marginal areas of the Alpine chain (the Alpine/Apennine junction and the Alps/Dinarides/Pannonian junction); (4) ascent along inherited lithospheric-scale faults, possibly explaining intraplate magmatism as observed in the Veneto magmatic province, which is located relatively far from the Alpine front.

Chapter 2 of Volume 3 summarizes the present-day knowledge on the external crystalline massifs, basement units deformed during the collisional stage of the mountain belt. In these massifs, the amount of shortening increases from south to north, as well as the P-T conditions of barrovian metamorphism. Each massif experienced a complex sequence of shortening: after slicing off the Dauphinois cover, ductile internal shortening affected the basement, before localization of deformation along basement thrusts resulted in "en bloc" exhumation of the massifs. Specificities between different massifs are described in detail and different hypotheses on the driving forces leading to the formation of the massifs are synthetized.

Chapter 3 of Volume 3 reviews the structure, age and stratigraphy of the northern foreland basin of the Alps, and it discusses the large-scale geodynamic processes underlying its stratigraphic architecture, its facies relationships and the provenance of its clastic material. The along-strike similarity versus change of the stratigraphy through time is synthetized and related to the drainage pattern of the basin. The historical evolution of inferred geodynamic processes is reconstructed over the past decades, thus showing that the stratigraphy and structure of the Molasse Basin may not only reflect thickening of the more internal part of the orogen, but rather the slab dynamics of the lower plate. Hence, subduction loads at deeper crustal levels rather than surface loads may exert the first-order control on basin subsidence.

Chapter 4 of Volume 3 reviews the currently available detrital thermochronological record of exhumation for the European Alps. It allows tracing and discussing temporal and spatial patterns of exhumation across the Alpine chain. The higher temperature white mica 40Ar/39Ar ages, if considered as cooling ages, depict phases of very rapid exhumation, on the order of 3 ± 1 km/Myr, in the Western Alps during the early Oligocene, and somewhat slower exhumation of ~1 km/Myr in the Central Alps during the late Oligocene–early Miocene. Detrital zircon fission-track data indicate a relatively steady exhumation signal since the late Oligocene for the Central Alps, whereas detrital apatite fission-track data emphasize regional differences in exhumation rates since at least Miocene times.

Chapter 5 of Volume 3 focuses on the northern deformation front of the Alps, that is, on the youngest, large-scale structure accommodating collisional shortening by the relative displacement of the orogen on its foreland basin. It shows that the structural style of this fault system varies along strike, with triangle zones being replaced by duplexes and basal décollements changing depth and the stratigraphic level. These variations are correlated to lateral changes in the mechanical stratigraphy, to differences in the links between these faults and their roots in the more internal parts of the orogen, or even to differences in the mechanical properties of the entire European Plate.

Acknowledgments

Romain Bousquet generously provided the vectorized version of his Tectonic Map of the Alps, used in most of the chapters in the three volumes of this book.

Part 1

Historical Introduction

1

Alpine Sections Through Time: A Mirror of Evolving Observations and Thoughts on the Tectonics of the Alps

Claudio L. ROSENBERG
ISTeP, Sorbonne Université,
UMR 7193, CNRS, Paris, France

1.1. Introduction

Cross-sections are the most commonly used tool to represent the deep structure of orogens, hence to constrain their three-dimensional anatomy. In the Italian publications of the 18th century, and of the first half of the 19th century (e.g. Villa and Villa 1844; Taramelli 1870), cross-sections were commonly termed "spaccato", meaning "broken up", suggesting that something is taken away from the landscape to show and disclose what would otherwise remain hidden. In the second half of the 19th century, the terms "taglio naturale" (natural section) and "taglio dimostrativo" (illustrative section) also appeared (e.g. Pantaloni et al. 2016) to describe and distinguish natural, steeply inclined surface outcrops from cross-sections that are constructed on vertical, planar surfaces, which are not visible in nature. The term "spaccato" has also been used to show geological structures on vertical, idealized

Geodynamics of the Alps 1,
coordinated by Claudio L. ROSENBERG and Nicolas BELLAHSEN. © ISTE Ltd 2023.

planar surfaces (Stoppani 1857; Omboni, 1879), but in such cases, until the 1870's, the authors specify that it is a "geological theoretical section". The term "taglio" is similar to the terms used in the German ("Querdurchschnitt", e.g. Ebel 1808), British ("section") and French ("coupe") literature, which expressed the idea of cutting across, hence also to disclose something, but by an artificial physical process. All these terms relate to the illustration of what is underneath the earth's surface, but not to what is above it.

In the 19th century, the term (geological) section was often associated with the adjective "ideal" (e.g. Studer 1834; Rogers 1836), when it illustrated structures that did not strictly reproduce what was visible in the landscape. Studer (1834) uses the term "Ideale Profile" for those Alpine sections in which he reduces the amount of 3D topographic effects that are observable in the field. The term "theoretical" section (e.g. Lory 1860; Favre 1867; Jaccard 1869) is often used to characterize sections in which stippled lines have been added to complete the structures of the eroded beds above the surface. The term "schematic sections" was first used by Termier (1903), when sketching orogen-scale sections that show both the inferred deep structure of the chain and the inferred eroded uppermost structures. Argand (1909) also mentions that his section is schematic, but he drops this adjective in 1924 when showing an even more simplified and schematic section of the same area of the Western Alps. From that moment onwards, the term "cross-section", generally without any other adjective, was used in Alpine literature.

The transition between landscape drawings, in areas characterized by well-exposed outcrops of sedimentary beds and/or tectonic structures, to cross-sections in a geological sense, takes place gradually through time. While the former ones simply reproduce a natural scenery, the latter focus on specific geological features of such sceneries, first documenting them by the addition of symbols linked to a legend in the drawings, and later by progressively removing all elements that are not strictly of geological nature. Eventually, cross-sections become fully independent of any specific point of observation of the author, hence showing constant thickness and dip of the beds throughout the section.

Modern sections through the Alps are constructed along ideal vertical planes that may integrate geological information from areas located tens of km away from the trace of the section plane (Argand 1911), and geophysical data imaging the structures of the mantle lithosphere (e.g. Giese et al. 1970) and the underlying asthenosphere (e.g. Lippitsch et al. 2013). This evolution took several centuries, and it is the aim of this chapter to show that it was marked by numerous, discrete steps.

How and why scientists came to accomplish such steps, both in terms of their innovative representation technique and their understanding of the tectonics of the Alps is discussed in the following.

Numerous papers traced the history of tectonic research in the Alps (e.g. Masson 1983; Dal Piaz and Dal Piaz 1984; Sengör 1989; Dal Piaz 2001, 2010; Trümpy 2001; Schaer 2010, 2011; Debelmas 2011), or the history of more specific aspects of Alpine tectonic research (Bailey 1935; Masson 1976; Greene 1982; Dal Piaz and Dal Piaz 1984; Trümpy 1991; Letsch 2014, 2017), or the contribution of individual Alpine geologists (e.g. Dal Piaz 1996; Trümpy and Lemoine 1998; Durand-Delga 2007; Dullo and Pfaffl 2019). In this chapter, the evolution of ideas concerning the structure and tectonics of the Alpine chain is also discussed, but only through the analysis of cross-sections from the early 18th century to the end of the 20th century. Therefore, the present text does not accord a lot of space to great Alpine geologists whose scientific contributions are not based and illustrated by innovative cross-sections, as is the case for Eduard Suess, Emile Haug and Hans Stille. A discussion of their achievements for the advancement of tectonic ideas on the Alps can be found in Dal Piaz and Dal Piaz (1984) and Sengör (1989, 1990, 2014). Because the Alps have been so intensely studied in the past centuries and many important steps in the understanding of orogenic processes were accomplished there, the following history of Alpine cross-sections since the 18th century is very much representative of the evolution of cross-sections and tectonic knowledge worldwide.

1.2. Transition from landscape drawings to cross-sections: beginning of the 18th to the first half of the 19th century

Given the subtle nature of the transition from landscape drawing of geological outcrops and cross-sections, it is difficult to pinpoint the precise initiation of drawing of cross-sections. Natural outcrops of sedimentary beds are masterly painted in the Renaissance, as shown and discussed by Ceregato and Vai (2017), but the scope of these landscapes was not a geological one. The first paintings and sketches of natural landscapes depicting the structure of exposed beds, with the specific goal to understand their origin, were those of Marsigli (1705; Figures 1.1(a–c)), reproduced by Gortani (1930), depicting the cliffs around Lake Uri (Figures 1.1(b) and (c)) in the Helvetic nappes of northern Switzerland. Indeed, Marsigli points out in the text of one figure describing an open antiform, that the lower part of the structure consists of an artificial wall, thereby focusing the attention of the reader only on the folded strata of the depicted landscape. Marsigli also proposes a classification of mountain types based on their internal structure and illustrated in schematic cross-sections. He defines 10 different types based on the dip, the orientation and the folding of the beds. It was the first structure-based classification of mountain belts.

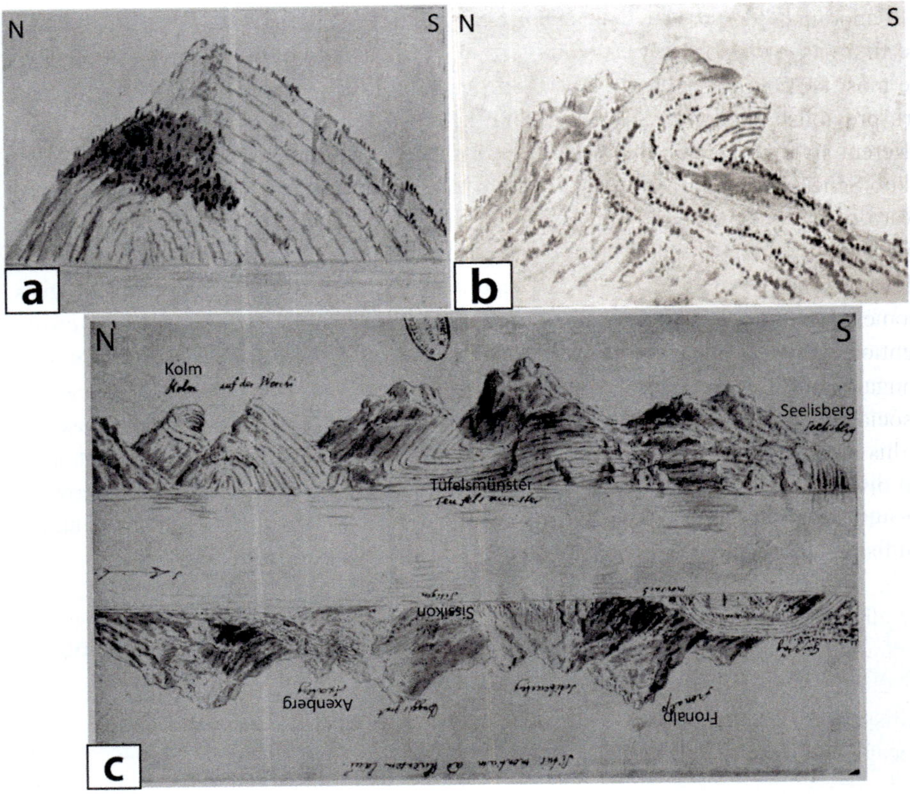

Figure 1.1. *Paintings of the shores of the Urnersee, eastern lobe of Lake Lucerne (Switzerland), by Marsigli (1705). Reproduced and modified from Gortani (1930). (a and b) Folded sedimentary beds. (c) The mountain flanks of each shore can be observed without any perspective-induced distortions by turning the picture upside down. Marsigli explicitly mentions folding of the beds in the legend to his figure, which is otherwise just the reproduction of an Alpine landscape. He writes: "Mons Kolm dictus ad lacum Uriensem, e stratis ita compositus ut vertex eius supra incurvatus quasi videatur, et stratorum positura quasi perpendiculariter arcuata videatur"*

One of his paintings (Figure 1.1(c)) shows the two sides of the lake that face each other, each of them painted as if they were observed from the opposite shore, hence providing the reader with an excellent, undistorted view of the structures and stressing the continuity of the structures from one shore to the other.

Panoramas focusing on the geometry of outcropping beds were rarely drawn in the first half of the 18th century, and interestingly, the ones that were published in the most famous books of that time are new drawings of the same outcrops of Lake Uri, previously shown by Marsigli (1705; Figure 1.1). They were redrawn using a different style by Johann Scheuchzer in 1708, and published by his brother Johann Jakob Scheuchzer in 1716 (Figure 1.2(a); Scheuchzer 1716). Johann Scheuchzer was aware of the original drawings by Marsigli (Vaccari 2001), but he made his own observations around Lake Uri (Koch 1952; Ellenberger 1990), and these are very relevant for the understanding of mountain building. Scheuchzer describes the geometry of the folded beds and their steep dip in the outcrops around Lake Uri. He mentions that sedimentary beds in the great planes (Flanders, Germany, Poland and Hungary) are horizontal and concludes (following Steno) that mountain building is associated with displacement and fracturing of the originally horizontal strata. Based on his observations around Lake Uri, Schechuzer also concludes that the geometry and orientation of the sedimentary beds controls the morphology of the mountainous topography. The drawings of Scheuchzer were modified and republished by Vallisneri (1715, 1726; Figure 1.2(b)), in his treaty on the origin of water sources. In 1732, Johann Jakob Scheuchzer, the brother of Johann Scheuchzer, published three new drawings of these same outcrops, drawn from a new perspective, to accompany a text on the Book of Genesis (Scheuchzer 1732), in which he discusses the origin of the Earth from the water. Finally, Moro (1740) again published the drawings of Vallisneri (1726) in his work on fossils exposed in mountainous regions. Hence, it appears as if the illustrations of folded beds around Lake Uri had a great impact on the naturalistic community of the first 18th century, and strangely, no attempts to reproduce drawings of sedimentary beds from areas other than that of Lake Uri are known from this period.

In a modern sense, the first cross-section in the history of Alpine research is the one of Giovanni Arduino (1758; Figure 1.3). He illustrates part of the Agno Valley (Italian Southern Alps) and shows for the first time the geometry of the beds and their lithological nature, using different symbols and short descriptions. A legend briefly describing the lithological units accompanies the section. Although the base of the section of Arduino is cut by a horizontal line (Figure 1.3), thus artificially interrupting the naturalistic character of his picture, many contacts between the lithological units are separated by irregular lines, representing the margins of cliffs that stand in front of other ones. Several lines at high angle to the orientation of the beds seem to represent topographic elements, such as little rivers (bottom, left-hand side of the figure). A small monastery is also shown at the southeastern end. All these features underline the fact that the figure was sketched in the field in front of the cliffs. However, the introduction of symbols (e.g. letters, and dots in the white pattern for unit R, continuous, parallel lines for unit P; Figure 1.3) transforms the section into something that is no longer the faithful reproduction of a landscape. It is now unambiguously the documentation and interpretation of a sedimentary tilted sequence.

The section of Arduino (1758) remained the most innovative one in the Alpine literature until the end of the 18th century. No other attempts to draw such sections are known. Although De Saussure's (1780) descriptions of geological outcrops and his thoughts about tectonic structures were very modern and innovative in his time, they were still based on classic drawings of mountain landscapes that cannot be distinguished from those of an artist who bears no specific interest in geology. Indeed, the excellent outcrop illustrations in De Saussure's book are not drawn by himself, but by a professional painter.

At the beginning of the 19th century, Ebel (1808) provided a new type of section, linked to a geological map (Figure 1.4). He shows three geological sections ("Querdurchschnitte") across the entire Alpine chain, one passing through the Gotthard Massif, one through the northeastern side of the Mt Blanc and one through the southwestern side of the Mt Blanc Massif, accompanied by a geological map of the entire Alpine realm. Given that no proper geological maps existed and that the topographic ones were rather poor (Ebel 1808), and that no cross-sections of the Alps had ever been drawn excepted for the small one of Arduino, both the idea and the achievement of Ebel's project must be considered as a scientific revolution and an immense piece of work. Indeed, the second attempt to depict the entire Alpine chain in one section would only be performed after one century (see below). Why did Ebel start such a project, instead of depicting smaller parts of the chain in more detail? He wished to show that rock units on the scale of the entire Alpine chain follow a specific order, which was still not known at that time. By providing one of the very first maps of the Alpine region (mainly Switzerland), and a series of sections, Ebel discloses a rational spatial pattern in the distribution of rock bodies:

> Nothing in nature is without order and law. Where these appear to be absent, it is only the fault of limited senses. The scouting eye and the investigating mind of the observers already found some great relationships and general laws of order in the terrible tangle of wild disorder, so that the mind is set in astonishment and comes to the certitude that a general order exists in the structure of the Earth.

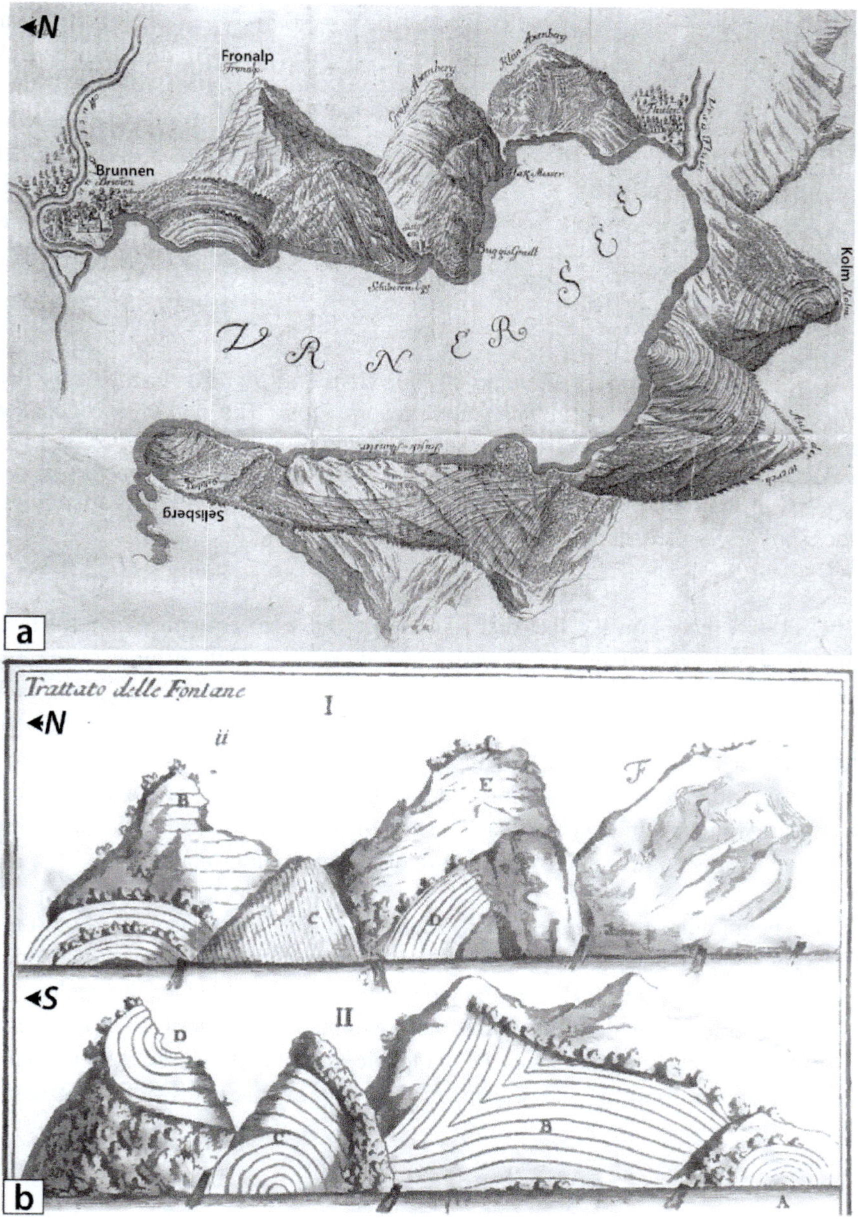

Figure 1.2. *Structures around Lake Uri (Urnersee). a) Scheuchzer (1716). The same approach of Marsigli (1705) is used for the perspective and expanded to the southern shore of the lake. b) Vallisneri (1726). The upper panel is the eastern shore, and the lower panel is the southwestern shore of Lake Uri, redrawn from Scheuchzer (1716)*

Figure 1.3. *Cross-section of Arduino (1758) from the Agnolo Valley (between Montecchio and Recoaro, near Vicenza, Southern Alps). The figure reproduces a Paleozoic-Tertiary sequence, exposed along natural, steeply dipping cliffs. The drawing does not merely represent a landscape, since different lithologies are distinguished and represented by symbols instead of their natural appearance. Symbols (letters) are described in a legend, not shown in Figure 1.3*

Ebel (1808) insists on the fact that rock bodies have a very conspicuous lateral extent in map view and questions their mutual, spatial relationships, even in a vertical direction: “The aim of these sections is to show the position of some rock bodies lying on top of others, and the steepening and deepening of some of the beds of the main rock body”. Indeed, by simply drawing what he observes, Ebel depicts the Cenozoic Molasse sediments lying underneath the Mesozoic ones in the northern parts of the Alpine chain (Figure 1.4). The age of these units was not known in his time, hence Ebel is not irritated by the nature of this anomalous contact.

At the beginning of the 19th century, mapping of the Alps was carried out on a very large scale. Detailed mapping on smaller scales progressively developed throughout the 19th century. Hence, the type of reasoning of Ebel on the geology of the Alps is also on the scale of the entire mountain belt. The map of Ebel provides a step forward, because he introduces 10 mappable units for the entire Alpine chain, whose contacts are more precisely mapped than in previous maps (e.g. Heitzmann 2008).

Figure 1.4. *N-S striking geological sections across the entire Alpine chain, modified from Ebel (1808). Upper panel: Mt Blanc section. Lower panel: Gotthard section*

Ebel's sections (1808; Figure 1.4) terminate at the bottom of the valleys, only showing what can be observed directly in the landscape and always showing a 3D view, representing more than a single "topographic steep plane" of the landscape in the sections. Although they locally show which lithological unit lies above the other, many of the contacts between different lithologies are merely apparent contacts between distinct topographic surfaces located one in front of the other. Topographic details are sketched in various parts of the section, always providing a 3D impression of the topography.

In the following two decades, cross-sections on smaller scales became very precise and commonly accompanied papers describing the geological maps of different Alpine countries. They document both the structure and the landscape surrounding the outcrops, as beautifully shown by Brongniart (1823; Figure 1.5(a)), where a 3D perspective and details of vegetation and urbanization are included. Even the shadows of the cliffs are shown in this section, showing that it is a drawing of nature as it can be seen from a specific place at a specific time of the day, rather than a construction from a geological map. Brongniart (1823) also shows 2D vertical sections (Figure 1.5(b)), where the different stratigraphic units are schematically depicted, but leave empty areas between some of the stratigraphic units, where no outcrops are visible. This indicates that the sections only show what the geologist sees in the field and nothing is added to that. Studer (1834) writes about his profiles: "In the sections, the outlines are as close to Nature as I could possibly draw them". This sentence indicates that Studer's attempt is to reproduce how things are as precisely as possible, and interpretative lines in the drawing would go against his goal. Interestingly, Studer (1834) presents two series of sections: one that he defines as "Profile" and another one that he defines as "Ideale Profile" (idealized sections; Figure 1.6), suggesting that the latter ones must provide some degree of abstraction with respect to the direct observation of the landscape. However, his "ideal profiles", in spite of their significantly smaller amount of topographic detail compared to the "profiles", still show several distinct topographic steep planes, set one in front of the other, thus creating some apparent contacts and discontinuities between different geological units (Figure 1.6).

This rigorous reproduction of what the eye sees in the field makes the geological interpretation more difficult. Studer (1834) seems to realize that geological interpretations need to filter topographic 3D effects, but he is not ready to give them up completely, and hence take a bigger step toward abstraction. Only in his "Geologie der Schweiz", in 1851, does he draw cross-sections that are entirely 2D in vertical planes. Most cross-sections until the end of the first half of the 19th century show such topographic elements (e.g. von Morlot 1847; von Cotta 1851).

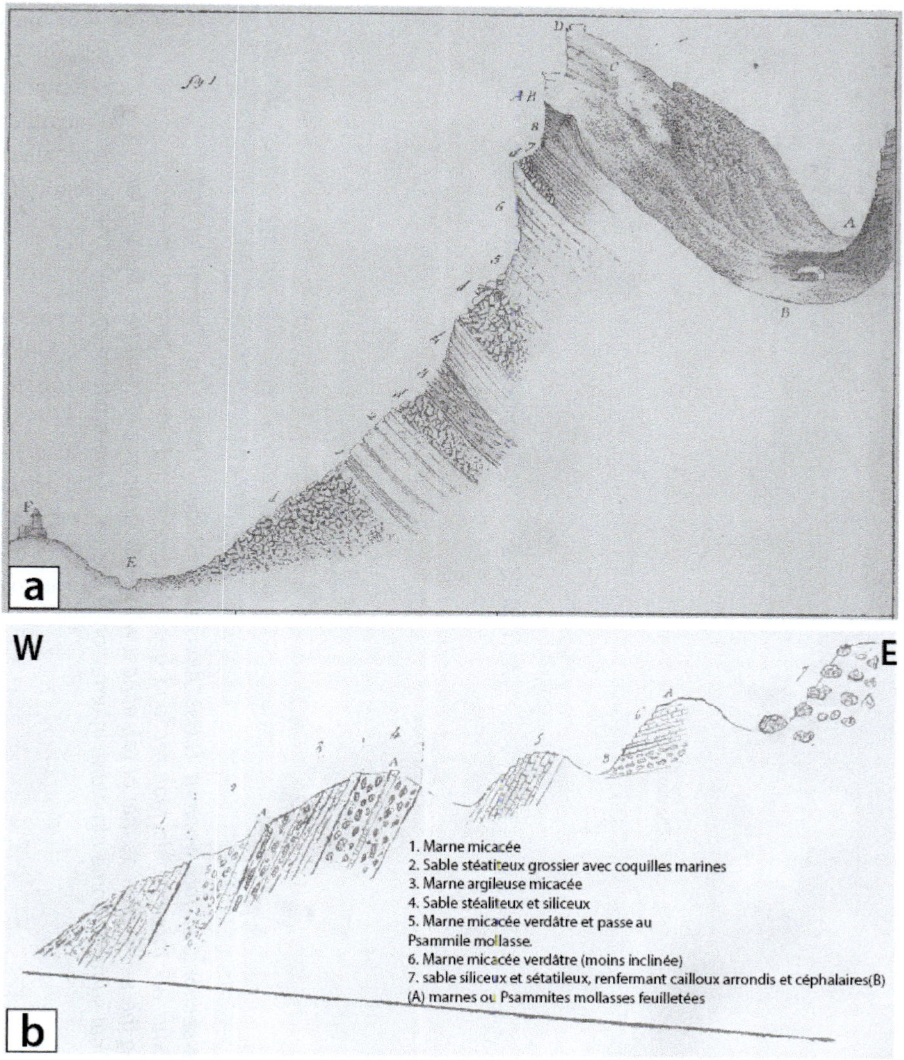

Figure 1.5. *Cross-sections after Brongniart (1823). a) Section through the Montagne des Fis, Servoz Valley (close to Chamonix, Western Alps). Note the details on 3D elements such as the little house on top of the hill and the shade of the cliffs. b) Cross-section of the Superga Hill (Torino). Note the discontinuous structure of the sedimentary beds, with "voids" where outcrops are not present*

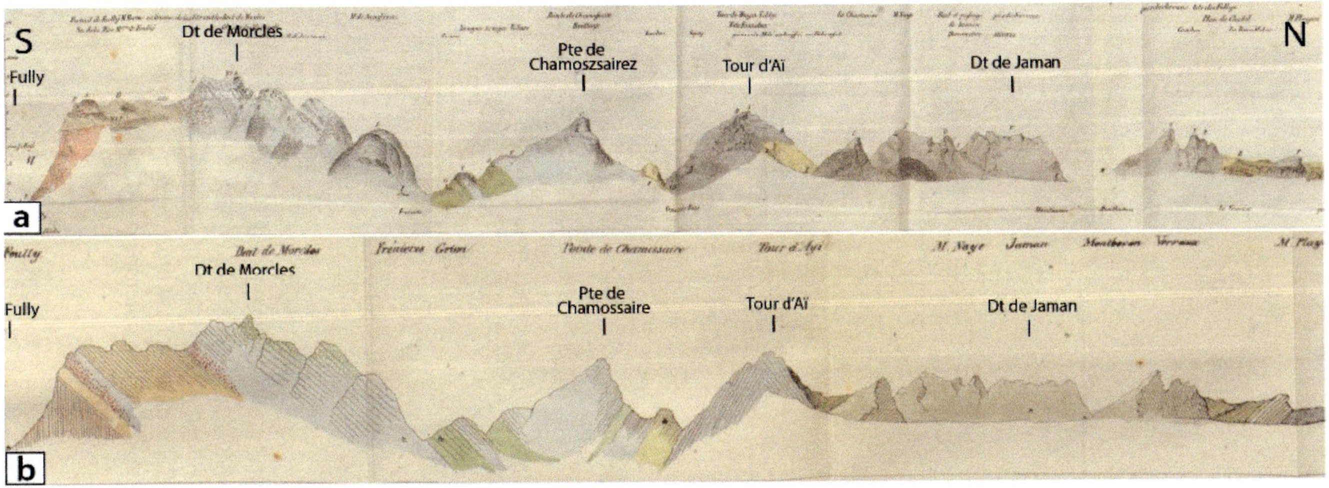

Figure 1.6. *Cross-sections of the western Swiss Alps, after Studer (1834). (a) Section termed "Profil", showing both 3D topographic features and geological units. (b) Section termed "Ideales Profil", representing the same area as that of (a), almost without 3D topographic features, but still far from a planar, vertical surface. Some structural information is shown, as the antiform at Tour d'Aï*

1.3. Cross-sections between 1840 and 1880: first attempts toward interpretative sections

Cross-sections representing a 2D vertical plane, in which sedimentary beds continuously fill the entire space between topographic surface and the horizontal "base line" of the section, have been published by Escher and Studer (1839, their Figures 3 and 4), Sismonda (1839; Figure 1.7) and Villa (1844). For the first time, these sections trace the geometry and location of beds, even across areas that are not exposed and visible at the surface and do not depict 3D topographic elements. Although not representing an Alpine example, some sections of Dufrénoy and Elie de Beaumont (1841) are very emblematic for that time, because they combine the 3D panorama-type of view of landscapes, with the 2D abstract illustration of geological structures, keeping them strictly separated in space, but as part of one and the same figure (Figures 4, 5 and 25 in Dufrénoy and Elie de Beaumont (1841)). This type of illustration creates a transition between the representation of natural landscapes and the construction of vertical, planar surfaces that are not directly observable in the field.

At the end of the first half of the 19th century, Alpine cross-sections looked close to what may be considered a cross-section in the modern sense of this term. The sections of Arnold Escher von der Linth (1848; Figure 1.8), published in Murchison (1848), as well as those of Murchison himself in the same manuscript (Figure 1.9), are definitely drawn as 2D, vertical and planar surfaces. As shown in the beautiful series of parallel sections of the Helvetic cover in the Säntis area, in eastern Switzerland (Figure 1.8), the orientation and geometry of the contacts, and the thickness of the beds, are represented in a way that is fully independent of 3D topographic effects and the location and point of view taken by the geologist while sketching the section. This allows for "lateral" correlations of beds and structures between different sections (stippled lines in Figure 1.8). The addition of stippled lines to show such a lateral continuity between sections had never been shown before.

The focus of the work of Murchison (1850) is mainly devoted to the determination of the age of stratigraphic units that he mapped in the Central Alps. However, as previously done by Escher von der Linth (1841), he probably notes that the Mesozoic Units lie above the younger Molasse beds in the northern part of his section (Figure 1.9), hence he draws a vertical fault, probably to avoid a large thrust. However, this fault is known to gently dip southward (Buxtorf 1914). Escher von der Linth (1841) shows a section of this contact (Figure 1.10), and clearly describes it as a thrust:

> [...] the well-known, abnormal height of the secondary carbonate units above the Tertiary Molasse, across all of Switzerland, is the consequence of a thrust of the Cretaceous Units (at least in eastern Switzerland) above the Molasse.

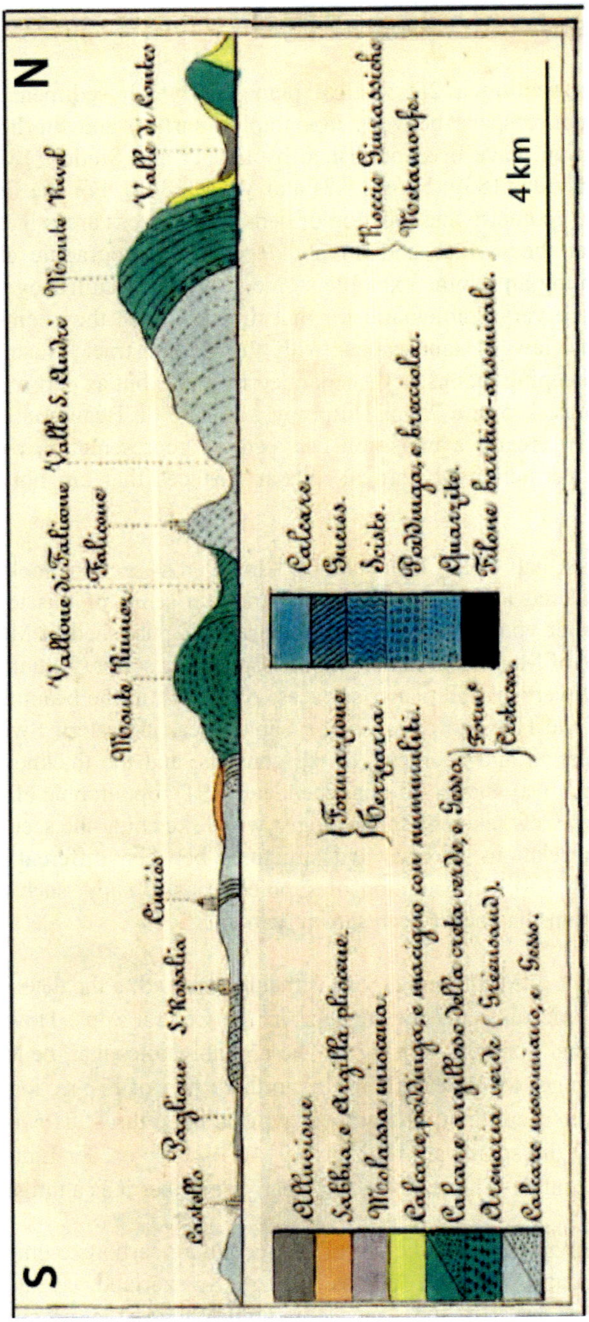

Figure 1.7. *N-S striking section, north of Nice (France, but "Regno di Sardegna in Terra Ferma" at that time). Adapted from Sismonda (1839)*

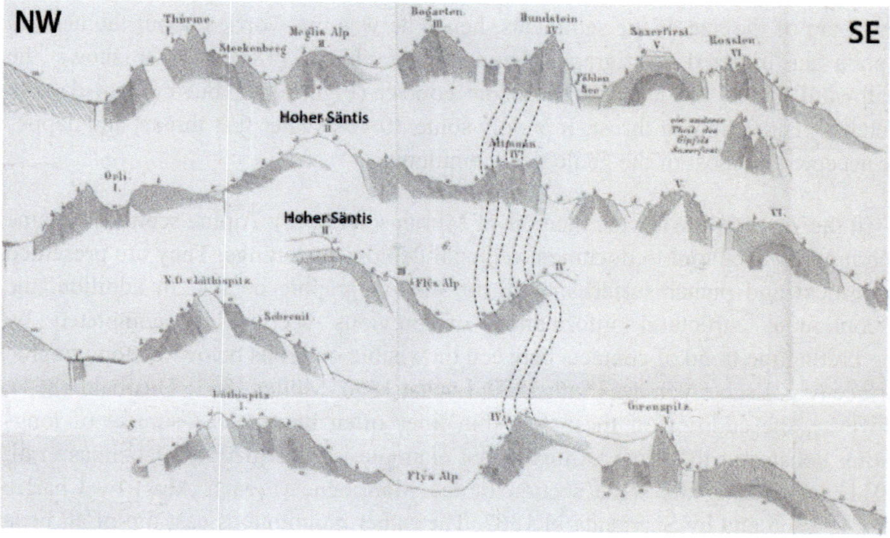

Figure 1.8. *Some of the serial sections across the Säntis Mountains, by Escher von der Linth, in Murchison (1850). Note the absence of any interpretations above the topographic surface. The lateral correlation of structures using stippled lines (accentuated here) was only added to the German edition of Murchison (1850). The English edition of 1848 does not show them*

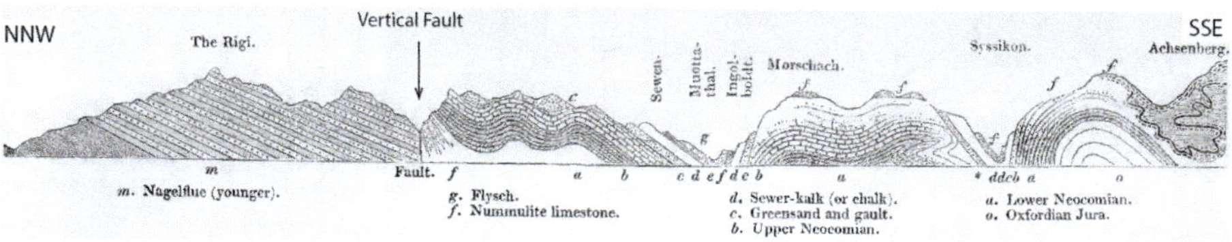

Figure 1.9. *Murchison's (1848) cross-section from the Molasse (The Rigi) to Lake Uri (Syssikon in the section). Inferred vertical fault marks the contact between Molasse and Helvetic nappes. See the text for further explanation*

Only the thrust plane itself is missing in the section (Figure 1.10). This may be the reason for ignoring his discovery in most studies on the history of tectonics and attributing the first descriptions of thrust faults to the early eighties of the 19th century (Callaway 1883; Lapworth 1883; Bertrand 1884; Geikie 1884). Ebel (1808) correctly represented the contact showing the Molasse below the Mesozoic (Figure. 1.4) at the beginning of the 19th century (Figure 1.4), possibly because he was not aware of the age of the sediments, hence he was not worried about the need to explain an inverted stratigraphic contact. Murchison (1848, 1850) shows the southward dip of the Molasse/Mesozoic contact (Figure 1.9), but did not dare to sketch a gently dipping thrust. It is only some 30 years later that thrusts and nappes are accepted by most of the geological community.

In the years following the memoir of Murchison (1848), Alpine sections became a commonly used tool to document regional geological settings. They are presented as vertical and planar surfaces, with no 3D topographic details. In addition, the discontinuous structural information of previous sections is completed by interpreting the trend of contacts between the visible outcrops below the topographic surface (e.g. Escher von der Linth 1850; Lachat 1858; Müller 1862; Theobald 1864). However, the addition of these few thin lines often became the subject of long-lasting debates at that time. An excellent example is described by Debelmas et al. (2011), comparing the same section of the Maurienne (French Alps) by Charles Lory in 1866 and by Sismonda in 1867. The rather monotonous east dip of all beds in the section is shown (by stippled lines below the surface) to be the result of tight, east-dipping folds in the first interpretation, whereas it is a monoclinal, unfolded sequence in the latter interpretation.

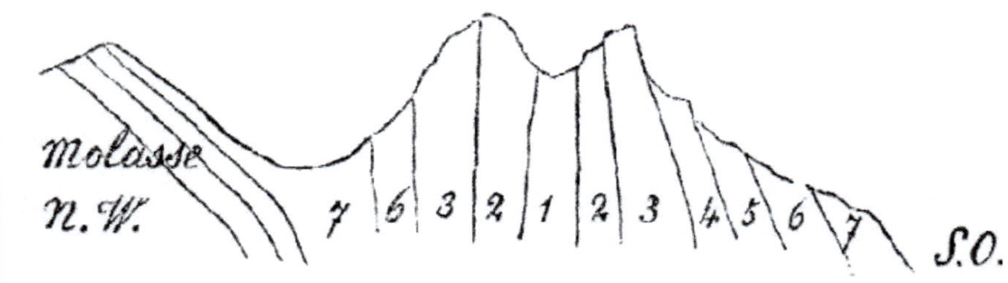

1. and 2. Upper Cretaceous; 3. Cretaceous; 4. Equivalent to Montagne de Fis;
5. and 6. Upper Néocomien; 7. Terrain Néocomien

Figure 1.10. *NE-SW cross-section showing the Molasse dipping at high angle below the subvertical Mesozoic cover (numbered units). The original figure illustrates the report of a presentation of Escher von der Linth, published by Studer and Desor (1841)*

The addition of interpreted lines to constrain the structures below the topographic surface observable in the field precedes interpretations of the same kind above the surface, which only appeared in the late 1960s (Favre 1867; Giordano 1869; Heim 1871). Interestingly, Favre (1867) shows some cross-sections with contacts extrapolated above the surface from the Mont Blanc area, stating that they are reproduced from his own early work from 1859. Indeed, Favre shows these interpretations above the surface for the first time in 1867, in a publication that recounts in great detail the debate on the Carboniferous terrains in the Alps, specifically in the locality of Petit-Coeur, close to Moutiers (France). Favre wanted to terminate a long-standing debate that lasted for decades on the structure of the "Petit Coeur" area, in the Western Alps. In order to clarify whether the subvertical and symmetrical distribution of sedimentary strata results from a synclinal or anticlinal fold, Favre presents two sections of the same structure, one with stippled lines that form an anticlinal hinge above the topographic surface, and one with stippled lines forming a synclinal hinge below the observable topographic surface. Thus, Favre (1867) oversteps the boundaries of the visible topographic surface for the sake of clarifying the structure of a complex area. Interestingly, Favre terms his old cross section of "Petit Coeur" published in 1859 without interpretative stippled lines (Favre, 1859) as "coupe classique", whereas he terms his new, interpreted version "coupe théoriqueé" (Favre, 1867). (For a historical overview, see Grandchamp (1997)).

Another significant advancement in the interpretation of cross-sections is the one provided by Felice Giordano (1869; Figure 1.11). In his sections across the Matterhorn at the Italian-Swiss border, Giordano dares to draw a very thin and light stippled line above the surface to underline the lateral continuity of the contact between his "talc-bearing gneisses" and the "calc-serpentinite formation" below. The second innovative aspect in the sections of Giordano concerns the drawing of structures down to the sea level, and possibly below, in areas where the valleys are still at altitudes above 1,500 m (Figure 1.11). Giordano does not draw any clear contact lines at such depth, but he writes the name of the geological unit with a folded shape that mimics the geometry of the contacts at depth (Figure 1.11). Thus, it takes over 100 years to arrive from the first illustrations of naturally occurring structures (Arduino 1758) to the latter ones that show some still very cautious and only indirect signs of the inferred extrapolation of exposed geometries both below and above the visible surface. Giordano dares to take this step for two main reasons: first he realizes the existence and importance of the lateral continuity of structures in spite of their interruption due to erosion (lithological similarity between M.te Cervino/Matterhorn and Dent Blanche, Figure 1.11), and second he is aware of the kinematic implications of such a lateral continuity. As described by Dal Piaz (1996), Giordano accurately assesses the large-scale geometry of three gently dipping units lying above each other (Figure 1.11): the "Formazione di Gneis Talcoso" lying above the "Formazione Calcareo-serpentinosa", which lies itself above the "Granito antico" (Monte Rosa basement, not visible in Figure 1.11). Because the intermediate unit consists of dolomites, quartzites and cargnieules, it is inferred to be Permo-Triassic, by analogy with other Alpine areas. But the presence of ancient granitoid gneiss in the "Formazione di Gneiss Talcoso" located at a higher structural level makes it difficult to infer a post-Triassic age for this unit, hence a normal stratigraphic sequence. Giordano (1869) is conscious that attributing an older age to the hanging wall unit would require its upwelling from below and its spreading out laterally: "[...] *overthrown in the form of huge nappes* (... *rovesciatosi poi in falde enormi*"), and in a second article in French "[...] *se serait épanché en nappes énormes*" (Giordano 1869b). In spite of this lucid kinematic concept, Giordano considers this process unlikely (Dal Piaz 1996), because it would correspond to a "*frightful overthrow*". Thus, Giordano precludes an older age of the upper unit, and interprets it as Cretaceous, denying the existence of nappes immediately after having coined their idea and name for the first time (Trümpy 2001).

In the same year, Jaccard (1869) also interprets the geometry of folded beds of the Jura Mountains with few stippled lines above surface and calls these sections "coupes théoriques". This shows that although these stippled lines complete the eroded parts of folds in a very obvious and intuitive way, Jaccard still wants to demark the difference between reproducing and interpreting the observable.

In spite of the publications of Lory (1867), Giordano (1869), Jaccard (1869), Gerlach (1871) and Heim (1871), sections were rarely interpreted above the topographic surface before the end of the 1870s. Only where folds are very clearly recognizable, but eroded in their hinges, would their eroded parts be represented locally by stippled lines, to document the continuity of the structures (e.g. Jaccard 1869; Gerlach 1871; Escher von der Linth 1878; von Gümbel 1878). Moreover, some geologists are seriously concerned about the lack of reality of such sections. Pfaff (1873) discusses the problems of documenting real (only observed) versus unreal structures (interpreted parts) in cross-sections. He takes the section of Favre (1867) as an example, probably the one that more boldly interprets the geometry of eroded structures at that time. Pfaff (1873) redraws the same topographic profile of Favre, showing those parts that can really be observed in the field, hence critically concluding that most of the section is not based on "real observations".

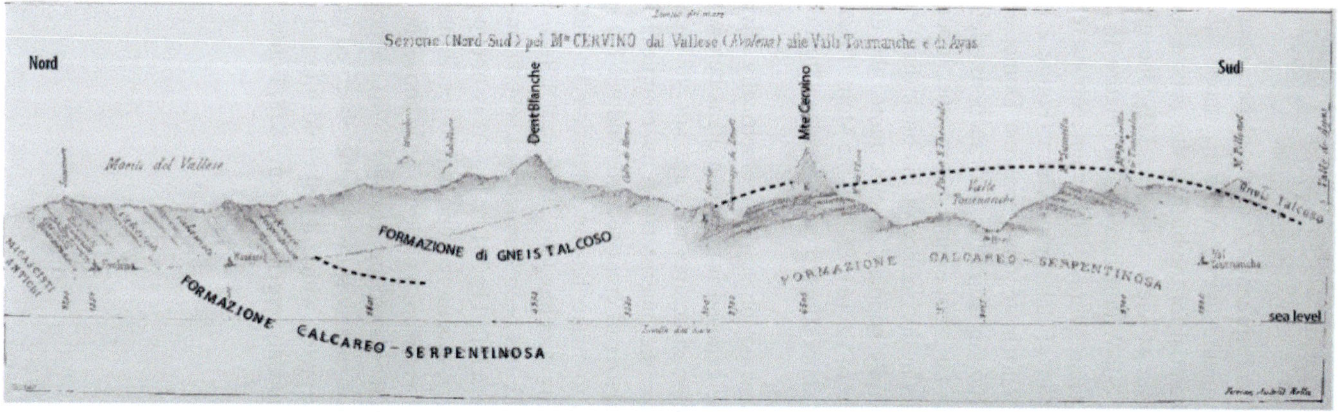

Figure 1.11. *N-S striking section across the Matterhorn (Cervino), and Dent Blanche, modified from Giordano (1869). Note the gently folded shape of the inscription "Formazione calcareo-serpentinosa", which points to the deep, folded structure of this unit, even below the sea level: an unprecedented interpretation and representation of the deep crustal structure. The stippled line (redrawn to make it more visible) along the base of the "Formazione di Gneis Talcoso" corresponds to the base of the Dent Blanche nappe*

1.4. Cross-sections in the last two decades of the 19th century: the first kinematic interpretations

In the last two decades of the 19th century, a new debate changed the way of thinking about tectonics and structure of the Alps: the nappe theory. The cross-sections of Heim (1878) are influential for the birth of this debate, and they are very innovative from a graphical point of view (Figure 1.12). One group of sections represents the structure of the Tödi-Windgällen Group (Figure 1.12(a)), nowadays famous for the outcrops of the Glarus thrust. The other sections represent the first structural synthesis of a large part of the Alpine chain (Figure 1.12(b)), from the Molasse Basin north of Lake Zürich, to the Lepontine Dome, south of Campo Lungo, hence covering almost 100 km length. In both sections, Heim interprets the structure of the eroded units far above and below the topographic surface (Figure 1.12(a)), but he mentions in the preface of his book that "everything being beyond the real observation is only stippled". Heim quotes the debate started by Pfaff (1873) on the cross-sections of Favre (1867) in the Mont Blanc area, and states very clearly that everything that is not stippled corresponds to "real observations".

The stippled lines of Heim are no longer the geometrically obvious and simple continuation of contacts that are exposed in different areas below the topographic surface. Heim interprets complex structures that are not yet understood by the geological community. Because the interpretation is complex, Heim needs to overstep the conventional boundaries of cross-sections and use the space above and below the observable surface to make his interpretation visible. One of these complex structures is the Glarus "double fold" (Figure 1.12(a)). The fascinating historical debate on this structure has been summarized several times (Heim 1908; Trümpy and Lemoine 1998; Letsch 2011, 2017), thus it is not re-proposed in the following. For the present subject, it is more important to recall that a discussion to explain the puzzling presence of Permian, Verrucano beds above the Tertiary Flysch starts several decades before the book by Heim (1878), hence the need for clarification is big. Heim dares to clearly show that the observed geometries below the surface can possibly be interpreted with a double fold inferred to lie above the surface, with hinges showing opposite vergence (Heim 1878; Figure 1.12(a)).

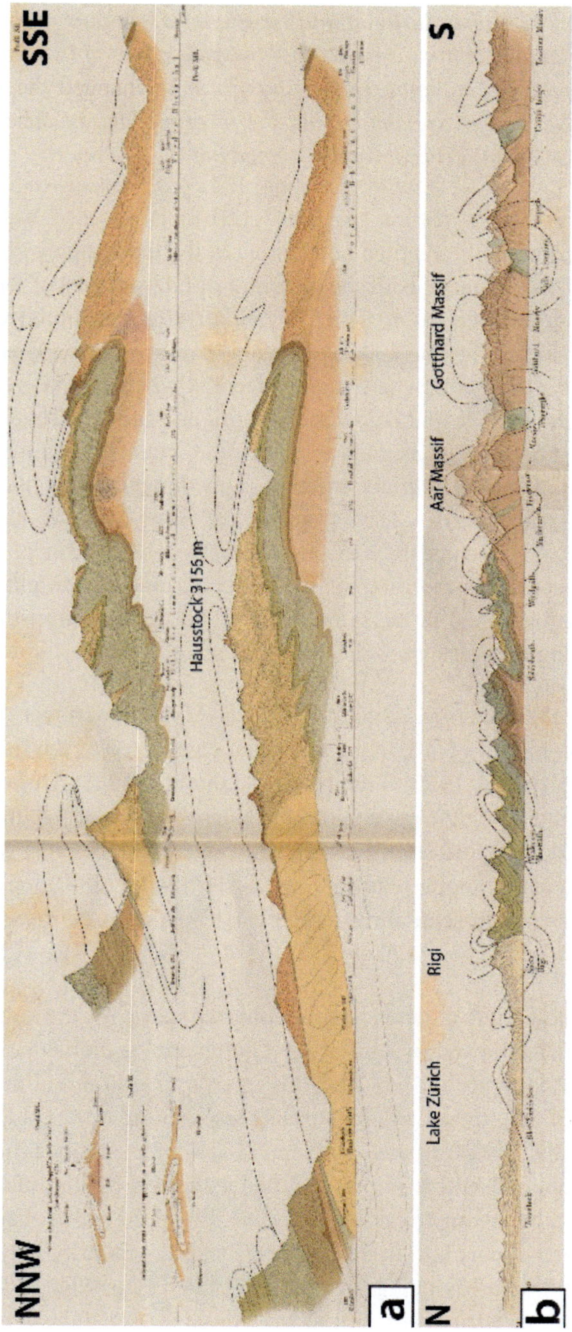

Figure 1.12. *Cross-sections from Heim (1878). a) Glarus Alps, showing the famous "double fold". b) Section striking from the Molasse Basin, north of Zürich, to the Lepontine Dome, showing a structural interpretation before the nappe theory*

COMMENT ON FIGURE 1.12.– *Shortening is only accommodated by folding, including the double Glarus fold. The Upper Jurassic units in the northern part of the section are interpreted to be in lateral continuity with the Mesozoic cover of the Aar Massif, the Gotthard Massif and the Lepontine Dome, thus forming a single folded layer throughout the chain. No thrusts are shown, but the inverted limb of the Mesozoic cover above the Molasse (S of Rigi) is severely thinned out.*

Concerning the larger scale section (Figure 1.12(b)), Heim also has a good reason to overstep the commonly accepted boundaries of the topographic surface: he constructs for the first time a synthesis of the structure of a larger part of the Alpine chain in one and the same section, in which a marker (Jurassic units) could be continuously followed all along the 100 km length of the section. This is only possible if the marker is shown both where it is eroded and where it is at great depth, hence at several kilometers above and below the topographic surface. This section is not only of great interest because of its graphic innovation, but also because it shows the state of understanding of the Alpine structure immediately before the general acceptance of the nappe theory. The structure of the Alps at this point (Figure 1.12(b)) is only represented by the basement/cover discordance, which is harmonically folded throughout the orogen. In the axial zone of the chain, folds affect the cover and the basement in the same manner. Not a single thrust or nappe accommodates shortening at this stage.

Bertrand (1884), without having visited the Glarus Alps, reinterprets the cross-sections of Heim (1878, Figure 11(a)), in a paper often considered as the first step of the nappe theory (e.g. Bailey 1935; Trümpy and Lemoine 1988; Durand-Delga 2010). After studying the geological maps of the Glarus area and Heim's cross-sections exhibiting the double fold (Heim 1878), he re-interprets the cross-section in a simpler and kinematically more plausible way. Bertrand (1884) replaces the double fold with a single, large-scale thrust (Figure 1.13(a)), thus defining the base of a nappe.

Bertrand (1881) interpreted inverted stratigraphic contacts by the presence of thrusts, and the approach followed by him in the Glarus area is emblematic for the historical evolution of ideas. He did not base his interpretation on additional field observations compared to those of his predecessors; on the contrary, his observational base is only the one filtered by Heim (1878) in his publication. On the one hand, Bertrand is guided by his impressive understanding of the link between kinematics and associated geometrical evolution and, on the other hand, by a synthetic capacity, which allows him to detect the structural similarity between two distinct areas. One of these areas being already interpreted (Gosselet 1879) as the result of thrusting: "I simply tried to apply to the Alps the so simple and rational explanation that Gosselet gave for the North (meaning the Bassin Houiller)" (Bertrand 1884). Bertrand's approach is purely structural. He explains in detail how the enormous thinning of the inverted fold limbs below the Glarus thrust (Bertrand 1884; Figure 1.12) results from simple shearing of layers that are not parallel to the shear plane. He interprets these as drag folds, resulting from displacement along the overlying thrust plane. Bertrand (1884) describes these folds with a term already coined at that time in French mining literature: "cran de retour". Based on his kinematic model, Bertrand rectifies the inferred finite geometry of geological units in the cross-sections of the Glarus Alps above the topographic surface, leaving the rest untouched because it is a "real observation", as already stated by Heim.

However, the interpretation of the Glarus thrust of Bertrand (1884) is still far from the present-day one. Although he recognizes the subhorizontal, top-to-the north displacement along the Glarus thrust, he summarizes his own interpretation as that of "a single fold instead of the double fold of Escher". He also states that "all glide phenomena could be brought back to folds". His interpretation remains that of a cooling Earth whose contraction is accommodated by a steeply inclined, folded area that can flow laterally, thus forming a recumbent fold of large amplitude, whose inverted limb evolves into a thrust.

These interpretations can be found again in his later cross-sections of the Mont Blanc-Mont Joly area (Bertrand and Ritter 1896; Figure 1.13(b)), where the Mesozoic cover is shown to be folded in a stack of recumbent, isoclinal folds of approximately 10 km of amplitude, all terminating in the sub-vertically folded basement-cover interface, whose folding amplitude is only of 1 km amplitude (Figure 1.13(b)). This area represents the inferred root of the nappes. These sections (Bertrand and Ritter 1896; Figure 1.13(b)) are the first to show the complete geometry of nappes, from their inferred root to their folded front, and more generally, they are also the first ones to provide such a detailed structural picture of an area above the topographic surface (Figure 1.13(b)). Bertrand and Ritter (1896) do not show this anymore with mere stippled lines, but with bold patterns, just like those that are conventionally used for the area below and at the topographic surface. Bertrand and Ritter (1896) term these sections as schematic, but they are based on sound assumptions and lateral projections, namely the idea that the Mont Joly area (line a-a′ in Figure 1.13(b)) represents a higher structural level compared to the Mont Blanc Massif (line b-b′ in Figure 1.13(b)).

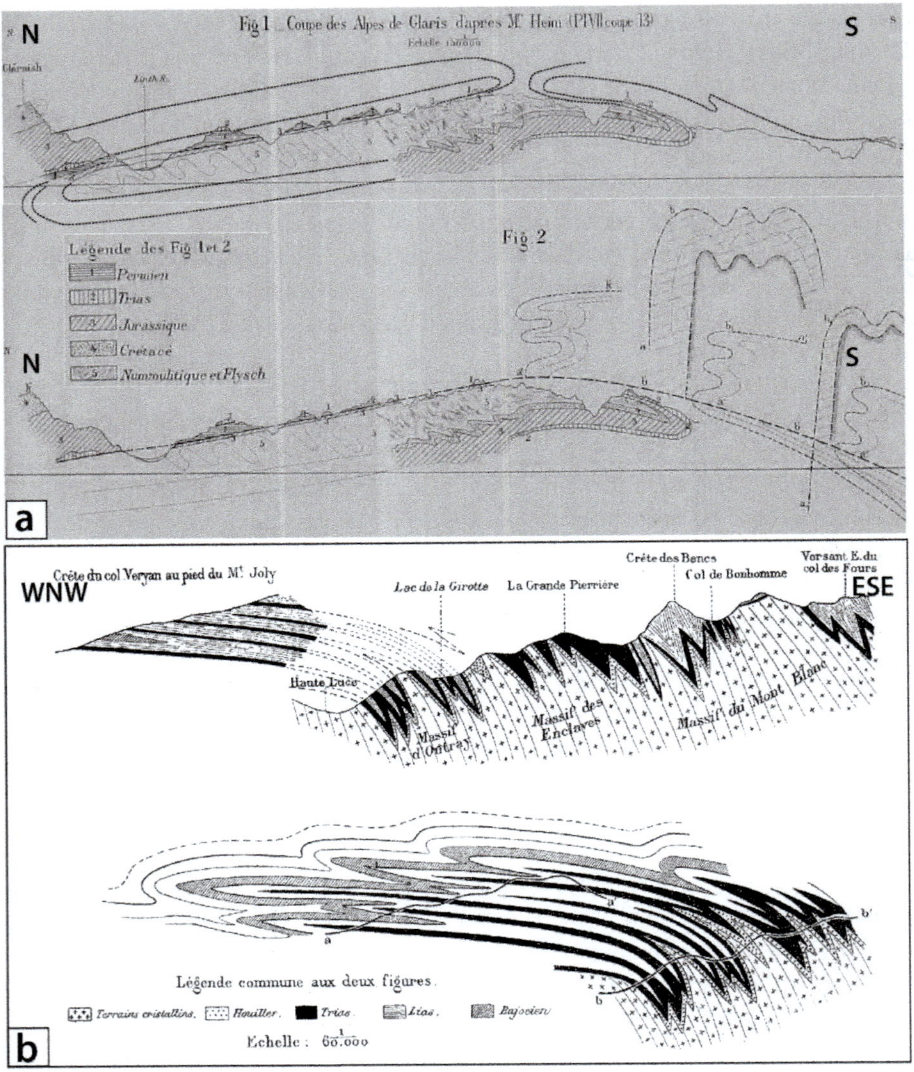

Figure 1.13. *Cross-sections of Marcel Bertrand. a) Re-interpretation of the cross-section of Heim (1878, Figure 11(a)). Bertrand deleted the southern half of Heim's section and replaced it by the trace of a large thrust that is now termed the Glarus thrust. Modified from Bertrand (1884). b) Cross-sections of the Mont Blanc-Mont Joly area, modified from Bertrand and Ritter (1896). Note the detailed representation of the area above the topographic surface and the low amplitude of the steep, basement-cover folded surface ("root zone"), compared to the cover nappes*

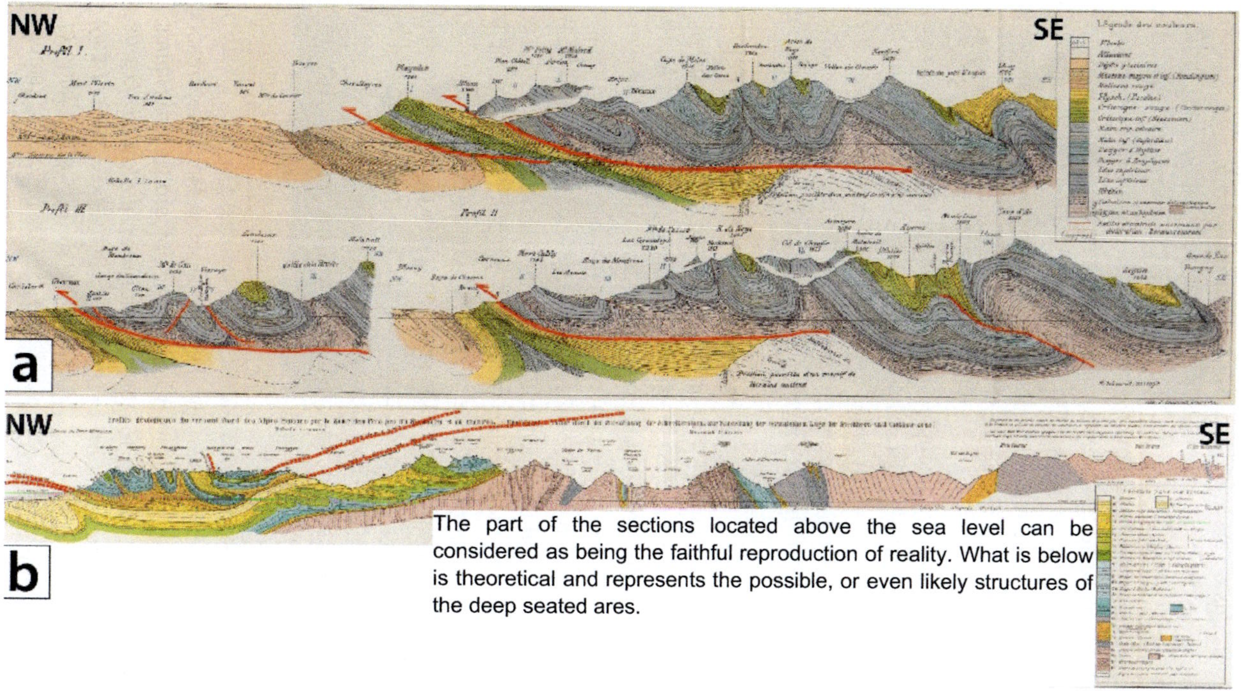

Figure 1.14. *Sections across the Préalpes. (a) Section (modified from Schardt (1893)) showing the Préalpes with a thrust at their base, lying above the Eocene Flysch. (b) Section (modified from Schardt (1898)) from Lake Geneva across the Préalpes and the northernmost Mont-Blanc Massif to the Austroalpine basement. Two major thrusts are shown: the basal thrust of the Préalpes and the overlying basal thrust of the Nappe de la Brèche. Both of them are shown far above the topography in the SE part of the section, suggesting that they must be rooted further SE*

The nappe theory and its representation in cross-sections mark a new step forward, with the description of the structure of the Préalpes by Schardt (1893, 1894). His numerous and considerate arguments are accompanied by a cross-section, which beautifully depicts the internal structure of the Préalpes resting on a thrust, above the Helvetic Flysch (Figure 1.14(b)). Schardt (1893) mentions that the construction of his sections is extended below the sea level "[...] against the common habit of doing it". Indeed, these sections represent the very first extrapolation and construction of structures at greater depth. On the figure showing the sections, Schardt writes explicitly that everything below the sea level should be considered as possible and probable, unlike what is above the sea level, which is ascertained.

Extending the depth of sections by a couple of kilometers should not be taken as a simple change in image formatting. Schardt (1893) increases the depth of his section, and explicitly justifies this act, in order to show for the very first time that the deeper level of the studied area is entirely different from the upper one exposed at the surface, hence to document the presence of a major thrust plane separating these two structural levels.

The structural approach of Bertrand (1884) only requires the sections previously drawn by Gosselet (1879) and Heim (1878), in which the thrust plane can be rooted in one and the same section. However, the observations of Schardt (1893, 1894) are of a different nature. What leads him to the concept of a nappe is both the geometric relationship between footwall and hanging wall, as in the case of Bertrand (1884), but also the different stratigraphy of a sequence compared to its surroundings. As he writes in 1894: "If the tectonics of the Chablais area do not show any relationship to their surroundings, this is even more the case when considering the sedimentary facies of the units that form this area". Indeed, the sedimentary facies of the Préalpes does not show the typical Helvetic, Mesozoic sequence; hence, it is recognized by Schardt as exotic compared to its Helvetic surroundings. This is the fundamental observation that leads him to assume that the Préalpes are far-traveled units originating from the southern part of the chain.

It was only in 1898 that Schardt constructed new sections (Figure 1.14(b)), interpreting the basal thrust of the Préalpes and the "nappe de la Brèche" (a nappe derived from the southeastern margin of the Briançonnais domain as thrust planes that continue southeastward, up in the air, for tens of kilometers). This construction is a big step forward in the representation of structures in cross-sections. These studies of Schardt (1893, 1894, 1898) increase the spatial dimension of cross-sections both toward deeper (not exposed) and higher (eroded) tectonic units. The lateral dimension also increases, as shown by the base of the nappes, which is still high above the surface at the right end of the section, hence rooted outside of the plane of the section (Figure 1.14(b)).

Schardt (1898) does not connect the large thrusts to any folds (Figure 1.14(b)). He is the first to decouple the concept of a nappe from that of a recumbent fold. In his description of the historical debate about his own idea of the nappes of the Préalpes, he states: "I am not talking about a recumbent fold, but about a glided nappe".

1.5. The beginning of the 20th century: first orogen-scale sections and the description of nappe piles

Geological mapping of the Alps was very advanced by the end of the 19th century. As summarized by Dal Piaz (2010), the first "geognostic" map of the Austrian empire was published in 1847. In 1887, the 21 sheets of the Geological map of Switzerland at the 1:100,000 scale were completed, including large parts of the northwestern Italian Alps. The geological map of the "Sardinia Mainland States", including the Western Alps of Savoy, Piedmont, and Liguria, was accomplished in 1866 by Sismonda, at the 1:500,000 scale, and at the 1:50,000 scale for the "Alpi Piemontesi", with 29 sheets that appeared between 1860 and 1879 by Baretti, Gastaldi and Gerlach (Campanino and Polino 2002). Hence, at the beginning of the 20th century, synthetic views of the Alps, from one side to the other, were made possible by the large number of existing geological maps.

Maurice Lugeon (1901) widens the horizon provided by Schardt (1898) by constructing a cross-section of the Helvetic nappe stack (Figure 1.15a) and showing that this nappe stack lies below the nappes of the Préalpes (Figure 1.15b). He constructs two parallel sections (Figure 1.15(a)), but shifts upward the one located further to the NE, in order to compensate for the northeastward dip of the basement and all the nappes above it (see the shift in the base line of A and B in the lower part of Figure 1.15(a)). By doing so, Lugeon (1901) achieves what is later performed by Argand (1909) by projecting structures parallel to fold axes into the plane of the cross-section. The result is the very first illustration of a stack of three nappes (Dent de Morcles, Diablerets and Wildhorn; Figure 1.15(a)) that are shown to lie below the basal thrust of the Préalpes. These sections represent the first descriptions and documentations of nappe stacks. In addition, because Lugeon was aware of the fact that the three Helvetic nappes (Figure 1.15(a)) lie below the nappes of the Préalpes, he constructed a schematic section showing both the Préalpes and the Helvetic nappes (Figure 1.15(b)). This impressive stack consisting of eight distinct cover nappes, and the observed ductile fold structures bring Lugeon (1901) to a lucid discussion on the depth conditions for the formation of the nappes, reflecting which units may have created the necessary pressure (temperature was not considered as the main factor) to induce ductility in the underlying nappes. Lugeon (1901) writes: “The nappes bearing helvetic facies [...] therefore postdate the prealpine nappes, and as a consequence, they very clearly formed at depth after, or maybe also during the displacement of these nappes of the Préalpes”. In contrast, concerning the formation of the nappes of the Préalpes themselves he states: “it is known that the poudingues were formed at the expense of the Prealpine nappes that cover them. Hence, thrusting appears to have taken place at the surface [...]”.

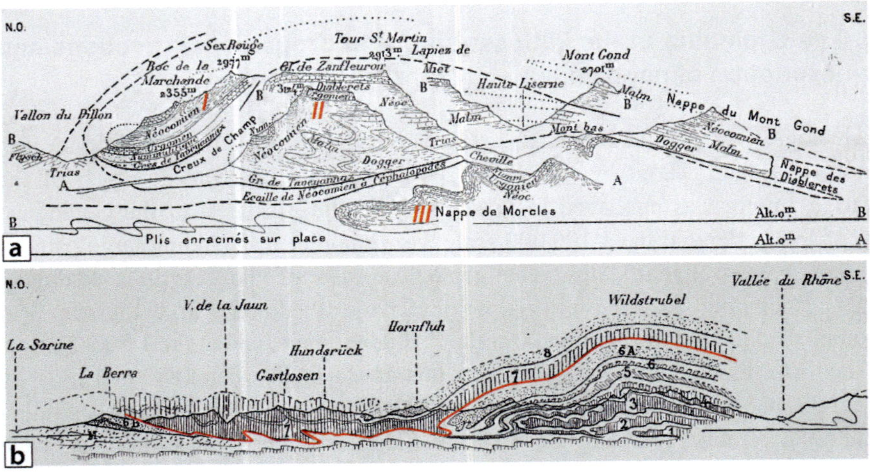

Figure 1.15. *Cross-sections after Lugeon (1901). (a) Section across the Helvetic nappes of western Switzerland. I: Mont Gond Wildhorn Diablerets nappe; III: Morcles nappe. (b) Schematic section from the Rhone Valley to the Sarine, crossing both the Helvetic nappes and the Préalpes. 1: Morcles nappe; 2: Diablerets nappe; 3: Mt Gond-Wildhorn nappe; 4–6: Nappes of the “Zone interne”; 7 and 8: Préalpes.*

In 1902, Termier (Figure 1.16(a)) published four sections across most of the Western Alps, from the Molasse basin to the “Zone du Piemont”. Only the upper plate of the orogen, which does not crop out in these sections located in the very SW of the arc, and the Italian Molasse basin are missing. They show that the orogen consists of a pile of six major nappes. The two uppermost nappes are suggested to be linked; hence, the structure consists of five orogen-scale nappes, only vaguely resembling the major tectonic (paleogeographic) units that are identified in present Alpine studies. The Schistes Lustrés nappe corresponds to the Liguro-Piemont Unit, and the Flysch and Nummulitique correspond to part of the Helvetic nappes, but the “Massifs granitiques” show one continuous basement unit throughout the chain, just as the “Terrains secondaires” (Figure 1.16(a)). The lowest nappe (number I in

Figure 1.16(a)) consists of "Terrains secondaires" that are rooted in the Pelvoux Massif (Figure 1.16(a)). All other nappes are rooted further east in a very large area of distributed folding, inferred to be the root zone of Alpine nappes (Figure 1.16(a)). A sixth and structurally highest nappe, not clearly described, must be rooted even further east, in an area not shown in the section (Figure 1.16(a)).

In 1903, Termier published a new set of five sections (Figure 1.16(b)), across the entire width of the Alps, two in the Eastern Alps, one in the Central Alps and two in the Western Alps, accompanied by a simplified, tectonic map of the Alps ("Essai à une carte structurale des Alpes"). These can be considered the first modern orogen-scale sections of the Alps and probably of a mountain range in general. Ebel (1808, Figure 4) did not illustrate the structure of an orogen, but merely the lithologies exposed at its surface. Termier (1903; Figure 1.16b) shows the continuity of the internal structure of the orogen crust, from one side to the other.

It is noteworthy that Termier did not start with the publication of one orogen-scale section, but rather with five at the same time. These sections are conceived and sketched during a short time, namely after Termier's field trip to the Eastern Alps and to the Tauern Window in 1903, where, accompanied by Friedrich Becke and Ferdinand Löwl, he understood that the structural position and age of the "Bündnerschiefer" (or "schistes lustrés") is exactly the same as the one that he already assessed in the Western Alps (e.g. Termier 1902), namely below the Austroalpine. Termier (1903) recognizes that the "Hochstegenkalk" (carbonates of the Helvetic cover, lying below the Schistes Lustrés) is not Paleozoic, but Triassic (it is now ascertained that it is Upper Jurassic: Kiessling (1992)) and he knows (based on the work of Franchi (1898); Frech (1901)) that the "Schistes Lustrés" are Mesozoic. Therefore, he understands that, like in the Western Alps, a Mesozoic cover sequence (Helvetic) is overlain by another Mesozoic series (Schistes Lustrés) that is eventually covered by a Paleozoic one (Austroalpine). Based on these observations, Termier is the first to interpret the structure of the Eastern Alps as a pile of nappes, as already inferred for the Western Alps.

Termier's (1903) section across the Western Alps (Figure 1.16(b)) is similar to the ones in his 1902 paper, but differs in at least one important aspect, which is the root of the uppermost nappes. The base of the Austroalpine, that is, the base of the "surface de traînage des Dinarides", is now clearly rooted in the southernmost basement as a line and no longer as a tightly folded region (Figure 1.16(b)). In contrast, all other nappes terminate, like in Termier (1902), in an area of steeply inclined, low-amplitude folds, reflecting his idea of nappe root. As he states in 1906: "Instead of land of the roots we could call it land of the folds, or even, following M. Lugeon, autochthonous land". His idea is that the root is shortened, but not displaced from its footwall. Only the upper part of the nappe is sheared and displaced northward along a flat-lying surface:

> Like the trees of the Rhône Valley under the irresistible push of the Mistral, like the smoke of an industrial country under the wind that folds it down and rolls it, the folds of the Alps were laid down northward towards the external part of the chain. And the force that laid them down was so full of energy to pile them up one on top of the other, to laminate them, stretch them, fragment them, so much and so well that the nappes formed in such manner up to 100, 120, maybe 150 kilometers away from their origin. (Termier 1903)

The sections of Termier (1903; Figure 1.16(b)) show several first-order innovative features and similarities between them, pre-announcing that tectonic processes must have been similar all along strike in the Alpine chain. These innovative features are as follows: (1) the Alpine chain consists everywhere of a pile of basement and cover nappes. Their geometry varies from one section of the Alps to the other, but the nappe sequence is the same everywhere on the very first order; (2) all sections show the "Surface de traînage des Dinarides", a very large thrust that is above the topographic surface in all sections and rooted in the "Dinarides" (the South-Alpine) basement. This surface corresponds to what is considered the base of the Austroalpine, that is, the base of the upper Plate in modern Alpine literature; (3) all sections show a south-verging fold belt in the southern, "Dinaric" part of the chain, in contrast to the rest of the orogen; (4) tertiary deposits (Molasse) on the European side of the chain are overthrust toward the N (in the Eastern and Central Alps) and WNW (in the Western Alps) over long distances (up to 60 km in the Eastern Alps) by the basement and its Mesozoic cover units. The amount of Cenozoic displacement dramatically decreases in the Western Alps.

Termier terms these sections as "schematic" and indeed they are far from depicting the geometry of the units as we know them from present-day reconstructions. The basement and its cover are affected by immense amounts of folds that are schematically drawn and, according to modern surveys, either not existing or of entirely different geometry. However, the sections do show a surface topography that links these Alpine structures to specific areas where they are truly exposed. Termier has extensive mapping experience in the Western Alps (e.g. Termier 1894, 1901), where he constructs numerous detailed and accurate sections, very different in style from those of 1902 or 1903. In the latter publications, Termier illustrates in one and the same figure something that is not visible as such in the field, but which synthesizes all field observations and interpretations.

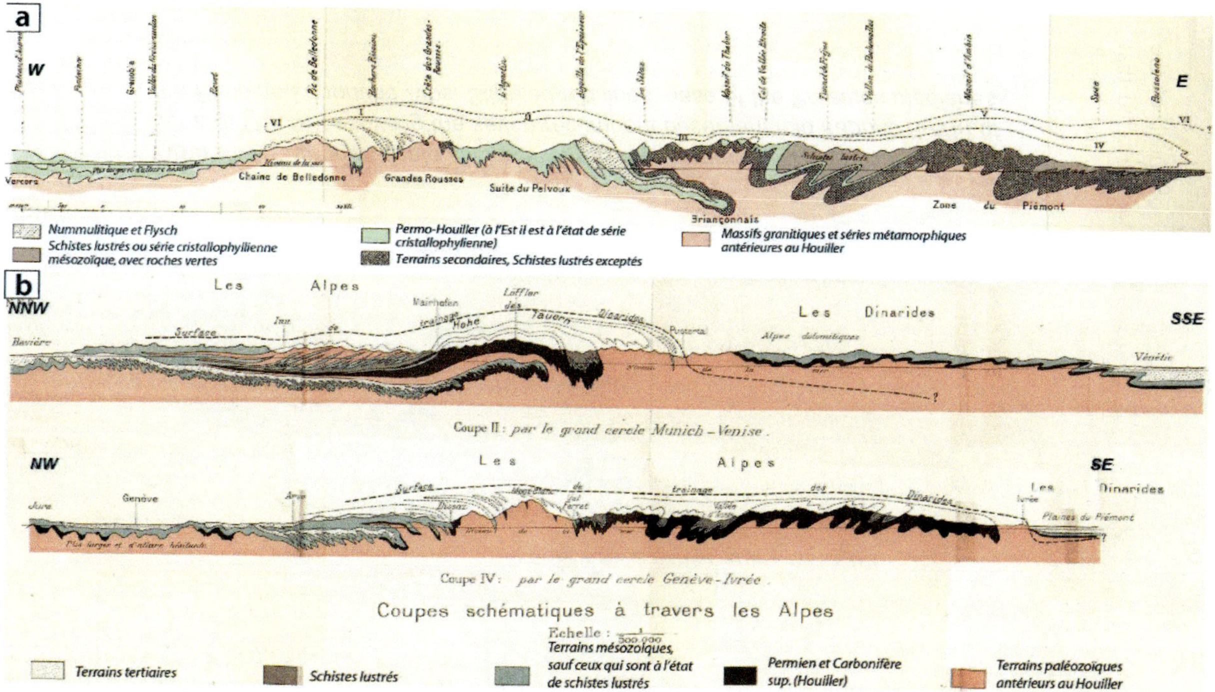

Figure 1.16. *Cross-sections after Termier. (a) Section across the W-Alps. The Alps are categorized into six major nappes (roman numbers) that are all rooted in a southeastward-located zone of distributed folding. Modified from Termier (1902). (b) Sections across the Eastern Alps (top) and the NW Alps (bottom). Note the presence of an "upper plate". Modified from Termier (1903)*

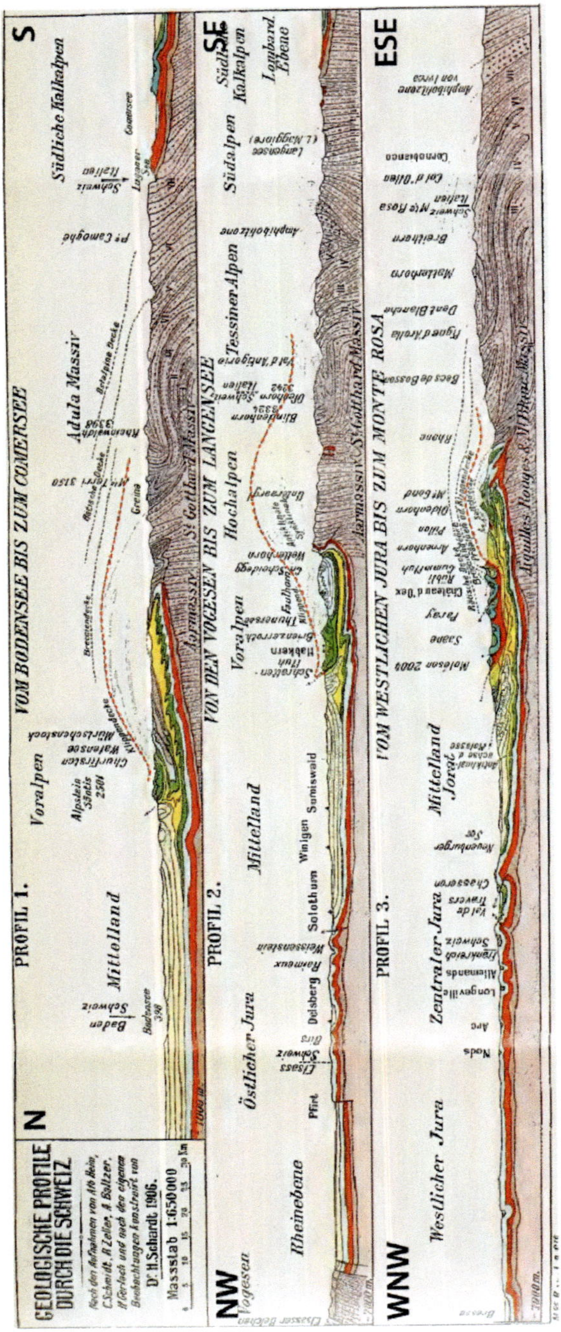

Figure 1.17. *Orogen-scale sections across the Western and Central Alps (from Schardt (1906)), showing the seven recumbent basement fold nappes (I–VII) of the Pennine Lepontine Alps. Stippled red lines: base of the Préalpes médianes*

Even if schematic, these sections show the very first-order and most important structural features of the Alpine chain. The work of Termier is a world-wide revolution in the field of tectonics. Indeed, it opens the way for rapidly increasing attempts to synthesize the structure of the entire chain in cross-sections.

Following the publications of Termier, several cross-sections showing the simplified nappe structure of the Alps were published (e.g. Steinmann (1906) for the Eastern Alps). Hans Schardt (1906; Figure 1.17) constructed three cross-sections to compare the orogen structure along the Central Alps, the NW Alps and the Western Alps on the 1:650,000 scale. The sections are interpreted down to a depth of –7,000 m and locally up to over 10,000 m. Alpine nappes and structures in these sections are shown with unprecedented detail on the scale of the orogen. Some areas are very precisely depicted, especially the cover nappes of the northern part of the Chain.

Other parts are not yet geometrically well constrained, such as the Southern Alps and to a certain degree the Lepontine nappe stack. However, for the very first time, the entire Alpine crust is shown to consist everywhere (except in the Southern Alps) of nappe piles, and the specific geometry of these nappes is also constructed and recognizable, even in the basement. Structures illustrated at that time are largely derived from sparse observations and fantasy, especially in the basement (Schmidt 1907). Some thrusts forming the base of nappes are projected upward to heights larger than 10,000 m (Figure 1.17). As later stated by Heim (1908): "Schardt has drawn the first Alpine section in which an entire folded part of the chain is thrust over a younger one". However, where and how these thrusts penetrate into the basement is still not entirely clear (Figure 1.17).

Understanding the structure of the Penninic basement nappes and that of the basement nappes in general is not achieved within a few years; it is the result of a long standing work of mapping, interpretation and re-interpretations (Schardt 1904; Figure 1.18). Schardt (1904) excellently summarizes the history of research in the Simplon area (Western Lepontine dome, Central Alps), where many studies were performed before the construction of the Simplon Tunnel. He reproduces 10 different interpretations performed between 1851 and 1903, showing how the inferred structure of the (almost) same, 20 km long, section, changes throughout the last decades of the 19th century (Figure 1.18). Constructing sections across these basement nappes at the end of the 19th century is no longer a mere reproduction of what the eyes see in the field, otherwise the cross-sections of different authors would be more similar to one another. Schardt (1904) states this clearly when commenting on the previous literature: "One cannot say that Gerlach was ahead of his time". Gerlach saw "what there was to be seen", and in Schardt (1898), commenting on the excellent work of Gillerin: "He only tried to assess with the greatest precision what is really visible". Indeed, the last decades of the 19th century are those where the main concern of geologists is no longer restricted to documenting what the eyes see (e.g. Studer 1834), but also and mainly to show what the mind adds to the eye. This explains why the 10 sections of the same area reproduced by Schardt (1904) are so different from each other (Figure 1.18). Schardt (1904) writes about the section construction of Lugeon (1901; Figure 1.18): “M Lugeon also applies the concept of recumbent folds to the gneiss of the Simplon [...]”. Hence, it is now a matter of conceptual interpretation, and no longer one of direct observations.

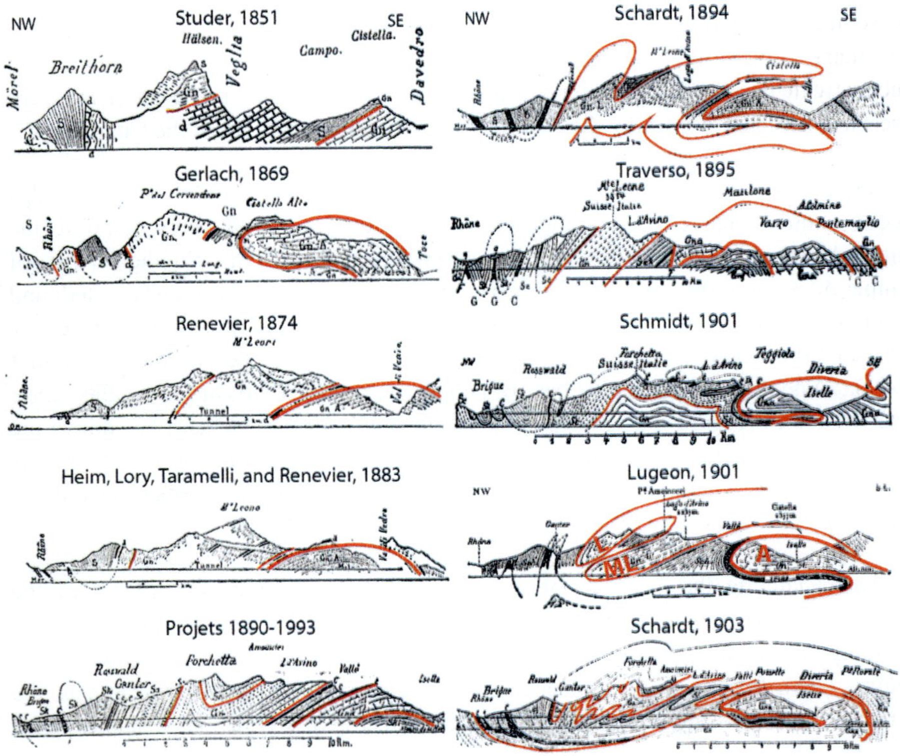

Figure 1.18. *Different interpretations the Simplon area, Western Alps, through the second half of 19th century. Thick red line: boundary of Antigorio gneiss (nappe). Thinner red line: boundary of gneiss units. A: Antigorio nappe; ML: Monte Leone nappe; L: Lebendun nappe. Modified from Schardt (1904)*

Indeed, it is with concepts of fold vergence and nappes in mind that Lugeon (1901; Figure 1.18) “solves” the problem of the Simplon structure, understanding that the basement synforms north of the Antigorio nappe are refolded anticlines, which represent the front of a higher nappe (Monte Leone and Lebendun). Hence, as anticlines, they are like the Antigorio recumbent fold (Figure 1.18) and separated from the latter by recumbent synclines. The result is a pile of gently dipping basement nappes: the first one ever described.

A significant improvement in the drawing of Alpine-scale sections is performed by Heim (1908; Figure 1.19), showing a "schematic attempt of illustrating the Alpine nappe stack along the Säntis-Chiasso line", in the Central Alps. The improvement in these sections is in terms of the geometry of the nappe stack that is more accurate than that of Termier (1903), Schardt (1906) and Schmidt (1907). The geometry of the External Aar and Gotthard massifs is quite similar to the one that is reconstructed in present-day studies (e.g. Schmid et al. 1996). The same is true for the nappe stack of the eastern Lepontine dome, showing very nice details like the back folds of the Suretta nappe (Figure 1.19). All these nappes are overlain by the Austroalpine ones that are vertically rooted in a region of very large N-S extent, between the Tambo nappe in the North and the Southern Alps (M.te Generoso) in the South, as inferred by present-day interpretations. The entire chain, including the Southern Alps, shows a northern vergence, instead of a south-directed one in the Southern Alps, as already pointed out by Termier (1903) and unanimously recognized by modern studies (e.g. Laubscher 1985). In fact, the section of Heim (Figure 1.19) indicates that the Southern Alpine basement represents the root zone of a large number of nappes of the orogen, which were all emplaced by top-to-north displacement. In spite of these inaccuracies, this section is the first to integrate the information of a vast area in map view into one and the same vertical plane, by lateral projection parallel to the main fold axes. In contrast to Termier (1903), the structures projected far above the topographic surface do not only consist of long sub-horizontal lines, but of quite accurately constructed nappe contacts, each with a specific size and geometry.

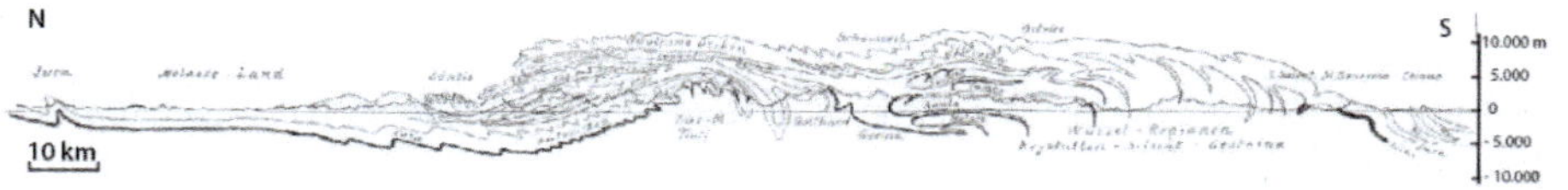

Figure 1.19. *Cross-section of the Central Alps, modified from Heim (1908). The structure of the entire Central Alpine nappe stack is shown here for the first time*

Heim's concept of nappes appears to be similar to that of Termier: they are squeezed out ("*herausgepresst*") from an area consisting of their roots, the latter being subvertical. Interestingly, after having denied their existence at the end of the 19th century, Heim states that the idea of the nappes

> is neither a hypothesis nor a theory, but a summary of observations. The theory, i.e. the true mechanical explanation of the processes and their causes, will be later deduced from additional and thorough observations.

The N-vergence of all nappes including the Southern Alps in the section of Heim (Figure 1.19) probably results from the well-constrained, north-verging fold geometries in the eastern Lepontine, in addition to the following observations reported by Heim: (1) nappes usually show a sedimentary facies that is characteristic of areas located further south compared to the underlying units, whereas the contrary is never true; (2) the small-scale geometry of the folds inside the nappes generally suggests a top-to-the-north shear sense; (3) if most nappes are displaced northward, it is mechanically very unlikely that some of them moved southward.

In 1909, Schardt presented several sections striking across most of the Western Alps (his Figure 4). Schardt shows the nappe pile, correlating their gently dipping parts in the west with their vertical roots in the east. As in the case of Heim (1908) for the Central Alps, the Internal Zone is now shown with its nappe structure, with roots remaining vertical down to 10,000 m and no longer showing their folded base close to the surface, as in Termier (1903), or a distributed basement folding as in Schmidt (1907).

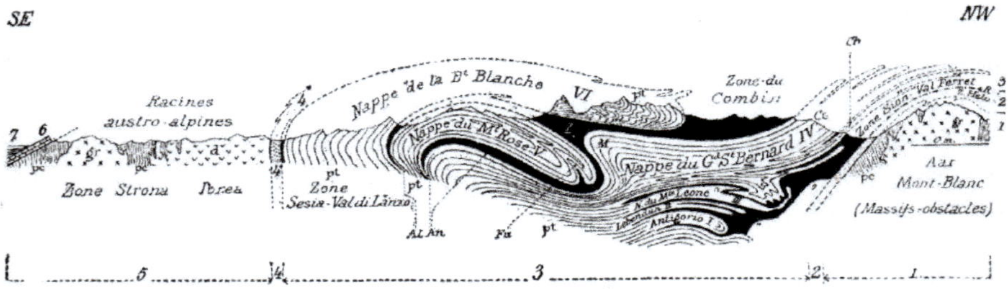

Figure 1.20. *Schematic cross-section of Argand (1909), of the Pennine Alps, from Aar/Mont Blanc Massif to Ivrea-Strona zones, through the I-VI Pennine fold-nappes*

In the same year, Argand published his first orogen-scale, "schematic" section (Figure 1.20), striking from the Ivrea-Strona Zone in Piedmont (Italy) to the External Massifs (Aar-Mt Blanc), and crossing the Matterhorn and Monte Rosa. Although the topography of the Matterhorn is clearly recognizable, his unifying term Mont Blanc-Aar Massif given to what he calls the "Massif-obstacle" at the NW margin of the section shows that Argand constructed a general external massif and not a specific one. The scale of the section is not provided, but based on the topographic features the entire section appears to represent a vertical distance of nearly 20 km, hence significantly more than previous sections. The nappe geometries of Argand (1909) are very accurate and do not differ significantly from his later sections and the ones of the present-day literature.

Argand (1909) discusses the "déroulement" (retro-deformation) of the cross-section, without yet showing it. His reasoning is in terms of paleo-geographic position of the nappes based on their structural levels in the constructed section, that is, the higher the structural level, the farther southeastward the corresponding unit before Alpine stacking. His approach is very quantitative for his time: he estimates 50 km displacement along the base of the Dent Blanche nappe, and in 1916, he even explains that retro-deformation can be performed by volume balancing.

In 1911, Argand shows a block diagram of the Northwestern Alps (Alpes Pennines, from the Simplon-Tessin dome to the Dent Blanche nappe), showing a tectonic map at the surface and a NW-SE cross-section on the side (Figure 1.21). The section is largely constructed by projecting nappe contacts from different parts of the map into the vertical plane of the cross-section, parallel to the fold axes that dip gently southwestward and quite homogeneously in the latter area. As a result, the geometries and depths of the structures in cross-section are the result of the projection of numerous surface structures that are not lying in the same vertical plane. The section is "the best representative" of this broad area, but not the exact (precise) representation of the geology along a specific vertical plane. Lateral shifts of the section trace in this area would not show any changes in the cross-section.

In the same year, Ampferer and Hammer (1911; Figure 1.22(a)) depicted the entire structure of the Eastern Alps, along an N-S section. Their paper has the explicit title: "Cross-section through the Eastern Alps, from the Garda Lake to the Allgäuer Alps" (translated from the original German). It is 183 pages long and its orogen-scale section, in color, documents in detail some 200 km horizontal length and a vertical extent usually attaining a maximum of a few 100 m only. As a result, the proportion of this cross-section, which is presented in a fold out of 11 pages all horizontally bound together, is rather unusual. Their very detailed legend consists of 58 lithological and stratigraphic units, ordered by age. No attempt is performed to simplify them into a smaller number of tectonic units. No correlations, aimed to provide a geometrical and structural continuity between one part of the section and the other, are shown.

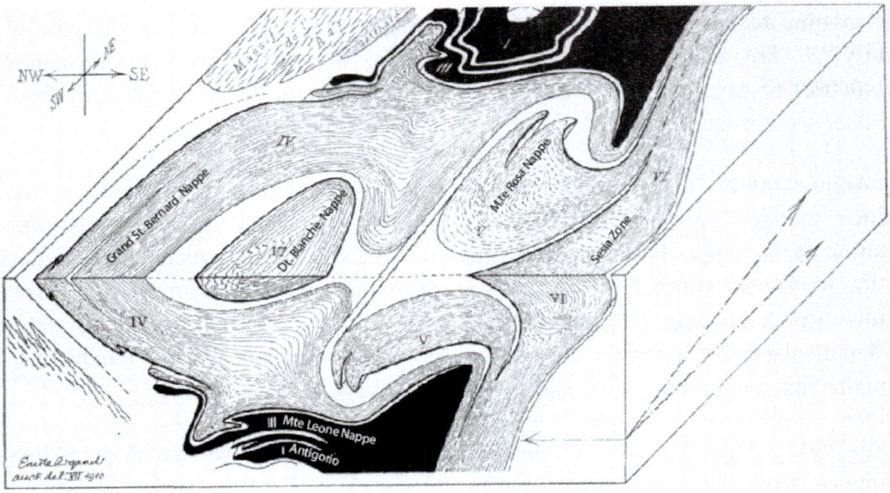

Figure 1.21. *Block diagram of the Western Alps, in which the Lower Penninic fold-nappes (I–III, black), exposed in the Tessin dome, are projected downward to the west, parallel to fold axes; modified from Argand (1911)*

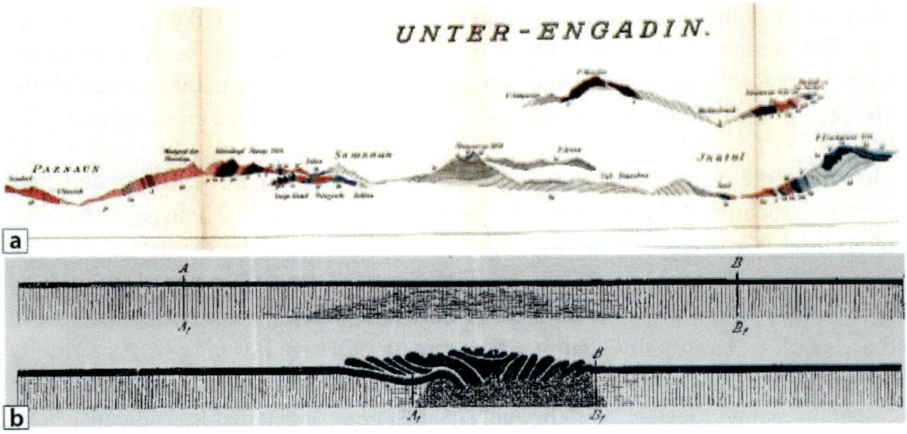

Figure 1.22. *a) Segment of cross-section through the Eastern Alps, after Ampferer and Hammer (1911). b) Schematic section illustrating how shortening accommodated by nappe stacking in the upper crust needs to be compensated by shortening ("Verschluckung") at depth (see the text for explanation)*

This section does not attempt to achieve a synthetic representation as Argand (1909, 1911) did, and underlines that at this time it was an exceptional representation.

The E-Alpine section of Ampferer and Hammer (1911) does not provide newer insight into the structure of a mountain chain. However, in addition to this section, their paper shows some extremely schematic drawings that do synthesize the structure of the Alps, but to such a degree that, geometrically speaking, nothing of the sections described in their long manuscript can be recognized in the latter pictures (Figure 1.22(b)). These schematic sections, which do not appear to be more specifically related to the Alps than to any another orogen, are the base for the first discussion on the quantitative relationship between shortening in the upper and lower (middle?) crust. Ampferer and Hammer (1911) infer that shortening in the cover, as can be deduced from the cross-sections, is by far more important than in the basement:

> If we imagine the nappe of the younger strata retro-deformed to its original position, then we get a 2-3 times wider band as that of the retro-deformed young basement folds, even if the thickness of the basement beds is considered as very small.

As a result, Ampferer and Hammer (1911) show and discuss for the first time that a nappe pile in the upper crust may possibly be compensated by the disappearance of middle and lower crust at deeper levels ("mass suction towards depth"; "absorption of the deeper zones"; "huge sinking"). This discussion allows the authors to use for the first time the term "Verschluckungszone" (swallowing zone), which is classically considered to be the ancestor of the term subduction zone.

While these brave attempts to depict the structure of the entire Alpine chain within cross-sections are making progress, the construction of detailed and precise sections of specific segments of the Alpine Chain continues to be remarkably refined. Without being able to cite them all, we should recall those of Buxtorf (1908) for the Jura Mts, Sander (1911) for the Tauern Window, Lugeon (1914) for the Aar Massif and its cover, Dal Piaz (1912) for the eastern Southern Alps, Heim (1919) and Staub (quoted in Heim 1922) for the Eastern Central Alps and Western Eastern Alps.

1.6. Argand and the invention of crustal plate tectonics (1916–1924)

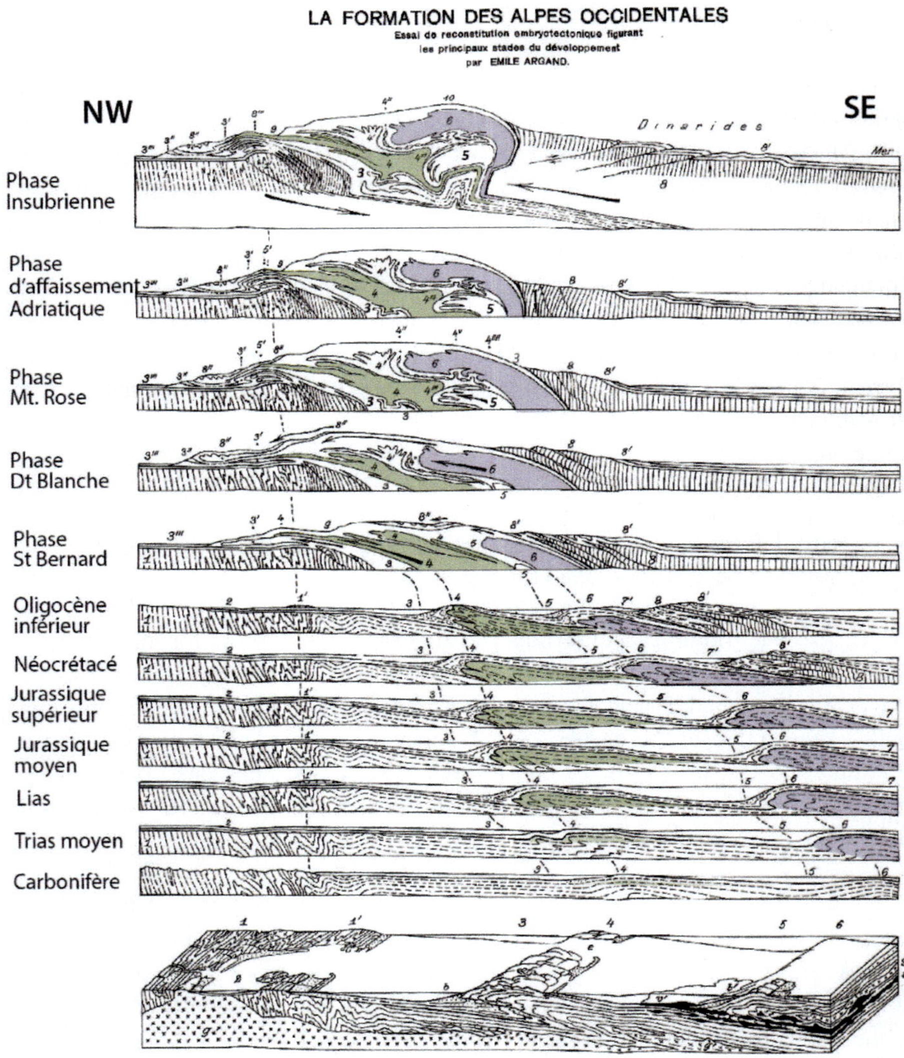

Figure 1.23. *Cross-section through the Western Alps and its retro-deformation, modified from Argand (1916). Two nappes are colored to facilitate comparison of their geometric evolution through time. The purple nappe (6) corresponds to Sesia-Dt Blanche, and the green nappe (4) corresponds to the St Bernhard nappes*

In 1916, Argand published his seminal paper "Sur l'arc des Alpes Occidentales", describing the geometry of the first-order structures of the Western Alps and interpreting both their kinematics and mechanics. Without giving any arguments on this subject, Argand explains in the first paragraph of his article that the structure and orientation of mountain chains depend on three major factors: first, on the orientation of pre-existing obstacles: the hercynian basement, forming the External Massifs, termed the hercynian hemicycle; second, on the rheological characteristics of the nappes (their "plasticity"); and third, on the intensity and duration of the applied "effort" (the force of convergence), which influences the plasticity. A fourth factor, but of subsidiary importance, is the angle of convergence between the continental blocks.

For the first time in the history of Alpine Geology, Argand performed a retro-deformation of a cross-section (Figure 1.23), and of the orogen in map view (or at least a reference line of the orogen), as he consciously notes: "the evolution of a tectonic object has never been depicted in such a way, which combines its large size and complexity". Besides the great quality of his construction, his section (upper panel of Figure 1.23, "Phase Insubrienne") brings major advancements in the understanding of Alpine Tectonics: (i) The "Dinaric" (South-Alpine) block only shows modest shortening accommodated by south-verging thrusts, which form in the most recent deformation phase ("Insubric"). As a whole, due to the very small offsets of the thrusts, it has a rather rigid behaviour compared to the Penninic nappe stack that accommodates shortening by ductile folding and adapts to the shape of the "Dinaric" block. This is what modern literature describes as indentation of the upper plate (Dinaric block) during Alpine collision. (ii) Between Europe and the Dinaric block, a nappe stack forms, characterized by very ductile, mainly isoclinal, recumbent fold nappes. In the later stages of tectonic evolution, these nappes become folded, thus accommodating additional convergence between the European and the Dinaric blocks. These structures correspond to what is termed "post-nappe folds" (Milnes 1974) and inferred to represent syn-collisional shortening that overprints the subduction nappe stack. Most of this shortening, in the structural evolution sketched by Argand, is accommodated in the immediate vicinity of the indenter (Dinaric block), where the nappes become verticalized and thinned by vertical stretching. This area was classically defined as the root zone of the nappes (e.g. Termier 1903). In contrast to Bertrand (1884) and Termier (1903), Argand shows for the first time that the steeply oriented roots of the nappes are not a primary structure, but rather the result of late refolding of previously formed, gently dipping nappes. (iii) The lowermost unit of the nappe stack, immediately above the Europe-derived basement (number 3 in Figure 1.23), forms the core of an orogen-scale antiform and continues to the southeast, below the lowermost Dinaric Units. This unit 3 corresponds to the lower crust of the Penninic zone that is shown to underthrust (subduct?) below the Dinaric block. The same result is shown by seismic interpretations of the nineties (e.g. Polino et al. 1990; Schmid et al. 1996), and corroborated by geophysical studies in the past 20 years. It is amazing that Argand is able to infer such a structure without any geophysical data or the knowledge of subduction. (iv) The external massifs represent antiformal basement culminations that result from high angle, N-vergent basement thrusts, as suggested by modern reconstructions (e.g. Boutoux et al. 2016), except for some minor gently dipping thrusts in the northern flank of the massif. (v) Two arrows pointing in opposite directions (upper panel of Figure 1.23) indicate that the southeastern "Dinaric" block moved toward the NW on top of the European block. An arrow pointing to the SE is depicted on the European block. The arrows are subparallel to the main contact between these two blocks, and they are drawn on the least deformed parts of the two continents, clearly indicating the movement of the entire continental blocks, not the one of individual nappes (see also Figure 1.24). This is the very first illustration of one continental margin thrust on top of the other as a result of convergence. From a kinematic point of view, the youngest stages depicted in the retro-deformed sections of Argand (Figure 1.23) are not significantly different from those that are still published since the discovery of Plate Tectonics.

The structural evolution of the Alpine nappe stack of Argand shows that nappes represent the amplification of recumbent folds that nucleate on preexisting, hercynian "ge-anticlines". Stretching of their lower, reverse limbs leads to the formation of discrete thrusts, hence of fold nappes in their hanging walls (e.g. evolution of unit 6, Dent Blanche in Figure 1.23). This evolution is very different compared to that suggested by Bertrand (1884), Termier (1903) and Schardt (1909). The latter authors consider a nappe as the lateral flow of vertically folded units due to the drag of orogen-scale thrust ("traîneau écraseur") at the top of the folded area, which is consequently termed the root zone of the nappe. It is also the first time that displacement, implicit in the geometry of the cover nappes, is somehow balanced with respect to displacements accommodated within their basement.

After 1916, Argand does not construct any new Alpine cross-sections. He does not add more details to his section or construct any new ones in other parts of the Alpine chain. This task is continued by many other Alpine geologists like Heim (1919), Staub (quoted in Heim 1922) and Lugeon (1932). Heim (1919) constructs an accurate section of the Penninic nappe pile, including the base of the Austroalpine, up to 20 km height. In this section and even more in those of Staub (1924) in the Eastern Alps, the influence of Argand (1911, 1916) is very visible, maybe even too visible for Argand's taste (see Sengör and Bernoulli (2011, pp. 903–904)).

In 1924, Argand published his seminal work ("La tectonique de l'Asie"). As stated by Lugeon (1940), the book title does not fit its content: "One imagines that it will only be about that old continent, whereas it is about the entire Earth". Indeed, Argand (1924) shows there a simplified and more schematic version (Figure 1.24) of his Alpine cross-section of 1916. The only structural elements that are still recognizable are the backfold of the Monte Rosa nappe and the smaller amplitude antiform of the external massif. The major changes compared to 1916 consist of the addition of the "sima" to the "sal"-crust and the inclusion of a discontinuous horizon of basic rocks inferred to be related to the "sima" all along the contact between Penninic and underlying European units (Figure 1.24). These sima-derived lenses correspond to what is presently termed ophiolitic units. Argand shows that they exist in all orogens, and that they are always located in a similar position along the interface of converging continents (Argand 1924). Both the over- (Austroalpine nappes) and underthrust (European nappes) margins are shown to thin out severely before wedging out completely (Figure 1.24), as inferred today from seismic imaging of continental margins (e.g. Boillot et al. 1995). The schematic character of this section of Argand shows that, in this phase of his research, he no longer tries to define the precise anatomy of the Alpine orogen, but rather to compare it with other Cenozoic orogens to understand what the first-order similarities are, and hence the general process of orogeny. Based on his sections (Argand 1924 and Figure 1.24), these similarities are as follows: (1) thrusting of one, thinned continental margin on top of the other, similar to a "flat subduction" of one margin; (2) intrusion of sima from the mantle along the plate interface; and (3) folding of a nappe stack between continental blocks.

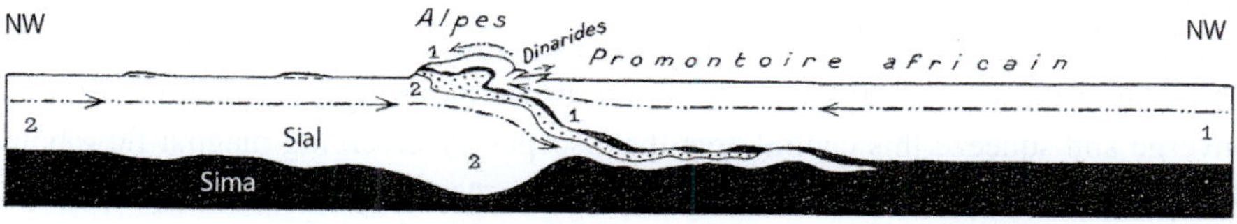

Fig. 24, Argand 1924

Figure 1.24. *Cross-section of the Alps, modified from Argand (1924). The section is reduced to its first-order structures and taken from a series of figures used by Argand to compare sections of several orogens, each of them showing only the very first-order, orogen-scale structures: number 1 corresponds to the "Promontoire Africain", Adria in the present-day terminology. Number 2 corresponds to Europe*

1.7. Alpine cross-sections after Argand and before Plate Tectonics (1924–1960)

In 1934, for the 50th celebration of the Swiss Geological Society, a geological guide of Switzerland was published, in which Lugeon presented a series of five orogen-scale sections (one of which is shown in Figure 1.25(b)) through the Swiss Alps, all constructed and redrawn down to –10,000 m, after Buxtorf, Gagnebin, Alb, Heim, Oberholzer, Preiswerk and Staub. These sections nicely show the nappe structure of both basement and cover, differentiating the European, Penninic and Dinaric basements. Almost 50 years later, in the classical book "Geology of Switzerland" (Trümpy et al. 1980), a major reference on the Alpine chain for decades, three orogen-scale cross-sections were constructed by Spicher (Figure 1.25(a)). These sections are interpreted down to –7,000 m and no significant differences can be assessed compared with those of Lugeon (1934). The time interval between 1930 and 1980 does not bring any significant improvements to Alpine orogen-scale sections. The structure of the chain has been well known since the 1920s and additional studies only refine it to the second order.

An exception to the latter trend comes from very few scientists that start to imagine and image the deeper parts of the chain (Heim 1927; Kraus 1932, 1954; Kober 1942, 1955), but their illustrations are rather schematic, lacking both physically sound models to assess what may be expected at depth and more importantly, geophysical data to provide robust constraints on such deeper structures. Heim (1927) attempts to include the new knowledge derived from gravity surveys into a schematic section showing a subsided nappe pile, due to the effect of isostatic compensation. Kober (1942; Figure 1.26(a)) shows that the Alpine orogen is underlain by a stratified area going from magma at the base of the crust, to migmatite in the middle crust and to granite in the upper crust. Africa and Europe converge and squeeze this central area that escapes by diverging magma flow both laterally at depth and upward, where it leads to the crystallization of granites.

Kober (1955) modifies the latter models showing a series of three sections across the Western, Central and Eastern Alps, interpreted down to 70 km depth. The latter section is shown in Figure 1.26(b). In contrast to Kober (1942), this section is simplified, and schematic, but strongly inspired by the known structure of the Alpine nappe stack. The novelty concerns the deeper part of the section, imaging the 30–70 km depth range, and showing how northern and southern thick continental blocks (from 5 to 35 km depth below the foreland) gently and symmetrically bend below the nappes, overlying two thick zones of magma that also bend below the orogen.

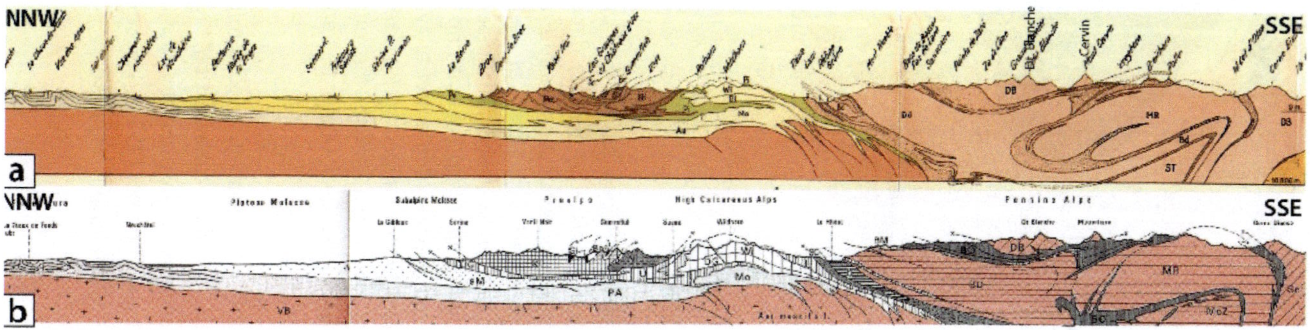

Figure 1.25. *Cross-sections of the NW Alps. a) Section constructed and compiled by Lugeon (1934) on the base of sections of Arbenz, Argand, Buxtorf, Gagnebin, Heim, Lugeon, Oberholzer, Preiswerk and Staub. b) Section following exactly the same trace of upper section a). After Spicher, in Trümpy (1980)*

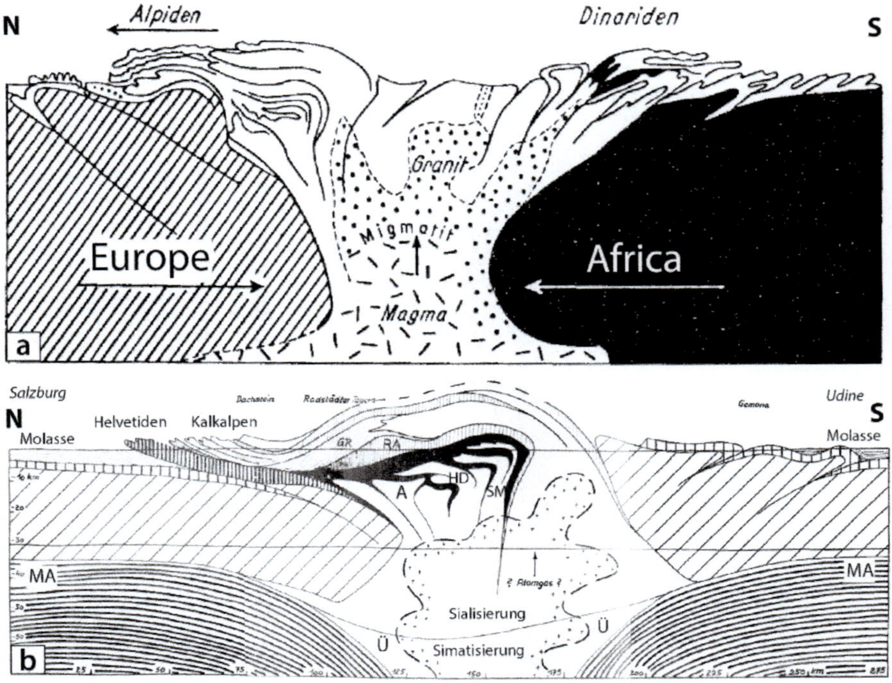

Figure 1.26. *Speculative models of the deep structure of the Alpine orogen in the time between Argand and Plate Tectonics. (a) Section through the Alpine orogen, after Kober (1940). The convergence between Africa and Europe is accommodated within a very large area characterized by magma at depth and a transition to migmatite and granite toward higher crustal levels. (b) Eastern Alpine cross-section, after Kober (1954). The crust (ca. 35 km thick) floats on an inferred "Magma Zone" (MA in b) that symmetrically warps downward in a sort of "Verschluckungungszone" of Ampferer type. The asymmetry of the nappe stack corresponds to an apparent plunge of Europe below Adria, which only affects the crust. A: Ankogel nappe; HD: Hochalm nappe; SM: Sonnblick–Modereck Dome; Ü: transition zone*

Neither the magma zones nor the continental blocks touch each other, because they are separated, respectively, by the "Simatization" and "Sialization" zones (Figure 1.26(b)). This cross-section is inspired by the conceptual models of Ampferer (1906) and Ampferer and Hammer (1911), rather than by observations of nappe structures, which are only schematically depicted.

1.8. Alpine cross-sections after Plate Tectonics

1.8.1. *Geophysical interpretations*

The 20th century brought a wealth of new geophysical data that was integrated in cross-sections. At the beginning of the 20th century, gravity measurements were performed in Switzerland (e.g. Niethammer 1904). The resulting mass deficit is already discussed by Heim (1921), who refrains from interpreting it in terms of differences in sial versus sima composition of the magma, inferred to lie at depth, below the chain. Heim explains that it is the result of thickening the crust by folding (Heim 1927). In the following decades, gravity anomalies are interpreted on the basis of different hypotheses concerning the density of rocks at depth, hence an image of the inferred "Sial/Sima" (crust/mantle) boundary below the Alps, showing its synformal structure with a maximum depth between 30 and 50 km, is provided (Gassman and Prosen 1948). From the end of the 1950s onward, seismic experiments (wide angle reflection) provide images of the deep structure of the Alpine orogen (e.g. Closs and Labrouste 1963; Berkhemer 1968; Giese 1968; Choudhury et al. 1971) and innovative Alpine sections are published by the geophysical community, interpreting the geometry and depth of the Moho in cross-sections (Figure 1.27). Contour maps of the Moho (e.g. Berkhemer 1968; Labrouste et al. 1968; Choudhury et al. 1971) and iso-velocity lines of the crust in cross-sections (e.g. Giese 1967, 1968; Giese et al. 1970; Choudhury et al. 1971; Perrier 1973; Müller et al. 1976) widen the image of the orogen from approximately 20 (e.g. Argand 1916) to over 60 km depth (e.g. Giese et al. 1970). Serial, radial sections (Giese 1968; Choudhury et al. 1971; Figure 1.27) illustrate for the first time the superposition of two Mohos in the Western Alps, or, more precisely, the superposition of two crust-mantle transition zones. The proper mantle, characterized by the 8 km/s velocity, is still shown as one continuous surface (Figure 1.27).

In spite of some differences in the absolute values of the Moho depth, most models show it as a single and continuous surface (Figure 1.28), which passes from a depth of approximately 30 km below the external massifs to almost 50 km in the Internal Zone (e.g. Berkhemer 1968) or almost 60 km below the Lepontine (Müller et al. 1976). Some models show that the Moho surface is folded by two open synforms, the eastern one located below the Ivrea body (Berkhemer 1968; Miller et al. 1982; Figure 1.28(a)), or by a single synform whose hinge, corresponding to the thickest part of the Alpine crust, is located below the Lepontine in the Central Alps and below the "Internal Zone" in the Western Alps (Giese 1968; Giese et al. 1973; Müller et al. 1976; Müller et al. 1980; Müller 1984; Figure 1.28(b)).

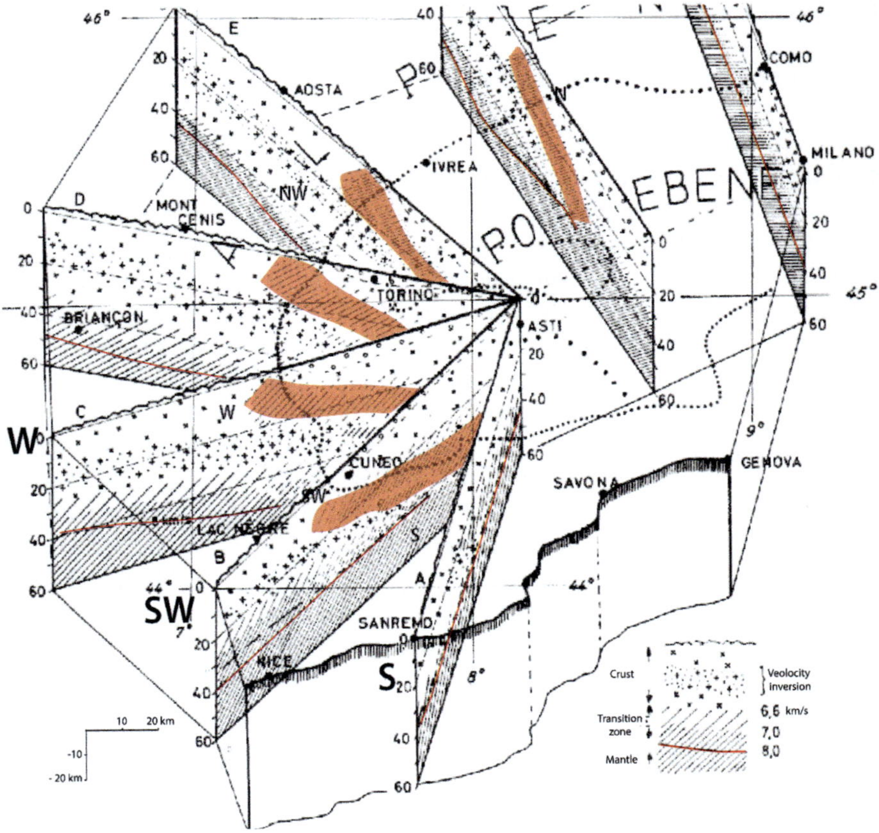

Figure 1.27. *Block diagram with cross-sections perpendicular to the arc of the Western Alps, showing the geometry of iso-velocity lines of seismic wave propagation. Orange shading: upper transitional zone. Modified from Giese (1968)*

Giese (1968) and Giese et al. (1968) showed for the first time in the Western Alps that the Ivrea body at depth represents the transition between lower crust and mantle that it is continuously linked to the Alpine (Adriatic?) mantle located at depth, further east. However, their serial sections (Figure 1.27) still show a single, continuous Moho surface below the Alps, from which the Ivrea slice penetrates into the crust. Giese (1967) notes that the Ivrea body lies above sialic rocks, which he thus interprets as a "tectonic slice or an intrusive body" (later described as a mafic-ultramafic body based on petrographic observations).

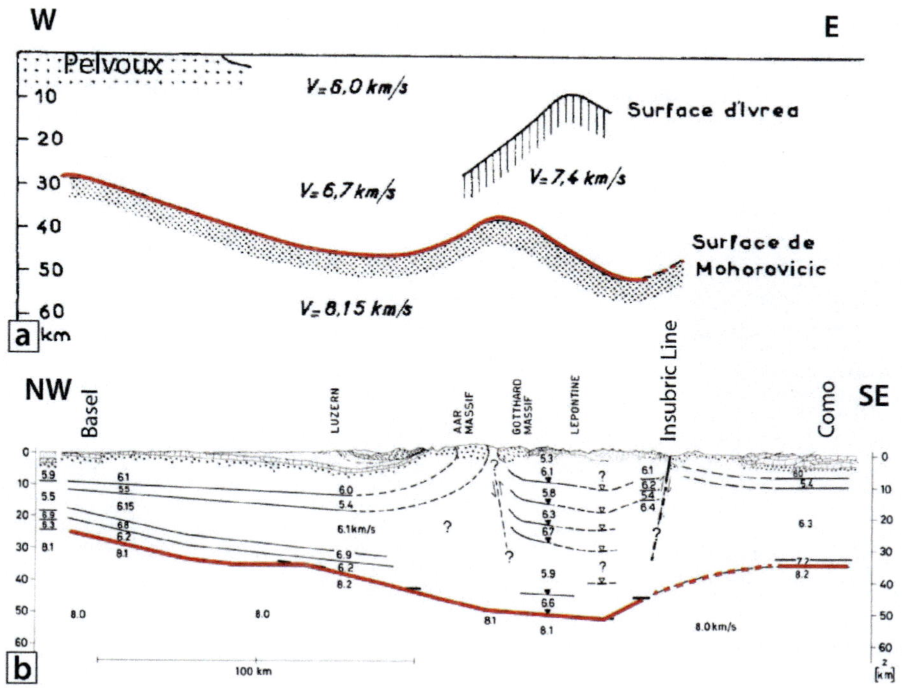

Figure 1.28. *Alpine Moho across the Alps. a) Section across the Pelvoux External Massif, showing a continuous Moho surface building two synforms, and a high velocity anomaly at higher crustal level termed the Ivrea surface. Modified from Berkhemer et al. (1967). b) Section across the Central Alps, illustrating the geometry of iso-velocity lines in the crust, and the Moho surface. A single Moho (red line) forms a large synformal structure whose hinge is located below the Lepontine Dome. Modified from Müller et al. (1980)*

The cross-section of the Western Alps of Giese et al. (1970; Figure 1.29) is the first one to integrate geological and geophysical data in a single section. The superposition of two distinct crust-mantle transition zones is still visible. The Ivrea body is shown as the termination of a larger, gently E-dipping body (Figure 1.29) consisting of crust-mantle transitional units, and this body lies above and is subparallel to the European Moho, even very far to the ESE (above the Monferrato, hence above the Apennines frontal thrust). This superposition is explicitly interpreted by Giese et al. (1970) as a thrust of the Adriatic mantle and overlying crust on top of the Penninic crust: “The Southern Alps-Po Plain with its crust and (crust-mantle) transition zone, and even with parts of its upper mantle must have been thrust over the simatic substratum of the Penninic Units". However, the same authors (Choudhury et al. 1971; Giese et al. 1973) continue to show Moho maps of the Alpine realm with continuous contour lines from the European to the Adriatic plates.

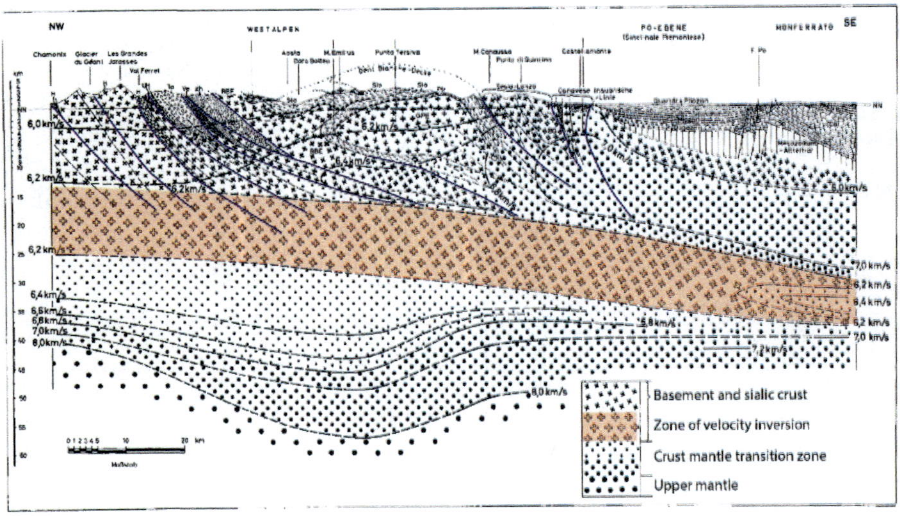

Figure 1.29. *Cross-section of the Western Alps integrating geological, tectonic and seismic data, modified from Giese et al. (1970). Note that major thrusts (blue lines) all terminate in the upper part of the "Zone of velocity inversion"*

Giese (1978) and Giese and Morelli (1978) finally present maps where the European and the Adriatic Mohos are distinguished, and where cross-sections show two distinct Moho's instead of two crust-mantle transition zones (Figure 1.30). One Moho is thrust over the other, leaving a crustal volume in between them, as indicated by the velocity inversion zone. The base of the upper mantle dips eastward and steepens upward, where it joins the Insubric Line. More detailed cross-sections and Moho maps are shown by Giese et al. (1982). They state that "it is unrealistic to assume a crust/mantle boundary running through an Alpine orogen without any interruptions", meaning that there must be a thrust plane (subduction interface in modern terms) that accommodates shortening within the mantle. In spite of the 20 years that passed since the formulation and first acceptance of Plate Tectonic processes, these figures and ideas are still in contrast with those of most other groups working on the deep structure of the Alps (e.g. Müller 1980; Miller et al. 1982) that suggest a continuous and rather symmetric synformal deflection of the Moho below the Alps (Figure 1.28(b)). The latter studies relate the symmetry of the Moho structure to an inferred symmetric type of subduction, as previously postulated by Ampferer (1906), Ampferer and Hammer (1911) and Laubscher (1969).

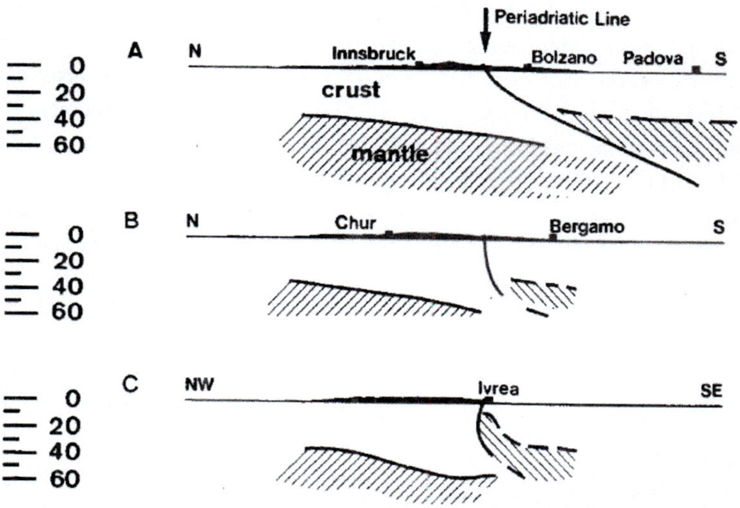

Figure 1.30. *Cross-sections showing the crust-mantle boundaries in the Eastern (A), Central (B), and Western Alps (C), after Giese (1978). Oblique-line pattern: mantle*

1.8.2. *Geological interpretations*

While the geophysical community tried to assess the deep structure of the Alpine crust-mantle transition based on the newest results of gravimetric and explosion seismic studies, geologists attempted to image the deep structure of the Alps in cross-sections that integrate the new concepts of orogeny based on Plate Tectonics. These studies resulted in lithospheric-scale cross-sections that image the Alps down to several hundreds of kilometer depth (Laubscher 1970, 1974, 1983; Dal Piaz et al. 1972; Hawkesworth et al. 1975; Dietrich 1976; Figures 1.31 and 1.32), thereby reducing the details of the crustal nappe stacks to only few schematic lines (e.g. Laubscher 1970, 1971, 1974, 1983; Hawkesworth et al. 1975; Dietrich 1976). The basis for all these sections is more conceptual than observational, the concept being that of subduction.

Laubscher (1970; Figure 1.31) was the first to sketch the evolution of the Alpine orogen, in which convergence of two plates is accommodated by subduction. It is a symmetric subduction of both plates, recalling the concepts of Ampferer (1906) and Hess (1962), and terminating by a break off of the lithospheric root. These schematic sketches show for the first time how the Alpine crust detaches from the subducting lithosphere, thus becoming accreted to the orogen. The paper illustrates the lithospheric-scale, structural evolution of the Alps in a series of cross-sections, from the Jurassic to the Present. After the kinematic evolution reconstructed by Argand and the similar ones (e.g. Staub 1924) that were inspired by him, Laubscher (1970) is the first to re-propose the structural evolution of the Alps, in a series of time slices, based on the new conceptual model of Plate Tectonics. Laubscher (1970) is conscious of the speculative character of these cross-sections that are based on a theory, not on observations. He states:

> [...] I left the subduction zone in the lithosphere in the south, without changes (compared to previous stages); there are no specific reasons for that; a shift to the north, for example through rolling of the plate hinges is absolutely possible.

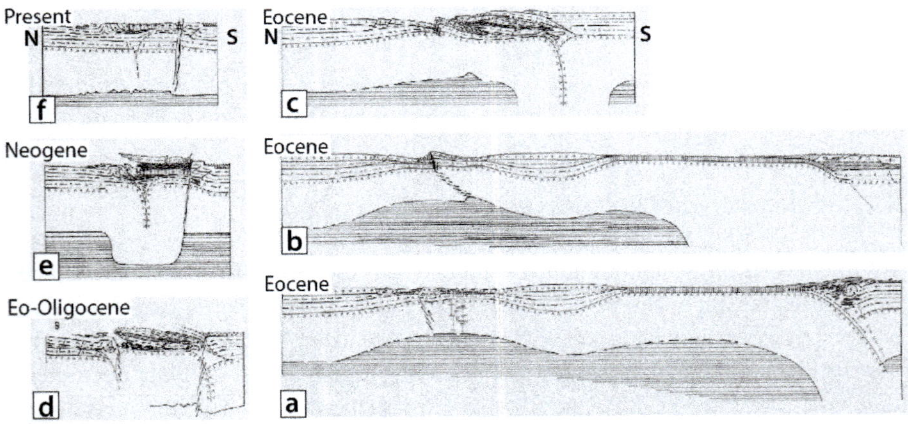

Figure 1.31. *Cross-sections illustrating the deep structure of the Alpine lithosphere and its evolution through time (from a to f). Modified from Laubscher (1970)*

Hence, the subduction zone shown in Figure 1.31 is drawn in one specific place, but it could just as well be elsewhere below the orogen. The model is well ahead of the geophysical data that could constrain it.

The theory of Plate Tectonics applied to the Alps also provides a refinement to the nappe concept. In the reconstruction of Argand, nappes result from the amplification of recumbent anticlines (Figure 1.23) and the entire crust is affected by these folds. In contrast, the "Plate Tectonics definition" of a nappe given by Laubscher (1970) is that an (Alpine) nappe is a crustal unit that detaches from a subducting lithospheric plate and becomes accreted to the orogenic wedge, building it at the same time.

Although the different phases of Argand are linked very strictly by geometric continuity, those of Laubscher are not. His concern is no longer the geometrical evolution of the nappes themselves, but the link between the nappes and the deeper lithospheric structure through time. The latter work focuses on the location of subducted Alpine slabs in relationship to the zones of deformation affecting the crust. Although the slabs represent a mere concept not yet physically linked to geophysical observations, deformation of the crust is well constrained at that time. The scarce data on the geometry of the Moho and the deeper mantle creates lots of space for interpretations and speculations. Laubscher (1970; Figure 1.31) discusses and sketches different possible locations for the inferred Cretaceous subduction below the Alps. The "late" localization of shortening below the External Massifs being well constrained, Laubscher suggests to relate it to a later, also symmetric subduction zone that develops below these massifs in the Neogene, as intracontinental European subduction (Figure 1.31(e)). A breakoff of the lithospheric root leading to a general isostatic uplift is also postulated by Laubscher (1970; Figures 1.31(e) and (f)). These concepts are amazingly modern. However, the cross-section representing the present-day structure (Figure 1.31(f)) becomes schematic to a degree that the Alpine structure cannot be recognized any more.

Following the first attempts of Laubscher to reconstruct the evolution of the Alpine lithosphere in cross-sections within the framework of Plate Tectonics, a new approach is undertaken by Dal Piaz et al. (1972; Figure 1.32) on the crustal scale. Their fundamental step forward is the ability to sketch specific nappes (Sesia-Lanzo and Monte Rosa) and show their kinematic evolution through subduction and collision. Their paper links the structure, the high-pressure metamorphic assemblages of these nappes and the radiometric ages based on the first age dating of that time (e.g. Hunziker 1970). In spite of the unprecise ages, setting these different data sets in the conceptual frame of Plate Tectonics, allows them to show for the first time the evolution of the western Alpine nappes throughout subduction and exhumation in cross-section (Figure 1.32).

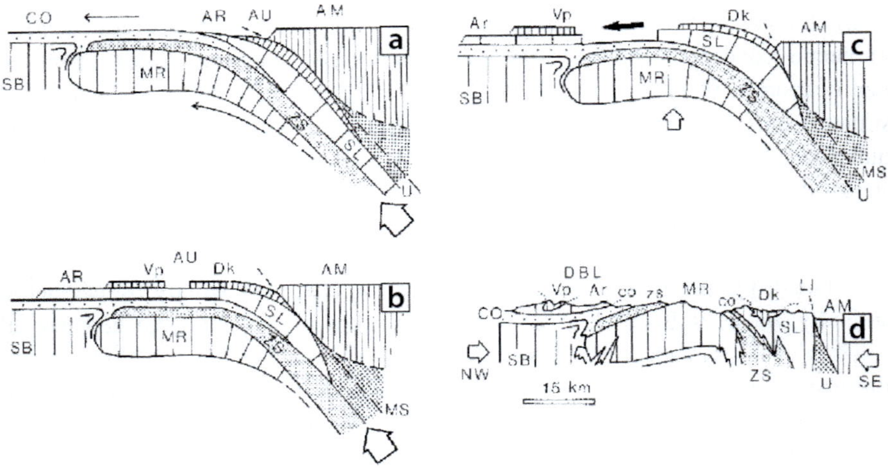

Figure 1.32. *Cross-sections illustrating the kinematic evolution of the western Alpine nappes through subduction and collision, modified from Dal Piaz et al. (1972). a) Subduction; b) exhumation; (c and d) collision. AM: Southern Alps; Co: Combin Zone; DBL: Dent Blanche Nappe; MR: Monte Rosa nappe; SL: Sesia Lanzo; ZS: Zermatt-Saas Zone*

The mismatch between the clarity of the Plate Tectonic theory and the lack of geophysical data providing insight into the deep structure of the Alps characterizes most of the papers that attempt to show the evolution of the Alpine lithosphere in the 1970s (Figure 1.33). A paper by Mattauer and Tapponnier (1978), explicitly about Plate Tectonics in the Alps, is very emblematic for that period. The authors sketch the evolution of the Western Alps through time, starting with Cretaceous subduction, based on a series of lithospheric cross-sections (Figure 1.33(b)). These show the subduction of oceanic crust, its partial exhumation by nappe stacking and erosion, the folding of these nappes after stacking, the bi-vergent thrusting of the continental crust, just as in modern interpretations, but they terminate this temporal evolution of the Alps at 40 Ma (Figure 1.33(b)). No lithospheric section of the present-state of the Alps is shown. The sections are not constructed by retro-deforming the present state, as the only one that can be really constrained. Paradoxically, the present state is the only unknown element of the temporal evolution shown by Mattauer and

Tapponnier (1978), in which the past is better constrained than the present thanks to a conceptual model: subduction followed by collision. The paper of Dietrich (1976; Figure 1.33(a)) on subduction-related magmatism of the Eastern Alps also shows the tectonic evolution of the Alps in a series of cross-sections starting at the Permian-Jurassic boundary and ending in the Oligocene. As in the case of Mattauer and Tapponnier (1978), no section of the present state is drawn. The reconstructions of Laubscher (1970, 1974, 1983) may show a section representing the present-day structure of the Chain, but the latter is less constrained than the ones showing the older stages of evolution.

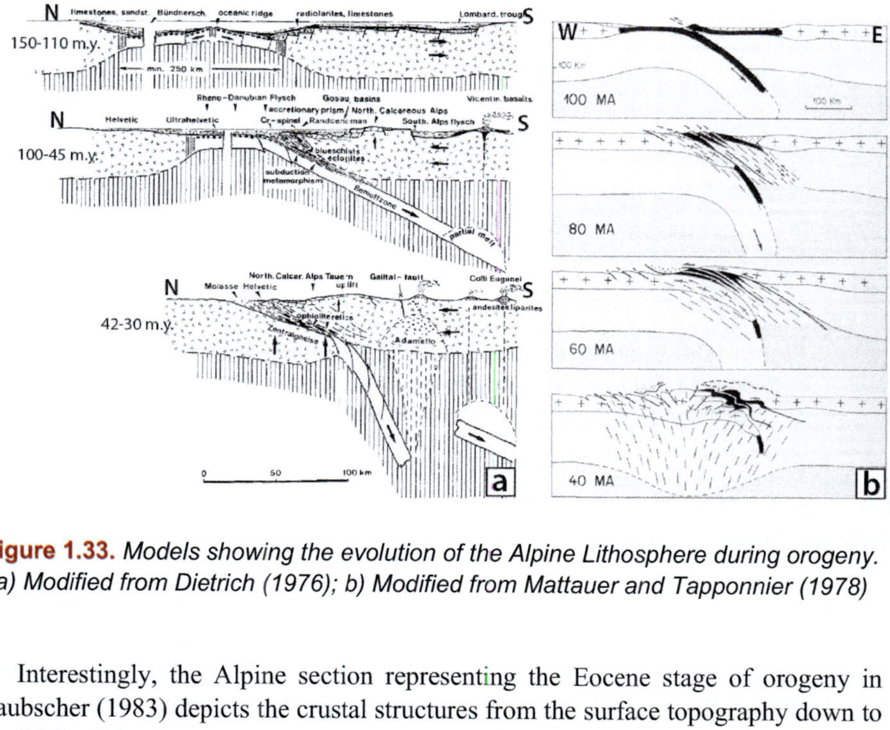

Figure 1.33. *Models showing the evolution of the Alpine Lithosphere during orogeny. a) Modified from Dietrich (1976); b) Modified from Mattauer and Tapponnier (1978)*

Interestingly, the Alpine section representing the Eocene stage of orogeny in Laubscher (1983) depicts the crustal structures from the surface topography down to the Moho (his Figure 16c), whereas the section illustrating the present state of the chain (his Figure 17) only shows the crustal structures of the upper third of the crust and a white surface below, showing that nothing is known about it! Similarly, alternative models to Plate Tectonics of that time (van Bemmelen 1973) also show the deepest part of the lithosphere in cross-sections with great detail throughout the Tertiary, but restrict their illustrations to the upper 8 km of the crust when it comes to the present.

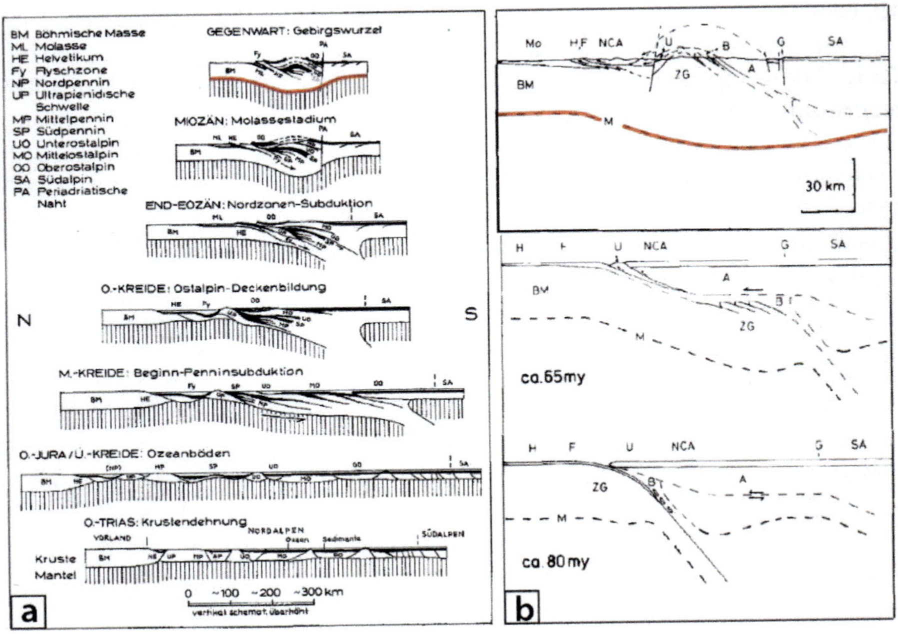

Figure 1.34. *Evolution of the Alpine crust and mantle from subduction to present. Red line: inferred present-day Moho. a) Modified from Tollmann (1980); b) Modified from Hawkesworth et al. (1975)*

1.9. Integrating deep geophysical imaging with surface structures (NFP20; Ecors-CROP)

Another breakthrough in the advancement of tectonic investigations in the Alps is marked by the integration of new geophysical results from the big trans-Alpine seismic campaigns of the beginning of the 1990s (ECORS-CROP in France and Italy, NFP20 in Switzerland). These new data are incorporated in cross-sections, in addition to the surface data (e.g. Nicolas et al. 1990; Pfiffner et al. 1990; Polino et al. 1990; Schmid et al. 1996), thus filling the gap that still exists throughout the 1970s. This step only precedes the one of seismic tomography, which images the "very" deep structure of the orogen, with subducted slabs down to 300 km depth (Lippitsch et al. 2003). The debates and type of imaging initiated by all these works are still ongoing at present.

1.10. Discussion and conclusion

1.10.1. *Changing style of cross-sections as a function of changing interpretations of the Alps*

The historical evolution discussed above traces a strong link between the advancement in the understanding of Alpine tectonics and the changes in the representation style of cross-sections. Why and how do cross-sections evolve from the illustration of the observable to its interpretation? Part of the answer lies in the nature of the object: outcrops and their representation in sections are not complete objects as fossils or plants. They are part of a larger scale entity that is never completely observable, and needs to be constructed. Advancements in geological understanding, hence interpretations, are themselves partly integrated, graphically, in the plane of the sections, requiring a continuous change in the style of representation.

The 18th century shows the evolution from paintings of outcropping sedimentary beds in Alpine landscapes (Marsigli 1705), only differing from artistic paintings by their purpose, to cross-sections that represent what the scientist's eyes are capable of discerning in the natural landscape (Arduino 1758). Which elements of the landscape are represented in the latter sections and which ones are not is based on specific criteria established by the scientist. Only by making such conscious choices, can the contacts between different lithological units be clearly depicted in figures. Symbols are added to the illustrations of natural outcrops to provide the reader with the key to geological interpretation. This change is the very first step toward a continuous increase in abstraction that will characterize the representation of geological structures in cross-sections. Marsigli (1705; Figure 1.1) illustrates a landscape, whereas Arduino (1758; Figure 1.3) illustrates an interpretation of the landscape.

In the first part of the 19th century, the evolution of cross-sections becomes more and more interpretative, in spite of some scruples, manifested by the adjectives "ideal" or "theoretical" that are added to the term cross-section. The example of Studer is emblematic for this approach. His sections of 1834 are either 3D representations showing all details of the 3D topographic surfaces of the outcrops (Figure 1.5), or what he terms "idealized" sections, where 3D effects of the topographic surface are reduced, but still clearly present. Eventually, in 1851 he draws a series of sections that are entirely bi-dimensional, as also performed by Sismonda (1839), Villa and Villa (1844), Lory (1846), Escher von der Linth (1848) and Murchison (1850). Hence, by the beginning of the second half of the 19th century, cross-sections are no longer reproducing landscapes, as observed from a specific point of view, but they are ideal, vertical, planar surfaces, in which the geological units are shown as if they were always continuously outcropping over their entire section, and sedimentary beds are shown with constant thickness. Therefore, although very closely related to the natural valley flanks that they represent, cross-sections no longer illustrate them in the naturalistic sense. This change takes place much earlier in the United Kingdom, where White Watson (Ford 1960) and later John Farey, inspired by the ideas of William Smith on correlating fossil-bearing units over large distances, present detailed maps and sections as early as 1808 (e.g. Ford 1967). These sections show laterally continuous beds in vertically interpreted planes, along more than 100 km length. The large lateral extent of stratigraphic units, as observed during mapping, provides the information to draw such interpreted cross-sections, but more importantly the geological maps allow us to construct cross-sections without drawing them from the visible landscape. Such maps of the Alpine realm only start to be published in the 1860s (Campanino and Polino 2002).

In the second half of the 19th century, when the valley flanks are sketched as if they were uninterruptedly outcropping along 2D, vertical planes, and the inferred geometry of beds both above and below surface topography (e.g. Favre 1867; Giordano 1869; Heim 1878; Bertrand 1884; Bertrand and Ritter 1900) are also represented, cross-sections start to document what is physically not visible in the natural landscape. This graphical change takes place at a time when different tectonic interpretations of the same area start to appear, and as stated by Schardt, we cannot only show what "the eyes see". The wish to clarify long-standing debates finally moves scientists to draw geological contacts below and above the observable topographic surface to show explicitly how different parts of a section are correlated. But it is also the illustration of new concepts on the structure of the Alps that requires the space of the section to be widened: the very first ideas of nappe tectonics need larger vertical dimensions to illustrate the structure of the nappe pile (Giordano 1869; Schardt 1894, 1897; Figures 1.11 and 1.14).

The integration of parallel sections representing different structural levels in one and the same plane (Bertrand and Ritter 1896) is the first step toward the precise illustration of structures that are located well above the surface (Figure 1.13(b)). A different technique is used and explained by Argand at the beginning of the 20th century, with cross-sections that are constructed by lateral projection parallel to fold axes (Argand 1911; Figure 1.21). These sections integrate the geological information of areas that are exposed many kilometers away from the trace of the section itself. Hence, one and the same section represents a large area. It certainly is not the faithful representation of structures on a vertical plane striking along a specific trace, but it is the most representative illustration of the structures of one area, and this over vertical distances (over 20 km) that could not have been attained without such lateral projections.

The evolution of cross-sections is not only characterized by enlarging their upper and lower portions. At the beginning of the 20th century, they are the tool to present a synthetic view on the structure of the Alps (Termier 1902, 1903). In his paper "The synthesis of the Alps", Termier shows in a series of cross-sections that a relatively simple structure describes the architecture of the entire orogen, and this structure consists of a sequence of major nappes accommodating shortening between two rigid continental blocks. The structures themselves, in terms of geometry and rooting of the nappes, are mostly wrong, but the idea that the basic structure of the Alpine orogen can be shown and understood by sketching only few thrusts out of millions of existing structures remains true at present. Nowhere in the pre-1902 publications of Termier, is an attempt to construct orogen-scale sections found. The latter do not result from progressively increasing the size of his sections, nor by adding piece by piece individual segments of cross-sections to obtain a larger one, as in the case of Ampferer and Hammer (1911). It is Termier's understanding of the first-order tectonic relationships of the Western Alps and their analogies to the Eastern Alps that allows him to synthetize the structure of the Alpine orogen in 1903. As Termier (1906) states himself: "It is the assessment of the permanence all along the chain of one and the same structural plan". Previous existing examples of orogen-scale sections (Ebel 1808; Figure 1.4; or Rogers (1836), for the Appalachians) are performed in the style of Ampferer and Hammer (1911), as described above. The sections of Termier are world-wide, the first ones to reduce the picture of an orogen to its essential structures, namely its nappe stack.

Interestingly, Termier, as the first geologist to recognize that the structure of an orogen can be summarized into a stack of nappes, is, to say the least, highly skeptical about continental drift and the theory of Wegener (1912). Termier is a mobilist, but like many others he criticizes the physical process proposed by Wegner to explain continental drift. Termier believes in the "shifting of continental masses [...] to explain the formation of folded zones, that is to say the mountain chains [...]". But the larger shifts required to group all continents in the past is merely the "beautiful dream of a poet" (Termier 1925).

Argand (1924) continues to increase the depth of cross-sections, representing the base of the crust and the upper mantle (Figure 1.24). Like no one else before him, Argand strives for an understanding of the evolution of the Alpine crust, based on the assessment of its present-day geometry. Once this task is achieved, Argand devotes the rest of his studies to the understanding of the kinematic and mechanical processes that lead to such present-day structure. He does this by moving from the investigation of the Alps to other Cenozoic mountain belts (Argand 1924). At this stage, his Alpine sections become simplified to the most essential structures that can be observed in all other orogens too (Figure 1.24; Argand 1924), namely the geometry of superposition of two continental "blocks", the lateral termination of the lower, thinned out "block" within the underlying sima, the 1st order fold geometry of the orogenic wedge, the buried penninic units inferred to derive from the axial zone of the Tethys, and the discontinuously "intruded" sima along the interface of the continental blocks. All this brings Argand to state that "Africa is thrust over Eurasia and the Austro-alpine nappes form its most distal salient".

The drawing of Alpine sections, in the long period after 1924 and until the late 1960s shows a slowing down in scientific advancement compared to the preceding decades, in the sense that neither the style of the sections nor the first-order structures shown until the 1970s are significantly different than those published 50 years earlier. The cross-sections of Heim (1919) are very similar to those of the 1980s (Figure 1.25). This observation should not discredit the work of Alpine geologists of the mid-20th century, but it suggests that the first-order structure of the Alpine chain down to some 10–20 km is well assessed by 1920. The following decades do not provide the necessary geophysical data to extrapolate surface structures to greater depth, nor a new tectonic conceptual model to allow a soundly based rethinking of the geosynclinal type of orogeny. Both these elements appear at the beginning of the 1960s, with seismological experiments, on the one hand, and the theory of Plate Tectonics, on the other hand. From 1924 to 1970, only few Alpine geologists propose new ideas on orogeny and illustrate them with cross-sections (Figure 1.26). These are mainly the models of Kober (1942, 1955), partly, but not only, based on the symmetric "Verschluckung" (subduction) of Ampferer (1906). Their ideas are founded on logical reasoning and physical processes, but they have no observational basis to rely on: neither a direct one from geophysical data of the Alpine realm nor an indirect one like those underlying the theory of Plate Tectonics, which relies on geophysical observations, albeit far away from the Alps. It is probably for this reason that the models by Kraus (1932) and Kober appear to be the result of wild imaginations nowadays.

1.10.2. *Historical evolution of the nappe concept as shown by cross-sections*

The numerous works discussed above also illustrate the historical evolution of the nappe concept, whose contours are discontinuously, but uninterruptedly refined over more than a century. The very beginning of this idea may be in the observations and interpretations of Escher von der Linth (Figure 1.10), as described by Studer and

Desor (1841). Gerlach (1869, Figure 1.18) clearly maps and shows in cross-section the existence of an isoclinal recumbent fold in the Antigorio basement, and interprets it as the result of a thrust accommodating nearly 10 km of displacement, hence a nappe.

Bertrand (1884) re-interprets the sections of Heim (1878) based on the descriptions of thrusts of Gosselet (1879), thus suggesting that the stratigraphic inversion is not due to complex folding, but to a large, gently dipping thrust. His reasoning about this newly interpreted structure leads him to suggest that "[...] it is simply the hypothesis of a single fold replacing that of the "double" fold of Escher". And more generally,

> [...] there is a general rule, and the result of contraction of the globe due to cooling is not only the folding of the crust, but also the outflow from the center of the folded zone.

Hence, his view of nappes is that of a region shortening by folding, from which large recumbent folds pour out and glide over younger units, toward the foreland. His sections of the Mont Blanc area (Bertrand and Ritter 1896; Figure 1.13(b)) illustrate this very clearly; the cover nappes are all rooted in a zone of low amplitude subvertical folds, suggesting that folding of the basement is by far smaller than shortening of the cover. It is this inferred discrepancy that forms the main argument for Ampferer (1906) to suggest a sort of subduction process of the basement in order to balance such sections.

In contrast to the model above, Schardt (1894) clearly decouples the development and displacement of nappes from that of folds. Schardt notes the map-scale "cran de retour" of Gosselet (1879), that is, the drag fold that Bertrand (1884) "postulated all along the northern hillside of the Alps exists nowhere". Schardt (1894) also notes that "the entire sedimentary sequence, from the Triassic to the Cretaceous had been driven, without inversion, above the flysch and the Oligocene". These observations allow him to infer that "the formation of nappes can start by a simple thrust". Termier (1902) depicts every nappe of the Western Alps as being rooted in an area of distributed, low-amplitude folding, of the Internal Zone (Figure 1.16(a)). Only in his 1903 sections, does Termier root the "Dinaric thrust" in the basement of the Southern Alps, without a zone of folding, but also without a visible offset of this basement. However, all nappes below the Dinaric "traîneau écraseur" are still rooted in the above described areas of folding. The mechanical process behind these geometries is suggested to be the displacement of a higher nappe (Termier 1903), causing the underlying structures to parallelize to the basal thrust of the nappe.

In 1906, Termier made a clear distinction between two types of nappes: the first ones are folds which initiate being nearly vertical, and later rotate attaining the horizontal plane or even further. The second type is defined as

> a fragment of the Earth's crust detached from its original substratum, and transported, without significant folding, just by simple translation, under the effect of a tangential pressure, gliding on a frictional surface.

The important difference lies in the fact that the latter type is detached from its original substratum and folding does not play a major role. This second type of nappe corresponds to the "Dinarides" in the view of Termier (Figure 1.16(b)), who thinks that they are displaced northward above the rest of the Alps, thus causing themselves the re-orientation into recumbent and stretched folds in their footwall.

Ampferer and Hammer (1911) provide a mechanical explanation:

> Between these large, apparently rigid zones, comparable to the jaws of a vice, lies an extremely intensively compressed, deeply folded zone, from which a bunch of steep, ascending hinges was squeezed out and laid down northwards gliding down as stacked ice flows or lava outpours to the wide northern depression.

Argand (1916; Figure 1.23) eliminates the low-amplitude, distributed zones of folding from the roots of nappes in his sections. His nappes do not evolve in time from a vertical to a horizontal orientation, as suggested by some of his predecessors. His embryo-tectonic evolution shows that nappes are the amplification of large-scale recumbent folds that re-activate pre-existing, Variscan "ge-anticlines". Hence, the nappes initiate as gently dipping structures, and it is only in the mature stages of collision that they become refolded with steeply oriented roots. This view is not really modified until the establishment of the Plate Tectonic theory and the interpretation of Laubscher (1970), stating that nappes are crustal units that detach from a subducting lithospheric plate and become accreted to the orogenic wedge. However, even at this stage, the precise geometry of the rear part of Alpine nappes has not yet been described. Ramsay (1963) interprets the basement-cover interface geometry of the External Massifs in the Western Alps as pinched synclines produced by a competence contrast between cover and basement undergoing shortening. It was only around 1980 that overthrusting of basement was recognized as the cause of the apparently folded geometry of the basement-cover surface (Bravard and Gidon 1979; Thouvenot and Perrier 1980; Ayrton 1980; Beach 1981) and that the root of many cover nappes was identified in a basement thrust. A clear description and illustration of the root zone of Alpine nappes in cross-section was only published in 1981 by Beach. This work finally shows, with a series of detailed cross-sections, how the cover nappes of the External Zone in the Western Alps are rooted in the basement of the External Massifs, providing the picture of Alpine nappes that we still use nowadays.

1.10.3. *Plate Tectonics*

Plate Tectonics certainly represents a revolution for the interpretation of orogens, but besides the change of paradigm that it brings concerning the driving forces of horizontal displacements, it initiates an immense change in the way geologists look at the bulk structure of the Alpine orogen. As firstly shown by Laubscher (1970), the deep structure of the orogen no longer terminates at 5 or 20 km depth, but at >100 km, and this deep structure is first depicted following Plate Tectonics concepts, because geophysical data are not yet available. A new task of tectonicists is to relate deformation structures observed in the crust with conceptually derived structures describing the entire lithosphere. Laubscher (1969) states:

> Homo sapiens has been and still is much of a myth-maker; he cannot collect isolated data without employing his imagination. He must see facts as part of a pattern, of a story, of a myth [...] The favorite current myth in global tectonics is "ocean floor spreading", which has produced a quantitative first approximation to global kinematics and although imperfect and questionable, is a concept, better than previous ones [...].

Laubscher (1970) draws cross-sections according to this myth, which are mainly conceptually derived. It is interesting to note that, as a consequence of Plate Tectonics, for the first time in the history of Alpine research, a theory becomes the main guide for geological understanding and representations of the Alpine orogen. The maxim of Studer (1834) to sketch cross-sections as close as possible to nature is abandoned in favor of the visualization of a concept explaining plate displacements. This is also manifested in the numerous papers of that time that illustrate the past phases of Alpine orogeny with cross-sections, based on the theory of subduction and collision, however not showing the link between structures in present-day cross-sections of the Alps and in their retro-deformed stages (e.g. Laubscher 1970; Dietrich 1976; Tapponnier and Mattauer 1978; Figure 1.33). This approach is exactly opposite to the one chosen by Argand (1916), who takes the precisely constrained geometry of the present-day cross-section as a starting point and retro-deforms it. The more Argand goes back in time, the more his retro-deformation also becomes guided by a tectonic conceptual model. In his case, it is that of geosynclines. In contrast, the aforementioned studies start from a tectonic conceptual model of inferred Cretaceous subduction, from which they develop the younger history, albeit without succeeding to attain the present-day structure.

From the beginning of the 1970s onwards, geophysical studies slowly fill the gap between theory and observations concerning the deeper parts of the lithosphere. By progressively assessing the structure of the deeper parts of the lithosphere, the a priori conceptually defined deep structure of the orogen is replaced by structures based on natural (seismic) experiments and observations. This effort is still ongoing at present.

1.11. References

Ampferer, O. (1906). Über das Bewegungsbild von Faltengebirgen. *Jahrbuch der Kaiserlich-Königlichen Geologischen Reichsanstalt*, 56, 539–622.

Ampferer, O. and Hammer, W. (1911). Geologischer Querschnitt durch die Ostalpen vom Allgäu zum Gardasee. *Jahrbuch der Kaiserlich-Königlichen Geologischen Reichsanstalt*, 61, 531–710.

Arduino, G. (1758). Lithostratigraphic cross-section of the Agno Valley, near Vicenza (dated October 1758). In Biblioteca Civica, Verona – "Fondo G. Arduino", b. 760, IV.c.11. *Arduino... Due lettere sopra varie osservazioni naturali dirette al Prof. A. Vallisnieri.*

Argand, E. (1909). L'exploration géologique des Alpes pennines centrales. *Bull. Soc. Vaudoise Sci. Nat.*, 45, 217–276.

Argand, E. (1911). Les nappes de recouvrement des Alpes Pennines et leurs prolongements structuraux. *Beiträge zur Geologische Karte der Schweiz*, Neue Folge, XXXI Lieferung.

Argand, A. (1916). Sur l'arc des Alpes occidentales. *Eclogae geologicae Helvetiae*, 14, 145–191.

Argand, A. (1924). La tectonique de l'Asie. *XIII Intern. Geol. Congr.*, Liège, 171–372.

Ayrton, S. (1980). La géologie de la zone Martigny-Chamonix et l'origine de la nappe de Morcles. *Geologicae Helvetiae*, 73, 137–172.

Bailey, E.B. (1935). *Tectonic Essays – Mainly Alpine*. Oxford University Press, Oxford.

Beach, A. (1981). Thrust structures in the eastern Dauphinois Zone (French Alps), north of the Pelvoux Massif. *Journal of Structural Geology*, 3, 299–308.

van Bemmelen, R.W. (1973). Geodynamic models for the Alpine type of orogeny (Test-case II: The Alps in central Europe). *Tectonophysics*, 18, 33–79.

Berkhemer, H. (1968). Topographie des "Ivrea-Körpers" abgeleitet aus seismischen und gravimetrischen daten. *Schweizerische mineralogische und petrographische Mitteilungen*, 48, 235–246.

Bertrand, M. (1881). Failles de la lisière du Jura, entre Besançon et Salins. *Bulletin Société géologique de France*, 10(III), 114–127.

Bertrand, M. (1884). Rapport de structure des Alpes de Glaris et du Bassin houiller du Nord. *Bulletin Société géologique de France*, 12, 318–330.

Bertrand, M. and Ritter, E. (1896). Sur la structure du mont Joly, près Saint-Gervais (Haute-Savoie). *C. R., Acad. Sci. Paris*, 122, 289–293.

Boillot, G., Agrinier, P., Beslier, M.-O., Cornen, G., Froitzheim, N., Gardien, V., Girardeau, J., Gil Ibarguchi, J.I., Kornprobst, J., Moullade, M. et al. (1995). A lithospheric syn-rift shear zone at the ocean continent transition: Preliminary results of the GALINAUTE II cruise (Nautile dives on the Galicia Bank, Spain). *Comptes-Rendus de l'Académie des Sciences, Paris*, 321(IIa), 1171–1178.

Bravard, C. and Gidon, M. (1979). La structure du refers oriental du Massif du Pelvoux : observations et interprétations nouvelles. *Géol. Alpine*, 23, 33.

Büchi, U.P. and Trümpy, R. (1976). Bemerkungen zum geologischen profil längs der geotraverse Basel-Chiasso. *Schweizerische mineralogische und petrographische Mitteilungen*, 56, 589–603.

Buxtorf, A. (1908). Geologische Beschreibung des Weissenstein-Tunnels und seiner Umgebung. *Geol. Karte der Schweiz*, N.F. 21.

Buxtorf, A. (1914). Profile zur geologischen Vierwaldstädersee-Karte, 1:50'000. *Geologische Kommission der Schweizerischen Naturforschenden Gesellschaft*, Luzern.

Campanino, F. and Polino, R. (2002). Carta geologica delle Alpi Piemontesi (1860–1879 ca.). Fotomosaico al 1:380.000 dei 29 fogli al 1:50.000 rilevati da M. Baretti, B. Gastaldi, H. Gerlach, con la collaborazione di C. Bruno, L. Bruno e G. Michelotti. Regione Piemonte, Museo Regionale di Sci. Nat. di Torino, Litografia Geda, Nichelino.

Closs, H. and Labrouste Y. (1963). *Recherches séismologiques dans les Alpes Occidentales au moyen de grandes explosions en 1956, 1958 et 1960*. CNRS, Paris.

von Cotta, B. (1851). *Die Alpen*. T.O. Weigel, Leipzig.

Dal Piaz, G. (1912). Studi geotettonici sulle Alpi orientali. *Mem. Istituto Geologia Univ. Padova*.

Dal Piaz, G.V. (1996). Felice Giordano and the geology of the Matterhorn. *Atti Acc. Sci. Torino*, 130, 163–169.

Dal Piaz, G.V. (2001). History of tectonic interpretations of the Alps. *Journal of Geodynamics*, 32, 99–114.

Dal Piaz, G.V. (2010). The Italian Alps: A journey across two centuries of Alpine geology. In *The Geology of Italy: Tectonics and Life Along Plate Margins*, Beltrando, M., Peccerillo, A., Mattei, M., Conticelli, S., Doglioni, C. (eds). Journal of the Virtual Explorer, Electronic Edition.

Dal Piaz, G.B. and Dal Piaz G.V. (1984). Sviluppo delle concezioni faldistiche nell'interpretazione tettonica delle Alpi (1840–1940). *Soc. Geol. It., vol. giubilare 1*, Pitagora-Tecnoprint Bologna, 41–70.

Dal Piaz, G.V., Hunziker J.C., Martinotti G. (1972). La Zona Sesia-Lanzo e l'evoluzione tettonico-metamorfica delle Alpi nord-occidentali interne. *Mem. Soc. Geol. It.*, 11, 433–466.

Debelmas, J., Antoine, P., Arnaud H., Gidon, M., Kerckhove, C., Desmons, J., Fudral, S., Grandchamp, P. (2011). *L'exploration géologique des Alpes franco-italiennes*. Presses des Mines, Paris.

Dietrich, V. (1976). Evolution of the Eastern Alps: A plate tectonics working hypothesis. *Geology*, 4, 147–152.

Dufrenoy, A. and Elie de Beaumont, L. (1841). *Explication de la carte géologique de la France, Tome I*. Imprimerie Royale, Paris.

Dullo, W.-C. and Pfaffl, F.A. (2019). The theory of undercurrent from the Austrian alpine geologist Otto Ampferer (1875–1947): First conceptual ideas on the way to plate tectonics. *Can. J. Earth Sci.* doi: 10.1139/cjes-2018-0157.

Durand-Delga, M. (2007). Marcel Bertrand (1847–1907), honneur de la géologie française. *Travaux du comité français d'histoire de la géologie (COFRHIGÉO)*, 3rd series, t. XXI.

Durand-Delga, M. (2010). *Marcel Bertrand (1847–1907), génie de la tectonique*. Transvalor-Presses des Mines, Paris.

Ebel, J.G. (1808). *Ueber den Bau der Erde in dem Alpen-Gebirge zwischen 12 Längenund 2-4 Breitengraden*. Orel F¨ssli und Compagnie, Zurich.

Ellenberger, F. (1990). Johann Scheuchzer, pionnier de la tectonique alpine. *Comité français d'Histoire de la Géologie*, 3rd series, 4, 85–115.

Escher v. der Linth, A. and Studer, B. (1839). *Geologische Beschreibung von Mittel-Bündten*. Bayerische Staatsbibliothek, Munich.

Favre, A. (1859). Mémoire sur les terrains Liasique et Keupérien de la Savoie. *Mém. Soc. Phys. Hist. Nat.*, 15, 24–29.

Favre, A. (1867). *Recherches géologiques dans les parties de la Savoie, du Piémont et de la Suisse voisines du Mont-Blanc ?* Editions Victor Masson et Fils, Paris.

Franchi, S. (1898). Sull'età mesozoica della zona delle pietre verdi nelle Alpi Occidentali. *Bollettino R. Comitato Geologico d'Italia*, 29, 173–247 and 325–482.

Frech, F. (1901). *Géologie der Radstädter Tauern. Geolog. und paläontolog.* Abhandl. von Koken, Jena.

Ford, T.D. (1967). The first detailed geological sections across England, by John Farey, 1806–8. *Mercian Geologist,* 2, 41–49.

Gassman, F. and Prosen, D. (1948). Zur Interpretation des Schweredefizits in den Schweizer Alpen. *Eclogae Geologicae Helvetiae*, 41, 135–140.

Gerlach, H. (1869). *Die penninischen Alpen. Beiträge zur Geologie der Schweiz*. Neue Denkschriften Allg. Ges. Ges. Natf., Zurich.

Gerlach, H. (1871). *Das Südwestliche Wallis mit den angrenzenden Landestheilen von Savoien und Piemont*. Beiträge zur Geologischen Karte der Schweiz. Druck von Zürcher und Furrer, Zurich.

Giese, P. (1968). Die Struktur der Erdkruste im Bereich der Ivrea-Zone. Ein Vergleich verschiedener seismischer Interpretationen und der Versuch einer petrographisch-geologischer Deutung. *Schweiz. Mineral. Petrogr. Mitt.*, 48, 261–284.

Giese, P. (1978). Crustal structure of the Ivrea-Zone within the frame of the Alpine crustal structure. *Memorie di Scienze Geologiche*, 33, 51–57.

Giese, P., Günther, K., Reutter, K. (1970). Vergleichende geologische und geophysikalische Betrachtungen der Westalpen und des Nordapennins. *Z. deutsch. geol. Ges.*, 120, 151–195.

Giordano, F. (1869a). Sulla orografia e sulla geologica costituzione del Gran Cervino. *Atti R. Acc. Sci. Torino*, 4, 304–321.

Giordano, F. (1869b). Notice sur la constitution géologique du Mont Cervin. *Arch. Sci. Phys. Nat. Genève*, 34, 255–267.

Gortani, M. (1930). Idee precorritrici di Luigi Ferdinando Marsigli su la struttura dei monti. In *Memorie intorno a Luigi Ferdinando Marsigli*, Corbelli, A., Marsiliano, C. (eds). Nicola Zanichelli, Bologna.

Gosselet, J. (1879). Sur la structure générale du bassin houiller franco-belge. *Bulletin de la Société géologique de France*, VIII, 505–512.

Grandchamp, P. (1997). Les origines de l'affaire de Petit-Cœur ou la découverte du célèbre gisement de végétaux fossiles et de bélemnites replacée dans son véritable contexte. In *De la géologie à son histoire*, Gohau, G. (ed.). Comité des travaux historiques et scientifiques, Paris.

Gratier, J.P. and Vialon, P. (1980). Deformation pattern in a heterogeneous material: Folded and cleaved sedimentary cover immediately overlying a crystalline basement (Oisan, French Alps). *Tectonophysics*, 65, 151–180.

Hawkesworth, C.J., Waters, D.J., Bickle, M.J. (1975). Plate tectonics of the Eastern Alps. *Earth Planet. Sci. Lett.*, 24, 405–413.

Heim, A. (1871). Notizen aus den geologischen Untersuchungen für Blatt XIV der eidg. *Karte*, 241–262.

Heim, A. (1878). *Untersuchungen über den Mechanismus der Gebirgsbildung im Anschluss an die geologische Monographie der Tödi-Windgällen-Gruppe*. Wentworth Press, Basel.

Heim, A. (1908). Der Bau der Schweizer Alpen. *Neujahrsblatt der Naturforschenden Gesellschaft in Zürich*, 110, 3–26.

Heim, A. (1919). *Geologie der Schweiz*. Tauchnitz, Leipzig.

Heim, A. (1922). *Geologie der Schweiz*. Tauchnitz, Leipzig.

Heim, A. (1927). Die Gipfelflur der Alpen. *Neujahrsblatt der Naturforschenden Gesellschaft in Zürich*, 129, 3–25.

Heitzmann, P. (2008). Die ersten geologischen Karten der Schweiz 1752–1853. *Cartographica Helvetica*, 38, 21–36.

Hess, H.H. (1962). History of ocean basins. In *Petrologic Studies: A Volume to Honor A. F. Buddington*, Engel, A.E.J., James, H.L., Leonard, B.F. (eds). Geological Society of America, Boulder.

Hunziker, J.C. (1970). Polymetamorphism in the Monte Rosa, Western Alps. *Eclogae Geol. Helv.*, 65, 151–161.

Jaccard, A. (1869). Jura Vaudois et Neuchatelois. *Matériaux pour la carte géologique de la Suisse*.

Kiessling, W. (1992). Palaeontological and facial features of the Upper Jurassic Hochstegen Marble (Tauem Window, Eastern Alps). *Terra Nova*, 4, 184–197.

Kober, L. (1942). *Tektonische Geologie*. Gebrüder Borntraege, Berlin.

Kober, L. (1955). *Bau und Entstehung der Alpen*. Franz Deuticke, Vienna.

Koch, M. (1952). Johann Scheuchzer als Erforscher der Geologie der Alpen. *Vierteljahrschrift der Naturforschenden Gesellschaft in Zürich*, Bd 97, 191–202.

Kraus, E. (1932). Die Alpen als Doppelorogen. *Geol. Rundschau*, 22, 65–73.

Kraus, E. (1954). Neue Überlegungen zum Bau der Alpen. *Mitteilungen der Geologischen Gesellschaft in Wien*, 47, 141–158.

Lapworth, C. (1883). The secret of the Highlands. *Geol. Mag.*, 10, 193–337.

Laubscher, H.P. (1969). Mountain building. *Tectonophysics*, 7, 551–563.

Laubscher, H.P. (1970). Bewegung und Wärme in der alpinen Orogenese. *Schweiz. Mineral. Petrogr. Mitt.*, 50, 503–534.

Laubscher, H.P. (1971). The large-scale kinematics of the Western Alps and the Northern Apennines and its palinspastic implications. *Am. J. Sci.*, 271, 193–226.

Laubscher H.P. (1974). The tectonics of subduction in the Alpine system. *Memorie Società Geologica Italiana*, 13, 275–283.

Laubscher, H.P. (1983). Detachment, shear, and compression in the Central Alps. *Geological Society of America, Memoir*, 158, 191–211.

Laubscher, H.P. (1985). Large-scale, thin-skinned thrusting in the southern Alps: Kinematic models. *Geological Society of America Bulletin*, 96, 710–718.

Letsch, D. (2014). The Glarus Double Fold: A serious scientific advance in mid nineteenth century Alpine geology. *Swiss J. Geosci.*, 107, 65–80.

Letsch, D. (2017). Swiss contributions to mid-nineteenth century tectonic research: A step backwards or the prologue to the nappe tectonics revolution? In *History of Geoscience: Celebrating 50 Years of INHIGEO,* Mayer, W., Clary, R.M., Azuela, L.F., Mota, T.S., Wołkowicz, S. (eds). Geological Society, Special Publications, London.

Lippitsch, R., Kissling, E., Ansorge, J. (2003). Upper mantle structure beneath the Alpine orogen from high-resolution teleseismic tomography. *Journal of Geophysical Research: Solid Earth*, 108.

Lory, C. (1846). Etudes sur les terrains secondaires des Alpes dans les environs de Grenoble. Thesis, Faculté des Sciences, Paris.

Lory, C. (1860). *Description géologique du Dauphiné (Isère, Drome, Hautes Alpes)*. Part 1. Savy, Paris.

Lugeon, M. (1902). Les grandes nappes de recouvrement des Alpes du Chablais et de la Suisse. *Bull. Soc. géol. France* (4th series), 1, 723–825.

Lugeon, M. (1914). Les hautes Alpes Calcaires entre la Lizerne et la Kander (Wildhorn, Wildstrubel, Balmhorn et Torrenthorn). Explication de la carte spéciale No. 60, Fascicule I.

Lugeon, M. (ed.) (1934). Préface. In *Guide géologique de la Suisse, Fasc. I.* B. Wepf & Cie., Basel.

Masson, H. (1976). Un siècle de géologie des Préalpes : de la découverte des nappes à la recherche de leur dynamique. *Eclogae Geol. Helv.*, 69, 527–575.

Mattauer, M., and Tapponnier, P. (1978). Tectonique des plaques et tectonique intracontinentale dans les Alpes franco-italiennes. *C. R. Acad. Sci. Paris*, 287, 899–902.

Miller, H., Müller, S., Perrier, G. (1982). Structure and dynamics of the Alps – A geophysical inventory. In *Alpine-Mediterranean Geodynamics,* Berkhemer, H. and Hsü, K. (eds). American Geophysical Union, Washington, DC.

Milnes, A.G. (1974). Post-nappe folding in the western Lepontine Alps. *Eclogae Geologicae Helvetiae*, 67, 333–348.

von Morlot, A. (1847). *Erläuterungen zur geologischen Übersichtskarte der nordöstlichen Alpen. Ein Entwurf zur vorzunehmenden Bearbeitung der physikalischen Geographie und Geologie*. Braumüller und Seidel, Vienna.

Moro, A.L. (1740). *De' crostacei e degli altri marini corpi che si trovano su' monti*. Stefano Monti, Venezia.

Mueller, S., Ansorge, J., Egloff, R., Kissling, E. (1980). A crustal cross section along the Swiss Geotraverse from the Rhinegraben to the Po plain. *Eclogae Geol. Helv.*, 73, 463–485.

Müller, A. (1862). Geognostische Skizze des Kantons Basel und der angrenzenden Gebiete. *Matériaux pour la carte géologique de la Suisse*, 1.

Müller, S. (1984). Tiefenstruktur, Dynamik und Entwicklung des Mittelmeer- und Alpenraumes. *Vierteljahrsschrift der Naturforschenden Gesellschaft in Zürich*, 129, 217–245.

Müller, S., Egloff. R., Ansorge, J. (1976). Struktur des tieferen Uniergrundes entlang der Schweizer Geotraverse. *Schweiz, Mineral, Petrogr. Mitt.*, 56, 685–692.

Murchison, R.I. (1850). *Ueber den Gebirgsbau in den Alpen, Apenninen u. Karpathen, namentlich um einen Uebergang aus sekundären Gebilde in tertiäre darzuthun, u. über die Entwicklung eocener Ablagerung im südlichen Europa: Bearbeitet von Gustav Leonhard*. Verlag B. Müller, Stuttgart.

Nicolas, A., Hirn, A., Nicolich, R., Polino, R., ECORS-CROP Working Group (1990). Lithospheric wedging in the western Alps inferred from the ECORS-CROP traverse. *Geology*, 18, 587–590.

Niethammer, T. (1904). Relative Bestimmung der Schwerkraft im Nikolaitale. Dissertation, University of Basel.

Omboni, G. (1879). *Le Nostre Alpi e la Pianura del Po. Descrizione Geologica del Piemonte, della Lombardia, del Trentino, del Veneto e dell 'Istria, Appendice alla Geologia d'Italia.* V. Maisner e Compagnia, Milan.

Panza, G.F. and Müller, S. (1978). The plate boundary between Eurasia and Africa in the Alpine area. *Memorie di Scienze Geologiche, Istituti di Geologia e Mineralogia, Università di Padova*, 33, 43–50.

Pfaff, F. (1873). *Allgemeine Geologie als exakte Wissenschaft*. Verlag von Wilhelm Engelmann, Leipzig.

Pfiffner, O.A., Frei, W., Valasek, P., Staeuble, M., Levato, L., Dubois, L., Schmid, S.M., Smithson, S.B. (1990). Crustal shortening in the Alpine orogen: Results from deep seismic reflection profiling in the eastern Swiss Alps, Line NFP 20 East. *Tectonics*, 9, 1327–1355.

Polino R., Dal Piaz G.V., Gosso G. (1990). Tectonic erosion at the Adria margin and accretionary processes for the Cretaceous orogeni of the Alps. *Mém. Soc. Géol. France*, 156, 345–367.

Rogers, W.B. (1836). *Report of the Geological Reconnoissance of the State of Virginia*. De Silva, Thomas & Co, Philadelphia.

Rogers, H.D. (1858). *Geology of Pennsylvania; a government Survey*. J. B. Lippincott & Co., Philadelphia.

Sander, B. (1911). Geologische Studien am Westende der Hohen Tauern (Erster Bericht). *Denkschr. Akad. Wiss. Wien, math.-nat. Kl.*, 82, 257–319.

de Saussure, H.-B. (1870). *Voyage dans les Alpes. Précédés d'un Essai sur l'histoire naturelle des environs de Genève*. Samuel Foche, Imprimeur et Libraire du Roi.

Schaer, J.P. (2010). Swiss and Alpine geologists between two tectonic revolutions. Part 1: From the discovery of nappes to the hypothesis of continental drift. *Swiss J. Geosci.*, 103, 503–522.

Schardt, H. (1884). Etudes géologiques sur le Pays-d'Enhaut vaudois. PhD Thesis, University of Geneva.

Schardt, H. (1893). Coup d'oeuil sur la structure géologique des environs de Montreux. *Eclogae Geol. Helv.*, IV, 29–43.

Schardt, H. (1894). Sur l'origine des Préalpes romandes. *Archives des Sciences physiques et Naturelles de Genève*, 30, 570–583.

Schardt, H. (1898). Les régions exotiques du versant Nord des Alpes suisses (Préalpes du Chablais et du Stockhorn et les Klippes) : leurs relations avec l'origine des blocs et brèches exotiques et la formation du Flysch. *Bulletin de la Société vaudoise des sciences naturelles*, 34, 113–219.

Schardt, H. (1904). Note sur le profil géologique et la tectonique du massif du Simplon. *Eclogae Geol. Helv.*, VIII(2), 173–199.

Schardt, H. (1906). Die modernen Anschauungen über den Bau und die Entstehung des Alpengebirges. *Actes, St Gallen*, 308–346.

Schardt, H. (1909). Coup d'oeuil sur la géologie et la tectonique des Alpes et du Canton du Valais. *Bull. Soc. Murithienne du Valais*, XXXV, 246–354.

Scheuchzer, J.J. (1708). De montium structura. Letter sent to the "Académie des Sciences de Paris", copied in the manuscript H92 of the Zentral Bibliothek in Zurich; this contains (p.30-39) the Latin text of Orographia, part of the Helvetiae Stoicheiographia by Johann Jakob Scheuchzer.

Scheuchzer, J.J. (1716–1718). *Helvetiae Stoicheiographia. Orographia et Oreographia. Oder Beschreibung der Elementen, Grenzen und Bergen des Schweizerlands*. Bodmer'sche Truckerey, Zurich[1].

Scheuchzer, J.J. (1732). *Physique sacrée, ou histoire naturelle de la Bible.* Pierre Schenk and Pierre Mortier, Amsterdam.

Schmid, S.M., Pfiffner, O.A., Froitzheim, N., Schonborn, G., Kissling, E. (1996). Geophysical-geological transect and tectonic evolution of the Swiss-Italian Alps. *Tectonics*, 15, 1036–1064.

Schmidt, C. (1907). Ueber die Geologie des Simplongebietes und die Tektonik der Schweizeralpen. *Eclogae Geologicae Helvetiae*, 9, 484–584.

Sengör, A.M.C. (1989). Die Tethys: vor hundert Jahren und heute. *Mitt. Oesterr. Geol. Ges.*, 89, 5–177.

Sengör, A.M.C. (1990). Plate tectonics and orogenic research after 25 years: A Tethyan perspective. *Earth-Science Reviews*, 27, 1–201.

Sengör, A.M.C. (2014). Eduard Suess and global tectonics: An illustrated "short guide". *Austrian Journal of Earth Sciences*, 107, 6–82.

1 This consists of three volumes. See in particular Volume 1, pp. 111–114 and pl. I.

Sengör, A.M.C. (2016). What is the use of the history of geology to a practicing geologist? The propaedeutical case of stratigraphy. *The Journal of Geology*, 124, 643–698.

Sengör, A.M.C. and Bernoulli, D. (2011). How to stir a revolution as a reluctant rebel: Rudolf Trümpy in the Alps. *Int. J. Eart Sci. (Geol. Rundsch.)*, 100, 899–936.

Sismonda, A. (1839a). Osservazioni geologiche e mineralogiche sopra i monti posti tra la Valle d'Aosta e quella di Susa in Piedmonte. *Mem. R. Acc. Sci. Torino*, 1, 1–42.

Sismonda, A. (1839b). *Memoria sui terreni stratificati delle Alpi*. Stamperia Reale, Turin.

Staub, R. (1924). Bau der Alpen. *Beiträge zur Geologische Karte der Schweiz NF*, 52, 1–272.

Steinmann, G. (1906). Geologische Probleme des Alpengebirges. *Zeitschr. d. D. u. Ö. A.-V.*

Studer, B. (1834). *Geologie der westlichen Schweizer-Alpen, Ein Versuch*. Druck und Verlag Karl Groos, Heidelberg and Leipzig.

Studer, B. and Desor, E. (1841). Verhandlungen der Schweizerischen Naturforschenden Gesellschaft, II. *Geologische und Mineralogische Section*, Erste Sitzung, 53–73.

Suess, E. (1875). *Die Entstehung der Alpen*. W. Braumüller, Vienna.

Taramelli, T. (1880). Il Canton Ticino meridionale e I paesi finitimi. Spiegazione del foglio Duf. XXIV, Colorito geologicamente da Spreafico, Negri, e Stoppani per Torquato Taramelli. *Beitr. Geol. Karte Schweiz.*

Termier, P. (1894). Le Massif des Grands Rousses (Dauphiné et Savoie). *Bulletin des Services de la Carte géologique de la France et des Topographies souterraines*, 40, 1–118.

Termier, P. (1901). Carte et coupes géologiques des montagnes comprises entre Briançon et Vallouise (région des nappes briançonnaises). Map.

Termier, P. (1902). Quatre coupes à travers les Alpes franco-italiennes. *Bulletin de la Société Géologique de France*, 2, 411–433.

Termier, P. (1903). Les nappes des Alpes Orientales et la synthèse des Alpes. *Bull. Soc. Géol. France*, 3, 711–766.

Termier, P. (1906). La synthèse des Alpes. *Conference held in Liège on January 26, 1906, for students of Ecoles spéciales*. Librairie Polytechnique Béranger, Paris.

Theobald, G. (1864). *Geologische Beschreibung der in den Blättern X und XV des eidg. Atlasses enth. Gebirge von Graubünden*, Bern.

Thouvenot, F. and Perrier, G. (1980). Seismic evidence for crustal overthrust in the western Alps. *Pure Appl. Geophys.*, 119(163–184).

Trümpy, R. (1991). The Glarus Nappes: A controversy of a century ago. In *Controversies in Modern Geology*, Müller, D.W., McKenzie, J.A., Weissert, H. (eds). Academic Press, London.

Trümpy, R. (2001). Why plate tectonics was not invented in the Alps. *International Journal of Earth Sciences*, 90, 477–483.

Trümpy, R. and Lemoine, M. (1998). Marcel Bertrand (1847-1907) : les nappes de charriage et les cycles orogéniques. *C. R. Acad. Sci. Paris, Sciences de la Terre et des planètes/Earth & Planetary Sciences*, 327, 211–224.

Vaccari, E. (2001). Between hydrology and geology: The Italian debate on the origin of springs in the early XVIII century. *International Symposium OH2 "Origins and History of Hydrology"*, Dijon, May 9–11.

Vallisneri, A. (1715). *Lezione Accademica intorno all'origine delle Fontane, colle Annotazioni per chiarezza maggiore della medesima*. G.G. Ertz, Venice.

Vallisneri, A. (1726). *Lezione accademica intorno all'origine delle fontane, con le annotazioni per chiarezza maggiore della medesima, [...] Seconda Edizione con la Giunta di varie Lettere Dissertatorie, un'altra Lezione Accademica, Osservazioni, Ragioni, ed Esperienze nuove, dimostranti la verità del proposto Sistema, con la Risposta alle Obbiezioni del dottore N.N. compilata da G. Giorgi*. A. Bortoli, Venice.

Villa, A. and Villa, G.B. (1844). *Sulla costituzione geologica e geognostica della Brianza e segnatamente sul terreno cretaceo*. Spettatore Industriale, Milan.

PART 2

Recent and Present-Day Alpine Chain

2

What We (Possibly) Know About the 3D Structure of Crust and Mantle Beneath the Alpine Chain

Anne PAUL
Université Grenoble Alpes, Université Savoie Mont Blanc, CNRS, IRD, UGE, ISTerre, Grenoble, France

2.1. Introduction

In 1993, Kissling published a seminal paper that presented a review of available geophysical data on the structure of the crust and upper mantle under the Alps and an assessment of the uncertainties associated with these data (Kissling 1993). The abstract of this publication began with the following sentence:

> In the last five decades, the deep structure of the Alps has been probed by every geophysical method applicable, and the resulting amount of data is unmatched for any other orogen.

Indeed, the publication came after a decade of major deep reflection or refraction seismic surveys in the Central Alps (European Geotraverse: Blundell et al. 1992; Swiss NFP20 program: Frei et al. 1990) and in the Western Alps (ECORS-CROP: Nicolas et al. 1990). These large seismic data acquisition programs (complemented with a gravimetric survey for ECORS-CROP: Bayer et al. 1989) provided Kissling with the main argument that the available data set on the deep

coordinated by Claudio L. ROSENBERG and Nicolas BELLAHSEN. © ISTE Ltd 2023.

structure of the Alps was quite unique for a mountain range. These data were all related to crustal structure, internal structure by vertical seismic reflection and Moho depth for wide-angle seismic and refraction.

In the early 1990s, the structure of the upper mantle beneath the Alps began to be studied by teleseismic tomography, which is travel time tomography of waves emitted by distant earthquakes (epicentral distances > 20°) and recorded by permanent seismic stations (Babuška et al. 1990; Cattaneo and Eva, 1990; Spakman 1990, 1991). Critical examination of these results by Kissling (1993) led him to conclude that a slab of European lithosphere is dipping southwards in the mantle beneath the southern Central Alps and the Po basin. A few years later, the development of permanent seismological networks and the collection and distribution by ISC (International Seismological Centre, http://www.isc.ac.uk) of a global seismological bulletin with hypocenter locations and arrival time picks have allowed increasingly accurate arrival time tomography, such as that of Piromallo and Morelli (2003) for the Alpine–Mediterranean area. Their P-wave velocity anomaly maps between 100 and 250 km depth clearly show the fast anomaly associated with the subduction of the European slab in the mantle of the Western and Central Alps. The anomaly is much less clear under the Eastern Alps. The mantle transition zone (410–660 km depth) is characterized by fast P-wave velocities over most of the Alpine–Mediterranean region. Piromallo and Faccenna (2004) interpreted this result using geological data on Alpine convergence and kinematic reconstructions. They proposed that most of the fast-velocity anomalies detected in the mantle transition zone correspond to slab fragments detached from the Alpine subduction by an event occurring around 30–35 Ma.

The teleseismic tomography of the Alpine mantle by Lippitsch et al. (2003) showed for the first time a high-velocity anomaly with a steep northeast dip under the Eastern Alps. On the basis of this dip, they interpreted this anomaly as evidence of subduction of the continental lithosphere of the Adria microplate under Europe. Lippitsch et al. (2003) also confirmed the presence of south-eastward dipping slabs under the Western and Central Alps, evidence of subduction of the European plate under Adria. This tomography was the first to use all the seismological data from the permanent networks available in the region, even though not all of them were open. They carefully picked all arrival times in a homogeneous way, and they took into account lateral variations in crustal structure by applying crustal corrections calculated in a 3D model. Because of the quality and accuracy of the work carried out by Lippitsch et al. (2003), from the preparation to the analysis and then the inversion of the P-wave arrival time data set, this publication has become an important reference on the structure of the Alpine uppermost mantle.

Over the past decade or so, seismic imaging of the lithosphere has benefited from a dual revolution in data and methods. Advances in seismological instrumentation have led to the development of broadband seismometers, which today equip most of the permanent seismological networks installed in the Alps. Instruments of similar quality can also be deployed in large numbers on temporary experiments lasting from a few months to a few years to improve the spatial coverage of permanent networks (e.g. AlpArray, http://www.alparray.ethz.ch/; AlpArray Seismic Network 2015; Hetényi et al. 2018a). Last but not least, most of the high-quality data provided by these permanent and temporary stations are now open and accessible to all because of the European federation of seismological datacenters EIDA (European Integrated Data Archive, http://www.orfeus-eu.org/data/eida/). This data revolution now makes large volumes of high-quality seismological data readily available.

The other revolution concerns tomography methods. It is of course related to the development of the performance of computers, but it remains inseparable from the data revolution: the most modern tomography is blind without a large amount of high-quality data. Much more than just arrival times, we now know how to use the entire seismic record in all its complexity in terms of amplitude and phase to probe the fine structure of the subsurface. This is achieved by full waveform inversion (FWI) that theoretically gives access to all the elastic parameters, P-wave velocity, S-wave velocity and density, as well as other physical parameters such as anisotropy or attenuation (e.g. Tape et al. 2009). Finally, we even know how to probe the subsurface without illumination by an earthquake or a man-made source, using ambient seismic noise. Campillo and Paul (2003) and Shapiro and Campillo (2004) have shown that correlating long time series of noise records at two seismic stations results in the same signal as if a seismic source were placed at one of the two stations and the waves recorded at the other station. Each seismic station thus becomes a wave source for all the others. This method has great potential for seismic imaging in regions with low to moderate seismicity, which is the case for a large part of the Alps.

Since the review by Kissling (1993) and key publications by Lippitsch et al. (2003) and Piromallo and Morelli (2003) on the structure of the Alpine upper mantle, several temporary passive seismic experiments (as opposed to active seismic reflection or refraction profiling) have been carried out in the Alps, either along profiles transverse to the belt or in deployments of sensor arrays such as AlpArray (Figure 2.1). Seismic tomography work was also carried out using data from permanent networks. This work will be reviewed in the following parts of this chapter, starting with imaging of the crust including the Moho, which geometry is particularly complex in the Alps where European, Adriatic and Ligurian Moho overlap, before moving on to tomography of Alpine subductions.

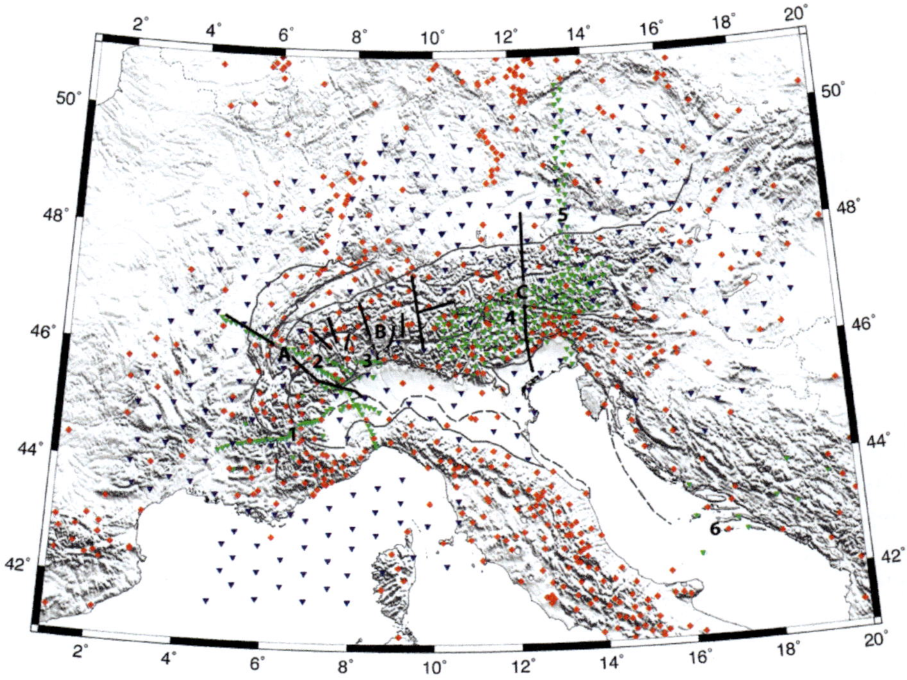

Figure 2.1. *Location map of seismic stations deployed in the Alpine region since 2012*

COMMENT ON FIGURE 2.1.– *Red diamonds: permanent networks; inverted triangles: temporary experiments; blue: AlpArray seismic network; green: (1) Cifalps, (2) Cifalps-2, (3) Ivrea, (4) Swath-D, (5) EASI and (6) CASE. Thick black lines show locations of controlled source seismic experiments: (A) ECORS-CROP, (B) NFP-20 and (C) TRANSALP. The main tectonic boundaries are shown as thin gray lines. Some of these deployments are recent and have not yet resulted in publications at the time of writing (Cifalps-2, Ivrea, Swath-D).*

At the time of writing, several temporary experiments including AlpArray have recently been completed in the Alps and the exploitation of their data is in progress. A further step toward more reliable and better resolved geophysical models of the Alpine crust and mantle is expected in the coming months and years.

2.2. Probing a very complex crust and crust-mantle boundary

The primary objective of most geophysical investigations of the crust is the crust-mantle boundary, or Moho because it is characterized by a generally abrupt petrological change between granulites of the lower crust (under continents) with P-wave velocity (Vp) of ~6.7 km/s and density (ρ) of ~2.7 × 10^3 kg·m^{-3}, and peridotites of the upper mantle with Vp ≈ 8 km/s and ρ ≈ 3.3 × 10^3 kg·m^{-3}. At a depth greater than the base of sedimentary basins, the Moho is thus the petrologic discontinuity that is the most accessible to active seismic profiling or passive seismology. Probing the bulk structure of the crust is more difficult due to generally weak seismic wave-velocity and density contrasts. This section begins by reviewing the most widely used geophysical methods for probing the crustal structure and for mapping Moho depth.

2.2.1. *A quick update on crustal exploration methods*

Refraction and reflection seismology, with the latter including near-vertical and wide-angle incidence, have been widely used because they are particularly effective for imaging strong velocity contrasts such as the Moho with excellent vertical resolution. They are rather expensive though and hardly compatible with current environmental constraints. As a result, the latest crustal-scale controlled source (or active) seismic experiments in the Alps are Transalp (Lüschen et al. 2006) in the Eastern Alps and the wide-angle experiments of the GéoFrance-3D Alps project in the Western Alps (Thouvenot et al. 2007).

Today, crustal imaging is mainly based on passive seismology, using earthquakes and noise recordings. Compared to controlled source seismology (CSS), the signals have a lower frequency and therefore a lower vertical resolution. However, a reasonably dense network of seismic profiles makes it possible to calibrate the depth of the interfaces by comparison with CSS data, while passive seismology provides 3D constraints and fills the gaps between CSS profiles. The most frequently used methods are described in the following paragraphs.

Receiver function analysis is often used to estimate the depth of the main velocity discontinuities in the crust, particularly the Moho, using the three-component records of teleseismic earthquakes (Vinnik 1977). When the teleseismic P-wave hits an interface from below, part of its energy is converted into an S-wave which is recorded on the horizontal components of the sensor a few seconds after the P-wave. Calculating receiver functions (RF) consists of enhancing the Ps waves (P-to-S converted waves at velocity discontinuities such as the Moho) in the P-wave coda. The arrival time difference between the P wave and a Ps wave converted at a velocity discontinuity depends on the average Vp and Vp/Vs ratio above the discontinuity and on the depth of the discontinuity. The amplitude of the Ps wave depends on the velocity contrast at the discontinuity. As in vertical-incidence seismic reflection, a depth migration is necessary to convert the arrival time difference between the P and Ps waves into depth. It is also required to place the conversion point at its exact position, due to the obliquity of the seismic ray. The simplest depth migration method, named CCP stacking (for common conversion point; Dueker and Sheehan 1997) assumes a flat-layered crust and requires a crustal velocity model. By calculating and migrating the RF at a series of seismic stations aligned along a profile, a CCP migrated section is obtained that in most cases shows the Moho depth variations along the profile. CCP sections are often displayed in a blue to red color scale, where blue corresponds to a velocity discontinuity of decreasing velocity with increasing depth, and red corresponds to a discontinuity of increasing velocity with increasing depth. Care must be taken to interpret these blue and red spots as interfaces and not as areas of high or low seismic velocity. Examples of CCP sections for the Transalp, Cifalps and EASI profiles will be discussed in section 2.3.

Local earthquake tomography (LET; Thurber 1983) consists of inverting arrival times of P (and S) waves generated by local earthquakes and recorded at a seismic array for absolute Vp, Vp/Vs ratio (if S wave arrival times have been picked) and hypocenter parameters (location, origin time). The LET method provides P-wave velocity images of the bulk crustal structure and not only velocity discontinuities. In the Alps, the LET may be inefficient in imaging the lower crust and the Moho because parameters are estimated between earthquake foci, which are mostly located in the upper, brittle crust and stations (located on the surface).

As outlined in the introduction, ambient noise tomography (ANT; Shapiro et al. 2005) has the potential to transform any seismic station into a source for other stations. The first step of ANT is to calculate the noise correlations between all station pairs available in the study region. If noise correlations are calculated over a time span long enough to ensure a sufficiently wide distribution of noise sources around the stations, they are dominated by the Rayleigh wave between the two stations. The second step is therefore a surface wave tomography. It involves measuring the group or phase velocity of the Rayleigh wave at different frequencies and inverting the resulting dispersion measurements to estimate a 3D S-wave velocity model under the station network. As for local earthquake tomography, ANT provides images of the bulk structure of the crust, in that case in S-wave velocity.

The forward modeling of gravity anomalies, in particular the Bouguer anomaly (onshore) is used to test crustal models derived from other geophysical methods, and geological and petrophysical data (e.g. Lardeaux et al. 2006). Inversion of gravity data can provide additional information provided that a suitable initial model is available. An important step forward will be provided in the next few months by the release of a new, homogeneous gravity database compiled from surface data across the Alpine area by the AlpArray gravity research group (Zahorec et al. 2021). Other methods exist, such as magnetotellurics (MT), which aims to estimate the conductivity of the subsurface, hence its fluid content; but it has not yet been used in the Alps, at least on a large scale.

2.2.2. *Crustal tomography at the scale of the entire Alpine belt*

2.2.2.1. *A review of Moho models from controlled source seismology, local earthquake tomography and RF*

The first 3D numerical model of Moho depth under the Alps was built by Waldhauser et al. (1998) from the results of seismic refraction and vertical-incidence reflection profiles. They systematically reviewed all the results, assigned to each reflector portion an uncertainty depending on the signal quality, and then they migrated these reflector portions in 3D before interpolating the data to determine the simplest and smoothest possible surface. Their Moho depth map presented in Figure 2.2(b) clearly shows the European Moho underthrusting the Adriatic Moho, which in turn underthrusts the Ligurian Moho. The locations of the vertical offsets between the three Moho surfaces are shown as dotted lines. It should be noted that there is little or no data available in the Southwestern Alps or in the eastern Po basin. The Ivrea body at the border between Adriatic Moho and European Moho under the Western Alps is not detected.

To complement the CSS data, Wagner et al. (2012) measured Moho depth in the Vp model of Diehl et al. (2009) derived from a LET. Since this LET uses a large number of arrival times of refracted Pn waves in the uppermost mantle, its resolution is fair down to the crust-mantle boundary. The Moho is approximated by the surface of isovelocity Vp = 7.25 km/s and the authors attribute to each depth measurement an uncertainty and a weight that depend in part on the velocity gradient. The interpolation procedure is identical to that of Waldhauser et al. (1998). The resulting map is shown in Figure 2.2(c). The main difference with the map of Waldhauser et al. (1998) in Figure 2.2(b) is the sharp rise of the Adriatic Moho to the top of the Ivrea body above the European Moho.

The densification of the permanent seismological networks has enabled the Moho depth maps to be progressively complemented with RF measurements (e.g. Lombardi et al. 2008; Piana Agostinetti and Amato 2009). Spada et al. (2013)

constructed a new Moho map for Italy by combining CSS and RF data in an interpolation similar to the one used by Waldhauser et al. (1998). They used the local earthquake tomographies of Diehl et al. (2009) and di Stefano et al. (2009) to verify the Moho depths obtained by RF and evaluate their uncertainties. The Alpine part of their map is shown in Figure 2.2(d). In the Western Alps, their Moho model is identical to that of Wagner et al. (2012) and essentially derived from the LET of Diehl et al. (2009), since CSS and RF data were scarce. On the other hand, it extends further south and includes the Ligurian Sea and the Corsica–Sardinia block. It also covers the Eastern Alps where it shows a Moho gap east of 12.5°E (gray-filled area in Figure 2.2(d)). This Moho gap is documented by the analysis of PmP waves (P waves reflected by the Moho at critical distance) recorded during the CELEBRATION2000 and ALP2002 CSS experiments (Behm et al. 2007). It is due to lack of data (i.e. PmP reflections) and/or poor data quality.

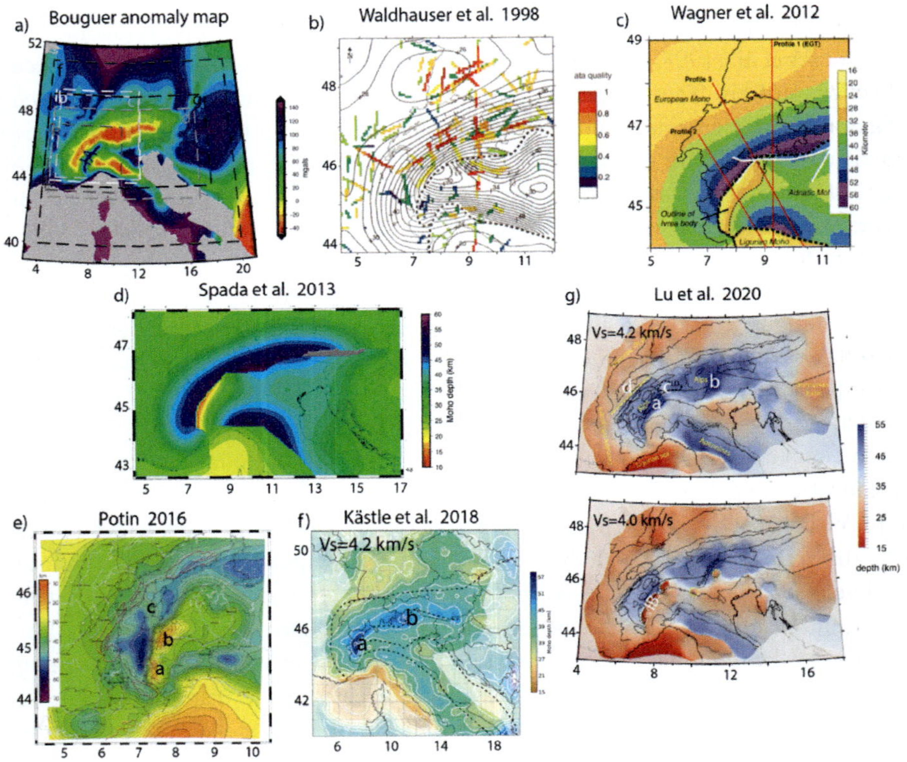

Figure 2.2. *Some depth maps of Alpine Moho, or proxy of Alpine Moho*

COMMENT ON FIGURE 2.2.– *(a) Bouguer anomaly map shown for comparison (World Gravity Map – 2012, Bonvalot et al. 2012). The areas enclosed by white, gray, black dashed lines and identified by letters (b–g) show the locations of maps displayed in Figures 2.2(b–g); IA: Ivrea anomaly. (b) Waldhauser et al. (1998): review and interpolation of seismic refraction and vertical-incidence reflection data; Moho depth mapped as contour lines; the colored line segments show the locations of the CSS data used, with color corresponding to data quality from poor (blue) to high (red); the thick black dotted lines show the locations of Moho steps. (c) Wagner et al. (2012): same CSS data as in (b) supplemented with segments of isovelocity surface Vp = 7.25 km/s from the LET of Diehl et al. (2009); the thin black dotted line is the outline of the Ivrea body. (d) Alpine part of the map in Spada et al. (2013): same as (c) supplemented with receiver function observations and CSS wide-angle data in the Eastern Alps; the gray-filled area is the so-called Moho gap of the Eastern Alps, where no wide-angle PmP reflection could be detected. (e) Potin (2016): depth of strongest vertical gradient in his Vp model computed by LET in the Western Alps. (f) Kästle et al. (2018): depth of isovelocity surface Vs = 4.2 km/s in the 3D Vs model calculated by ambient noise and teleseimic earthquake surface-wave tomography; main tectonic boundaries plotted as thick black dotted lines. (g) Lu et al. (2020): depth maps of isovelocity surface Vs = 4.2 km/s (top) and Vs = 4.0 km/s (bottom) in the Vs model calculated by ambient noise wave-equation tomography; main tectonic boundaries shown as black lines; IB: Ivrea Body; LD: Lepontine dome.*

The next Moho model shown in Figure 2.2(e) comes from the LET of Potin (2016) in the Western Alps. It is to date the best-resolved LET in the Western Alps because of the comprehensive database used, but it is not published yet. In the central area where the resolution is very good, the Moho depth is estimated from the strongest vertical Vp gradient (with Vp between 7.3 and 7.6 km/s). In more external areas, the map is obtained by averaging the depths of existing models, in particular that of Wagner et al. (2012). In the central region (5.5–9°E; 44–46.5°N), there are faster depth variations than in the very smooth models mentioned previously, including a two-block Ivrea body (labels “a” and “b” in Figure 2.2(e)) and an uprising of the European Moho under the Aosta region (label “c”). That particular geometry of the Ivrea body poorly corresponds to the Bouguer anomaly, which shows no evidence of the presence of two such differentiated blocks in the anomaly high of the Ivrea body (labeled “IA” in Figure 2.2(a)).

2.2.2.2. *S-wave velocity models from ambient noise tomography*

ANT has been applied several times since its emergence in the early 2000s to determine the crustal Vs structure under the Alpine arc. Stehly et al. (2009) used records from 150 permanent stations located mainly in the Western and Central Alps. This pioneering application to the Alpine region clearly demonstrated the ability of ANT to provide a spatially continuous 3D Vs model in a region where seismicity is not only moderate but also inhomogeneously distributed. Molinari et al. (2015a) used about the same stations as Stehly et al. (2009) in the Alps, and more stations in Italy. Their Vs model was derived from a nonlinear inversion of the group and phase velocities of the Rayleigh wave. This nonlinear inversion allowed them to explore the solution space in the vicinity of a priori models defined by synthesizing existing models, that is EPcrust (Molinari et al. 2011) for the European crust and MAMBo (Molinari et al. 2015b) for the Po basin. This inversion method led to a more robust crustal model than the linear inversion of Stehly et al. (2009). The Moho depth map is an improved version of the very smooth Moho model of Spada et al. (2013) that Molinari et al. (2015a) used as a priori model. The lateral resolution of this model remains however inadequate (100–200 km depending on the region) to add significant new constraints to the geological models.

Kästle et al. (2018) combined the Rayleigh and Love wave dispersion observations from ambient noise with dispersion data from regional and teleseismic records to compute a Vs model of the crust and upper mantle to 200 km depth. Measurements on earthquake records complement ambient noise data toward the low frequencies that are sensitive to deeper structures, from the base of the crust to the upper mantle. They used data from a larger number of permanent stations than previous publications, ensuring fairly good coverage of the entire Alpine arc. Like Molinari et al. (2015a), their 3D model of Vs is computed by stochastic inversion, but they explore a larger set of models with no a priori. The resolution of their model (= minimum size of the heterogeneities that the model is able to distinguish) is ~50 km in the upper and middle crust and ~100 km in the upper mantle, over all the Alps. The fast-velocity anomaly of the Ivrea body is as well detected as in the Vp model of Diehl et al. (2009). The depth map of the 4.2 km/s isovelocity surface, used as a proxy for Moho (Figure 2.2(f)) shows details that were not visible on the very smooth map by Spada et al. (2013) in Figure 2.2(d). The maximum depth of the European Moho is 55 km under the Western Alps, just in front of the Insubric line and the Ivrea body ("a" in Figure 2.2(f)), and 50 km in the Central Alps under the Giudicarie line ("b" in Figure 2.2(f)).

A further step in the number of stations used and thus the spatial resolution of the Vs model was taken by Lu et al. (2018) in an ANT of Europe using data from nearly 1,300 seismological stations, including the first six months of data from the temporary AlpArray network (Hetényi et al. 2018a). The Alps are particularly well covered by this Rayleigh-wave group-velocity data set, with an estimated resolution of 0.3° (~30 km) in the upper crust and 0.9° (~100 km) at Moho depth. The final Vs model resulting from a Bayesian nonlinear inversion is probabilistic in the sense that it includes the probability density on the values of Vs, but also on the depth of the interfaces including Moho. Lu et al. (2020) further improved the Vs model of Lu et al. (2018) in the Alpine region by inverting ambient noise phase dispersion data based on elastic waveform simulation. Unlike conventional ANT methods that rely on ray theory and apply a point-by-point 1D depth inversion of dispersion data for S-wave velocity, such a wave-equation tomography accounts for 3D and finite frequency effects. The resulting 3D Vs model is therefore more in line with the basic principles of seismic wave propagation. Figure 2.2(g) shows two depth maps of isovelocity surfaces Vs = 4.2 km/s (top) and Vs = 4.0 km/s (bottom) in the Vs model of Lu et al. (2020). The isovelocity surface 4.2 km/s (also used by Kästle et al. 2018; Figure 2.2(f)) is a good proxy for Moho everywhere except for the Ivrea body. The Ivrea body shows up on the 4.0 km/s depth map (dashed white line labeled "IB" in Figure 2.2(g), bottom) because it is made up of serpentinized Adriatic mantle whose S-wave velocities can be as low as 3.6–3.8 km/s (e.g. Reynard 2013). The 4.2 km/s map shows significant variations in the maximum depth of the European Moho along the arc, with two maxima below the Sesia zone in the Western Alps (52–54 km; "a" in Figure 2.2(g), top) and below the Giudicarie Line (54–56 km; "b" in Figure 2.2(g), top) in the central-Eastern Alps and a high point below the Lepontine Dome (40–42 km; "c" in Figure 2.2(g), top). The 4.2 km/s isovelocity depth map also shows a ~8 km abrupt step that closely follows the northwestern boundary of the external crystalline massifs of the Western Alps (yellow dashed line labeled "d" in Figure 2.2(g), top). This previously unknown structure coincides with the disappearance of the European Moho in the ECORS-CROP vertical seismic reflection section. Its rectilinear shape and SW–NE orientation suggest that it may be related to a Variscan lithospheric discontinuity. The Ivrea body is visible at 20–22 km depth on the 4.0 km/s map ("IB" in Figure 2.2(g), bottom), with a geometry that roughly corresponds to that of the positive Bouguer anomaly ("IA" in Figure 2.2(a)). The depth of the 3.6 km/s isovelocity surface (not shown here) is shallower than 10 km in some parts of the Ivrea body, in better agreement with the depth to the top of the high-velocity body observed with early refraction seismic experiments (Closs and Labrouste 1963).

2.2.3. *Crustal structure imaged along seismic transects across the Alps*

Seismic arrays provide 3D images of the lithospheric structure whose resolution is at best of the order of magnitude of the distance between stations, that is, ~ 40–50 km under most of the Alps (with the AlpArray seismic network). Geological structures are usually smaller in size. In order for the resolution of seismic tomography to tend toward the resolution of the geological map, sensor arrays must be made denser. For this reason, several dense temporary seismological experiments, with stations spaced 5–15 km apart, have been deployed along cross-chain profiles in the last ~ 20 years. The disadvantage of these experiments is that they usually only provide 2D sections of the subsurface. However, deploying sensor arrays a few kilometers apart over a significant area is still a logistical and, above all, a financial challenge. The Swath-D experiment of the 4D-MB project (Mountain Building in four dimensions, German component of AlpArray, http://www.spp-mountainbuilding.de) is such a sensor array (Heit et al. 2017, 2018). Note that 163 broadband sensors were installed in Austria and Italy from 2017 to 2019 in a rectangle of ~330 × 120 km^2 with a spacing of 12–15 km between stations (labeled "4" in Figure 2.1). The main objective of Swath-D is to study the assumed polarity flip of the subduction between Central and Eastern Alps and the Moho "hole" (Spada et al. 2013 and Figure 2.2(d)). Results are expected after the publication of this review.

2.2.3.1. *Recent results on the subduction channel of the Western Alps (Cifalps and AlpArray experiments)*

The temporary Cifalps experiment (Zhao et al. 2016a; see location in Figure 2.1, label "1") involved a profile of 46 stations with 5–10 km spacing across the Southwestern Alps from the southern Rhône valley (France) to the Po basin (Italy), and 10 offline stations to complete the coverage of permanent networks. The main objective of Cifalps was to study the structure of the Alpine subduction channel under the Dora Maira massif where the first mineralogical evidence of continental subduction was discovered by Chopin (1984) in the form of coesite inclusions in garnets. Coesite is a (U)HP/HT mineral formed by burying continental crust at a depth of 90 km or more.

Before Cifalps, two seismic experiments had been carried out in the late 1990s in the Southwestern Alps in the framework of the French GeoFrance-3D program. A temporary seismic array provided data for a LET focused primarily on the Ivrea body (Paul et al. 2001), which was the basis for a joint geological–geophysical interpretative crustal-scale cross-section (Lardeaux et al. 2006). A wide-angle experiment provided new data on the depth of the European Moho and failed to confirm the existence of the Briançonnais mantle flake proposed on the basis of the ECORS-CROP experiments (Thouvenot et al. 2007).

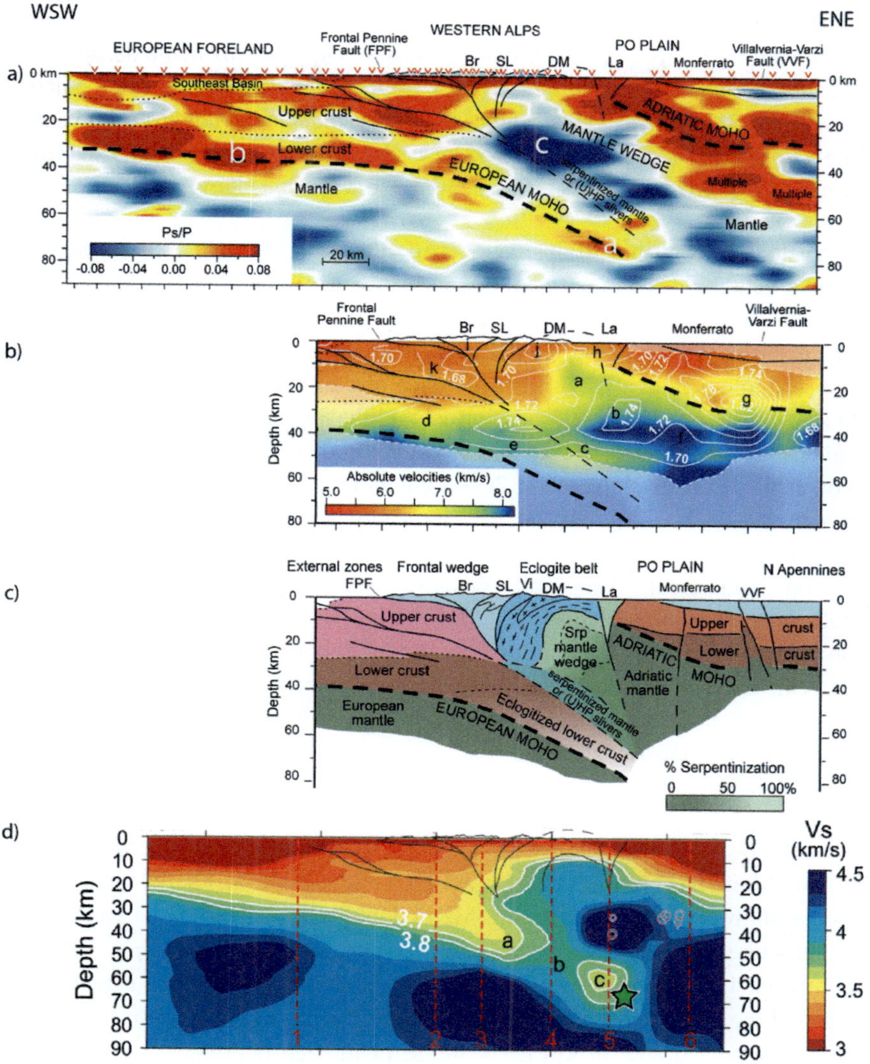

Figure 2.3. *Depth sections of seismic parameters along the Cifalps profile (Southwestern Alps, location in Figure 2.1, labeled "1"), and geological interpretation*

COMMENT ON FIGURE 2.3.– *(a) Result of receiver function analysis: common conversion point (CCP) section of Zhao et al. (2015) where the blue and red patches show converted waves on interfaces with velocity increasing (red) or decreasing (blue) with increasing depth; (b) result of local earthquake tomography (LET): P-wave velocity model of Solarino et al. (2018) and contours of Vp/Vs ratio (white lines); (c) interpretive geological section constructed from (a) and (b) (modified from Solarino et al. 2018); (d) result of transdimensional inversion of Rayleigh wave group velocity dispersion data: S-wave velocity model of Zhao et al. (2020); the small white circles are hypocenters of microearthquakes located in the vicinity of the section. Br: Briançonnais, DM: Dora Maira, FPF: Frontal Pennine Fault, La: Lanzo, SL: Schistes Lustrés, Vi: Viso, VVF : Villalvernia–Varzi Fault.*

Using the RF from the Cifalps data, Zhao et al. (2015) found P-to-S converted waves on a velocity discontinuity 75–80 km below the western edge of the Po plain (labeled "a" in Figure 2.3(a)) in the continuity of the European Moho, which is clearly visible further west ("b" in Figure 2.3(a)). They interpreted these signals as the first seismological evidence of the presence of the European Moho at a depth of 75–80 km, thus of subduction of the continental lithosphere under the Alps. A set of Ps conversions of negative polarity and high amplitude (blue color in Figure 2.3(a), labeled "c") is detected at a depth of 20–40 km under the Dora Maira massif and the western Po plain. The negative polarity indicates that signals are produced by a velocity discontinuity where the velocity decreases with increasing depth. Combining geological data and models, petrophysical data, gravity and seismological modeling, Zhao et al. (2015) proposed an interpretive crustal scale section that includes a thick suture zone under the Dora Maira massif (Figure 2.3(c)). This zone is characterized by velocities decreasing with increasing depth below the Ivrea body. It indicates that the European lower crust underthrusts the Adriatic mantle, almost vertical to the place where coesite was discovered by Chopin (1984).

The geophysical and geological model of the suture zone (or subduction channel) under Dora Maira has been refined by the LET of Solarino et al. (2018; Figure 2.3b). This paper shows that the (U)HP Dora Maira dome is, at a depth of 10 km, directly in contact with partially serpentinized peridotites of the Ivrea body (Vp ~ 7.5 km/s; Vp/Vs = 1.70–1.72; labeled "a" in Figure 2.3(b)). These serpentinites extend down to the top of the eclogitized European lower crust ("e" in Figure 2.3(b)). Toward the east and between 20 and 50 km deep, the serpentinized peridotites are in contact with the dry peridotites of the Adriatic mantle (Vp ~ 8 km/s; Vp/Vs > 1.74; labeled "b" in Figure 2.3(b)) from which they are separated by a quasi-vertical lithospheric and seismogenic fault marked by microearthquakes (red circles in Figure 2.3(d)). The region marked "c" in Figure 2.3(b) has lower Vp (Vp ~ 7.0–7.5 km/s) than the serpentinized peridotites of the Ivrea body. These could be more serpentinized peridotites than those of the Ivrea body and/or (U)HP continental rock slivers. At greater depth, the area marked "e" has higher Vp (~7.7 km/s) that is interpreted as eclogitized European lower crust. The seismic structure of the subduction channel can be seen here from the surface to the top of the eclogitized lower crust over a thickness of almost 50 km. The decrease in velocity between the Ivrea body (Vp ~ 7.5 km/s; "a" in Figure 2.3(b)) and the more strongly serpentinized peridotites (and/or (U)HP slivers) located below it (Vp ~ 7.0–7.5 km/s; "c" in Figure 2.3(b)) could explain the negative polarity of the Ps-converted waves in the CCP section ("c" in Figure 2.3(a)), provided that the S-wave velocities undergo the same change as Vp.

The objective of Zhao et al. (2020) was precisely to measure Vs in the subduction channel of the Western Alps. They calculated a probabilistic Vs model in the region of the Cifalps profile by transdimensional Bayesian inversion of the Rayleigh wave dispersion measurements of Lu et al. (2018). There is no a priori constraint (or very few a priori constraints) on the model in this type of inversion and all parameters, for example, the number of model layers, are estimated to best fit the data. The computations are much more cumbersome than in more classical inversions, but the velocity values resulting from such an inversion are more reliable than from other methods that require regularization of the inverse problem by ad hoc parameters.

The cross-section in the Vs model of Zhao et al. (2020) is shown in Figure 2.3(d). It shows a thick body of slow velocities (Vs = 3.6–3.8 km/s; labeled "c" in Figure 2.3(d)) at 50–70 km depth along the suture zone between the Adriatic mantle wedge above and the eclogitized lower crust and European mantle below. The decrease in Vs with depth under the Ivrea body is much clearer than in the Vp model of Solarino et al. (2018) whose resolution is not good beyond 50–60 km depth. Zhao et al. (2020) ascribe these very low velocities to serpentinites along the subduction channel based on laboratory velocity measurements and petrological and mineralogical data. The low viscosity of the serpentinites gives them an important role in the processes associated with continental subduction, both in the burial and exhumation of HP/HT metamorphic rocks (e.g. Guillot et al. 2009).

2.2.3.2. *Recent results on the Eastern Alps (TRANSALP and EASI experiments)*

The Eastern Alps have been traversed by two seismic profiles, TRANSALP and EASI. TRANSALP (1998–2001) was running north–south at 12°E longitude, and it crossed the Tauern Window between the Molasse Basin to the north and the Dolomites to the south ("C" in Figure 2.1). The experiment combined a vertical-incidence reflection seismic profile (Lüschen et al. 2006) and passive imaging with a temporary seismic array installed along the same transect (Kummerow et al. 2004). The CSS component of TRANSALP was the first reflection seismic profile in the Eastern Alps, and the only one to date. Installation of seismological stations along the CSS profile enabled a very interesting comparison between the line drawing of the CSS profile and the CCP section from the receiver function analysis (Figure 2.4(a)). Figure 2.4(a) shows that both methods image similar structures, with obviously much better vertical resolution for seismic reflection profiling. However, Kummerow et al. (2004) note a disagreement on the geometry of the Adriatic Moho, which is clearly horizontal at a depth of 40 km on the CCP section (red spot under the Dolomites in Figure 2.4(a)), whereas the line drawing of the CSS section shows a set of reflectors with an envelope that seems to dip northward and join the European Moho at a depth of ~55 km (black line segments in Figure 2.4(a)). They propose that the better vertical resolution of the reflection CSS profile allows the image of a pervasive deformation fabric in between the European and Adriatic Moho while the RF gives access to the Moho (as the velocity discontinuity between crust and mantle) on a large scale. In any case, the European lithosphere appears to subduct beneath the Adriatic lithosphere.

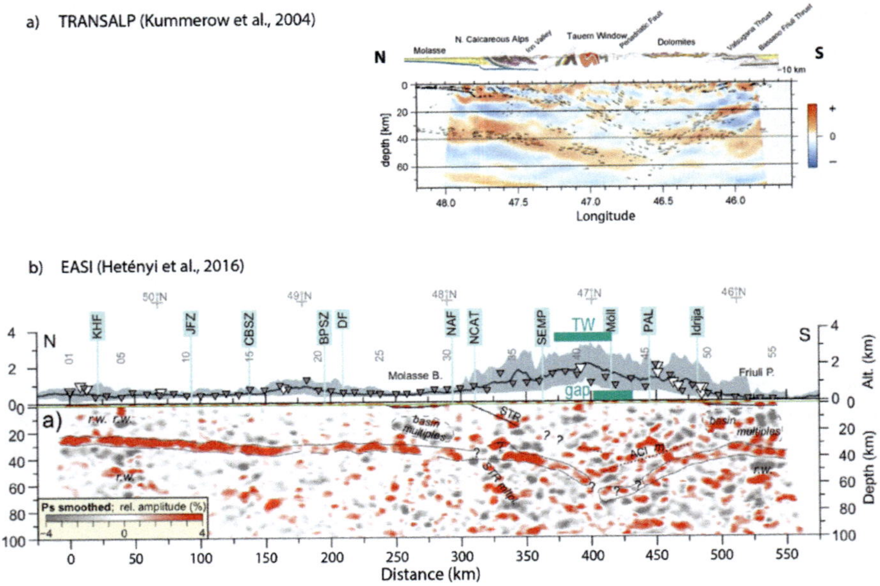

Figure 2.4. *Seismic imaging results across the Eastern Alps*

COMMENT ON FIGURE 2.4.– *(a) Transalp experiment (location in Figure 2.1, label "c"): line drawing of the vertical-incidence seismic profile (thin black line segments) and receiver function CCP section (modified from Kummerow et al. 2004); (b) EASI experiment (location in Figure 2.1): receiver function CCP section by Hetényi et al. (2018b). ACI: Adriatic Crustal Interface, BPSZ: Bavarian Pfahl Shear zone, CBSZ: Central Bohemian Shear Zone, DF: Danube Fault, JFZ:*

Jáchymov Fault Zone, KHF: Krušné Hory Fault, NAF: Northern Alpine Fault, NCAT: Northern Calcareous Alps Thrust, PAL: Periadriatic Line, SEMP: Salzach–Ennstal–Mariazell–Puchberg fault, STR: Sub-Tauern Ramp, TW: Tauern Window, r.w.: ringing waveform.

The EASI profile (AlpArray Seismic Network, 2014) is a long passive seismic profile located at 13.3°E longitude from the Bohemian massif in the north to the Adriatic coast in the south ("5" in Figure 2.1). It crosses the Moho gap documented in particular by Spada et al. (2013) (see section 2.1.2.1 and Figure 2.2d) south and east of the Tauern Window. The CCP section computed by Hetényi et al. (2018b) is shown in Figure 2.4(b). The Ps converted wave on the European Moho in the north is clearer and laterally more continuous than that of the Adriatic Moho in the south. It is hardly detectable in the suture area under the Tauern window (TW). Hetényi et al. (2018b) carried out a very detailed analysis of RF, including waveform inversion, to propose a model in which the Adriatic Moho underthrusts the European Moho to a depth of 70 km. They propose that the contact zone is characterized by low-amplitude converted waves (marked with "?" at 60–70 km depth in Figure 2.4(b)) produced by a transitional Moho where velocity gradually increases over ~20 km of thickness. Since this transitional Moho can only be detected by low-frequency (and long-wavelength) waves, it is invisible to seismic reflection, which may explain the Moho gap.

Qorbani et al. (2020) used ANT and part of the EASI dataset to compute a crustal 3D shear-wave velocity model in the area [46°N–48°N; 9°E–17°E] centered on the Tauern Window. As the model is restricted to a maximum depth of 40 km, it provides no information on the contact zone between the European Moho and the Adriatic Moho. The high velocities associated with the metamorphic rocks that form the Tauern Window at shallow depth extend to the bottom of the model at 40 km depth. A comparison is made between a vertical section through the Vs model along the TRANSALP line and a crustal-scale geological cross-section. The boundary between high and low velocities documented by the Vs section is in agreement with the geological boundaries assumed for the Tauern Window to 10 km depth, but it does not correspond to any geological boundary at 10–30 km depth. Qorbani et al. (2020) suggest interpretations for the Vs boundary, such as a post-nappe metamorphic front to the south of the Tauern Window. This example confirms that such tomographies have a great potential to extend geological maps at depth and in 3D. It also shows that geological models have to be reconsidered in the light of the results of seismic tomography.

2.3. Probing subduction slabs and mantle fabrics

2.3.1. *Subduction slabs*

As stated in the introduction, positive Vp anomalies of typical subduction slab geometries in the upper Alpine mantle have been known since the 1990s (Babuška et al. 1990; Cattaneo and Eva 1990; Spakman 1990, 1991). The tomographies of Piromallo and Morelli (2003) at the scale of the whole Euro–Mediterranean region and of Lippitsch et al. (2003) in the Alps were breakthroughs in terms of both the quantity and quality of the data used. The teleseismic tomography of Lippitsch et al. (2003) showed a northeastward dip of the fast-velocity anomaly under the Eastern Alps, while the fast-velocity slab dips southeastward under the Central and Western Alps. They interpreted these anomalies as traces of lower continental lithosphere, with the European lithosphere being the lower plate in the Western and Central Alps and the Adriatic plate the lower plate in the Eastern Alps. Interpretation of anomalies as continental rather than oceanic in origin is not based on a difference in the seismic image. The oceanic crust has a different structure (thickness, lithology and therefore seismic velocity) from that of the continental crust, which makes it possible to distinguish them by seismic tomography. This is not the case of the lithospheric mantle, which has similar properties for the two types of lithosphere. The interpretation of Lippitsch et al. (2003) was therefore based on the comparison between the length of the high-velocity slabs and the shortening values estimated by structural geology.

The change in subduction polarity along strike as well as the highly segmented aspect of Alpine slabs has been the targets of several recent tomographies. They have benefited from the development of permanent networks and the opening of databases. Upper mantle tomography has not benefited as much as crustal tomography from the methodological innovations related to the use of ambient noise correlations. We saw in section 2.1.1 that ANT can be used to perform surface-wave tomography without earthquakes. However, the empirical Green functions reconstructed by noise correlation have the same frequency content as the microseismic noise, so most of their energy is concentrated between a 5 and 30 s period. They are, therefore, only slightly sensitive to upper mantle depths. As Kästle et al. (2018) have stated, the surface wave spectrum can be extended to lower frequencies by using records of teleseismic events to illuminate the upper mantle, but the horizontal resolution of surface wave tomography remains limited. The most widely used method for imaging the upper mantle remains teleseismic tomography. The main limitation is vertical resolution, since teleseismic tomography uses arrival times of P (or PKP, etc.) waves that travel through the mantle in near-vertical rays. Its horizontal resolution depends on the station density. Input data of teleseismic tomography are most often travel time differences (residuals) between stations of an array measured for a set of teleseismic earthquakes. Therefore, the values of velocity anomaly (dVp/Vp) resulting from the inversion at a given depth refer to an unknown horizontally averaged Vp. They cannot be used to reconstruct the absolute Vp model.

FWI on a regional scale has not yet demonstrated its full potential in applications to the Alps (on a regional scale: Beller et al. 2018; on a continental scale: Zhu et al. 2015; Fichtner et al. 2018). In a recent review of published tomographies of the upper mantle under the Alps, Kästle et al. (2020) compares the results of Zhu et al. (2015) and Fichtner et al. (2018) with those of different teleseismic tomographies (their Figures S1 and S2 in the supplementary information file). Horizontal resolution is clearly not (yet) at the level of what can be achieved by teleseismic tomography, but new results are expected in the near future with AlpArray data (Paffrath et al. 2020).

We shall return in the following sections to a series of questions raised in Lippitsch et al. (2003).

2.3.1.1. *With or without slab tears?*

The tomography of Lippitsch et al. (2003; Figures 2.5(a) and (d)) shows on all depth slices from 90 to 240 km depth three well-individualized fast-velocity anomalies along the arc of size ~150 km × 150 km. At a depth of 200 km (Figure 2.5(d)), they are located under the western Po plain in the Western Alps, slightly west of the Giudicarie line in the Central Alps, and northeast of the eastern end of the Tauern Window in the Eastern Alps. The anomaly amplitudes are similar at 200 km depth; but at depths shallower than 150 km, the western anomaly is attenuated, and even positive at 120 km. The along-dip amplitude change in the Western Alps is clearly visible in the cross-section of Figure 2.6(a), while the dip of the deep high-velocity slab is ~50°. Lippitsch et al. (2003) interpreted this attenuation of the amplitude of the positive anomaly at 100–150 km as an indication of a breakoff of the European (Alpine) slab in the Western Alps. Subsequently, this hypothesis has been used in a significant number of publications, for example to explain the rapid exhumation measured in the Western Alps since 2 Ma (Fox et al. 2015), or the present fast uplift rates measured by GNSS in the highest regions of the Western Alps (Nocquet et al. 2016; Sternai et al. 2019).

Every teleseismic arrival time tomography published since Lippitsch et al. (2003) shows a continuous fast-velocity slab across the upper mantle of the Western Alps (Koulakov et al. 2009; Zhao et al. 2016b; Hua et al. 2017; see Figures 2.5 and

2.6(b) and (c)). It is also continuous in the tomography of Piromallo and Morelli (2003), as well as in the tomographies of the crust and mantle by FWI of Zhu et al. (2015) and Fichtner et al. (2018), even if an argument can be made about the low resolution of the latter two models. As mentioned earlier, teleseismic tomography has a better resolving power in the horizontal direction than in the vertical direction, due to the near-vertical incidence of rays from teleseismic sources. Horizontal resolution depends primarily on the spatial coverage of the station array and the distribution of earthquakes around the study region, that is, on the dataset used. In the Western Alps, the low vertical resolution problem is further enhanced by the presence of the Ivrea body at crustal depths. Its anomalously fast velocities (as compared to normal continental crust) generate arrival time advances that are difficult to separate from the time advances produced by the subduction slab in the mantle. To prevent the Ivrea body anomaly from leaking vertically and masking a possible slab breakoff, an optimal ray crisscrossing is required in the uppermost mantle, that is, a dense network of stations and a uniform distribution of earthquakes in back-azimuth. The second condition is never met in the Alps as very few events have back-azimuths between 100° (ESE) and 250° (WSW). In addition to data from the permanent networks, Zhao et al. (2016b) used data from the temporary Cifalps network, with an average interstation distance of 7 km and five stations located directly above the Ivrea body (see location in Figure 2.1, label "1"). They used finite frequency tomography, which is more in line with the physics of wave propagation since it takes into account the sensitivity of waves to the medium around the ray, at a distance depending on the frequency of the used signals. Conventional teleseismic tomography (known as ACH in reference to the authors of the reference publication, Aki et al. (1976) uses infinite frequency approximation and considers that arrival times are only influenced by media traversed by infinitely thin rays. Zhao et al. (2016b) avoid vertical smearing thanks to the station density along the Cifalps profile and, to a lesser extent, thanks to finite frequency tomography. Using synthetic examples, they demonstrate that, under the Cifalps profile, their tomography can reconstruct a fast-velocity anomaly located in the crust (akin to Ivrea body) with negligible leakage to mantle depths. If a slab breakoff did exist beneath the western Po Plain, I am convinced that it would be detected by the tomography of Zhao et al. (2016b).

Most tomographies have applied crustal corrections calculated in an ad hoc model, for example, the model of Waldhauser et al. (2002) for Lippitsch et al. (2003), to overcome the leakage problem of the Ivrea body anomaly in the mantle. Therefore, the result of the tomography is not only determined by the observations, but also by the a priori model.

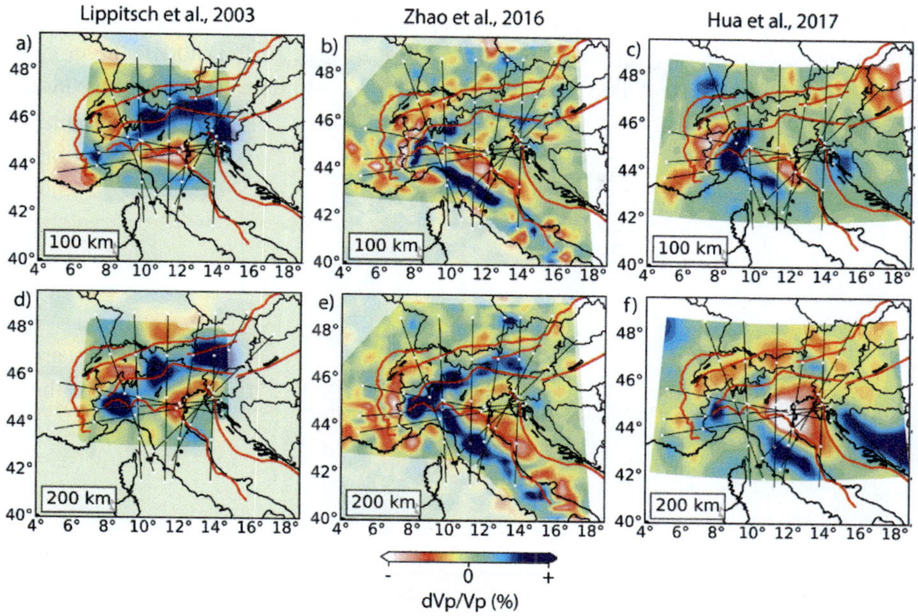

Figure 2.5. *Comparison of depth slices at 100 km (top) and 200 km (bottom) in three dVp/Vp tomographic models published for the Alpine mantle. a) and d): Lippitsch et al. (2003); b) and e): Koulakov et al. (2009); c) and f): Zhao et al. (2016b)*

COMMENT ON FIGURE 2.5.– *The boundaries of the color scales are ±5% for (a) and (d), ±3% for (b) and (e) and ±4% for (c) and (f). Thick red lines show the main tectonic boundaries. Thin black lines show the locations of cross-sections, three of which are shown in Figure 2.6. Modified from Kästle et al. (2020, supplementary information).*

Kästle et al. (2020) provides a comprehensive review of recent tomographic models of the upper mantle under the Alps. In addition to the body-wave tomographies mentioned above, they discuss the contributions and limitations of the surface-wave tomography of Kästle et al. (2018) that combines observations of Rayleigh wave phase velocities from noise correlations (for the crust, see section 2.1.2.2) and teleseismic recordings for the mantle. Surface-wave tomography has better vertical than horizontal resolution and can therefore complement body-wave tomography. The Vs model of Kästle et al. (2018) covers depths shallower than 200 km. They find a fast anomaly limited to the first 100 km depth in the Western Alps, whereas it extends to 150–200 km in the Central Alps. They also detect a fast mantle anomaly under the Po plain, which they ascribe to the Apenninic subduction. Other tomographies, however, show rather slow velocities along this same section under the Po plain. The surface wave tomography of Lyu et al. (2017) along the Cifalps profile finds fast velocities down to 200 km (base of the model) under two mini-arrays of the western Po plain, in contradiction with the results of Kästle et al. (2018).

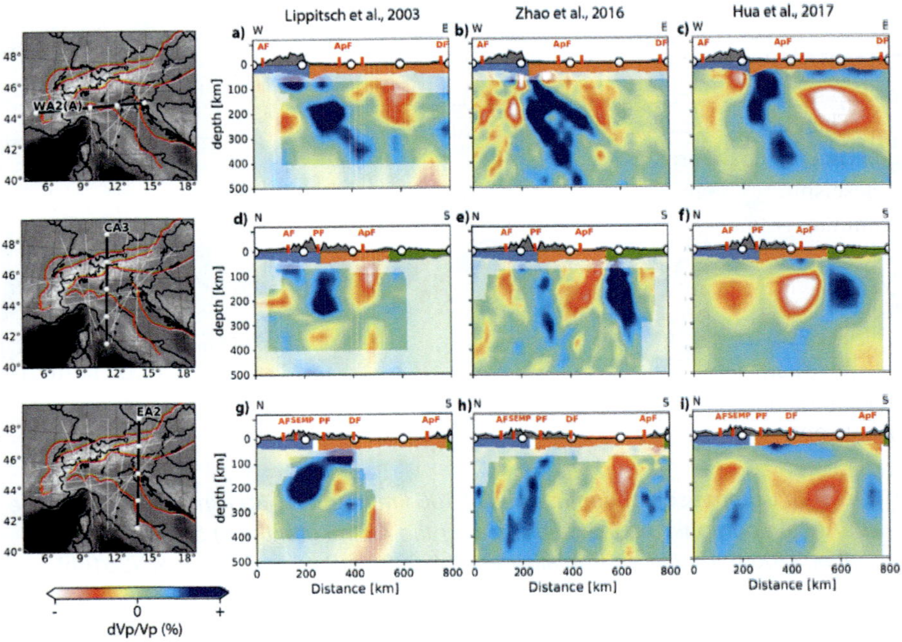

Figure 2.6. *Comparison of vertical cross-sections through the tomography models of (a, d, g) Lippitsch et al. (2003), (b, e, h) Koulakov et al. (2009) and (c, f, i) Zhao et al. (2016b). Locations of cross-sections are shown on the right: Western Alps (top), Central Alps and transition to Apennines (middle), Eastern Alps (bottom)*

COMMENT ON FIGURE 2.6.– *The boundaries of the color scales are ±5% for (a, d, g) and ±3% for (b–c), (e–f) and (h–i). Unresolved regions are masked with lighter colors. Crustal thickness profiles from Spada et al. (2013) are shown on top of each section, with blue color corresponding to European crust, orange to Adriatic crust, green to Ligurian–Tyrrhenian crust and white to the slab gap in the Eastern Alps. Red lines and annotations are the main tectonic boundaries: AF: Alpine Front, ApF: Apenninic Front, DF: Dinaric Front, PF: Periadriatic Fault, SEMP: Salzach–Ennstal–Mariazell–Puchberg fault. Modified from Kästle et al. (2020, supplementary information).*

The teleseismic full-waveform inversion of Beller et al. (2018) on Cifalps data detected a low Vs anomaly between 70 and 120 km depth that they propose to interpret as a trace of the breakoff of the Western Alps slab. This anomaly is more superficial than those of Lippitsch et al. (2003) or Kästle et al. (2018). Moreover, the slow anomaly is located at the edge of the European Moho detected by RF by Zhao et al. (2015). It is therefore more likely related to subduction of the European crust or to (U)HP rocks and/or serpentinites in the subduction channel (Zhao et al. 2020).

The AlpArray data will certainly provide decisive elements on the existence or absence of slab breakoffs in the Western Alps. The preliminary results of the teleseismic tomography of Paffrath et al. (2020) rather point toward a vertically continuous slab in the Western Alps as in the other parts of the chain.

2.3.1.2. *Relationship between the Alpine and Apenninic slabs*

The lack of continuity between the southwestern end of the Alpine slab and the northwestern end of the Apenninic slab is in line with recent kinematic reconstructions in the Mediterranean (e.g. Jolivet and Faccenna, 2000; Vignaroli et al. 2008; Dumont et al. 2012; Malusà et al. 2015). The Apenninic subduction retreated eastward from the end of the Oligocene, while the Adriatic microplate was still moving northward with respect to Europe, resulting in continental collision in the Alps (Malusà et al. 2016).

The two subductions of the European lithosphere under the Western Alps and the Adriatic lithosphere under the northern Apennines are very close and have opposite dips toward the East for the Alps and toward the Southwest for the Apennines (e.g. Piromallo and Morelli 2003). Imaging the deep structure at the intersection of two subductions of opposite dips and two arcuate mountain ranges is a real challenge for seismic tomography. The cross-sections of Figure 2.6(d) (Lippitsch et al. 2003) and Figure 2.6(e) (Koulakov et al. 2009) only display the southward dipping Alpine slab. The question of the relationship between Alpine (European) and Adriatic (Apenninic) slabs was addressed by Vignaroli et al. (2008) on the basis of the tomography of Piromallo and Morelli (2003), and the reinterpretation of geological data and paleotectonic reconstructions. Vignaroli et al. (2008) show an N–S section in model PM0.5 (Piromallo and Morelli 2003) across the Central and Ligurian Alps that display the south-dipping Alpine slab under the Central Alps, but no fast anomaly to the south. The two fast anomalies of the Alpine and Apenninic slabs seem horizontally disconnected on an NW–SE section. By combining this information with geological arguments and paleotectonic reconstructions, Vignaroli et al. (2008) proposes that the retreat of the Apenninic slab induced an outward toroidal asthenospheric flow from below the retreating Apenninic slab through the gap between the two slabs. This asthenospheric flow in turn forced the general westward retreat of the Alpine slab. The more recent teleseismic tomography by Giacomuzzi et al. (2011) shows that fast anomalies located beneath the northern Apennines merge with the fast anomaly beneath the Western Alps between 140 and 180 km depth beneath the Ligurian Alps.

Zhao et al. (2016b) showed that the two fast anomalies that remain distinct to 160 km coalesce from 180 km downward, where the north-dipping slab fragment at the northern end of the Apennine slab comes into contact with the European slab (Figure 2.6(f)). Zhao et al. (2016b) question the origin, Alpine or Apenninic, of this surprisingly north-dipping high-velocity anomaly observed at depths < 200 km along the Alps–Apennines transition zone (Figure 2.6(f)). Its location in the continuation of and at the northern end of the Apennine slab is more in line with an Apenninic origin. The northernmost tip of the Apenninic slab would then have been strongly overturned from a vertical to southwestward dip to a northward dip. Zhao et al. (2016b) rather favor an Alpine origin for the short north-dipping fast-velocity slab fragment because it is located beneath the Ligurian Alps that were structured by Alpine subduction (Malusà et al. 2015) and rotated counterclockwise by up to 90° since the Oligocene (e.g. Maffione et al. 2008). In this case, the southernmost tip of the European slab initially dipping to the east would have rotated to a northward dip while maintaining its dip angle during rotation.

Independently of the origin of the shallow slab fragment, the proposal by Vignaroli et al. (2008) of a toroidal asthenospheric flow at the northern end of the Apennine slab is incompatible with the close contact between the two slabs in the model by Zhao et al. (2016b). This issue will be further discussed in section 2.2.2 on seismic anisotropy.

2.3.1.3. *Northeast-dipping slab in the Eastern Alps*

Lippitsch et al. (2003) first imaged a gap between the high-velocity anomaly of the Central Alps and the one of the Eastern Alps, coinciding with a strong dip change from southeastward in the Central Alps to northeastward in the Eastern Alps (Figure 2.6(g)). They proposed that the Adriatic mantle lithosphere subducts northeastward beneath the European plate. This northeastward dip, though closer to vertical, was later confirmed by the tomographies of Koulakov et al. (2009; Figure 2.6(h)), Zhao et al. (2016b; Figure 2.6(i)) and Hua et al. (2017). The teleseismic tomography by Mitterbauer et al. (2011) displays a steeply to vertically dipping slab below the Eastern Alps, that they interpret as European lithosphere unlike Lippitsch et al. (2003) and other publications cited above. For Koulakov et al. (2009), the polarity of the slab is unclear, and both the Eurasian and Adriatic plates appear to be connected to the high-velocity body (Figure 2.6(h)).

An important question that can help us to understand the origin of the polarity reversal is its precise location along the arc in the tomographic studies that document it. In Zhao et al. (2016b), Figures 2.5(c) and (f) show that the high-velocity anomalies at 100 km and 200 km coincide with the Periadriatic fault (thick red line) west of longitude 13°E, while east of 13°E, the anomaly at 200 km depth is shifted northward by ~70–80 km with respect to the anomaly at 100 km depth. The slab polarity change from vertical to northeastward is therefore located at ~13°E longitude. Zhao et al. (2016b) interpret the northeastward dipping anomaly as Adriatic mantle lithosphere and they link it to the Dinaric slab. The less well-resolved tomography by Lippitsch et al. (2003) shows a broad gap between the Central Alps slab and the Eastern Alps slab, and a northward shift of ~70–80 km between the anomaly at 100 and 200 km (Figures 2.5(a) and (d)). The slab polarity reversal is located in the slab gap between 12°E and 13°E. Therefore, the change in slab dip between the Central Alps and the Eastern Alps is located at ~13°E longitude in the two publications that document it.

To summarize, all teleseismic tomographies since Lippitsch et al. (2003) show a vertical or steeply northeast-dipping high-velocity anomaly beneath the Eastern Alps east of ~13°E. This polarity change is confirmed by the receiver-function cross-sections along the TRANSALP line (12°E), where the European lower crust seems to underthrust the Adriatic lower crust (Figure 2.4(a)), and along the EASI line (13.3°E) where the Moho shift appears to be in the opposite direction, with Adria underthrusting Europe (Figure 2.4(b)). To move forward on the issue of the nature of the high-velocity slab, European or Adriatic, seismic tomography results should be compared with geological data, such as shortening estimates (e.g. Kästle et al. 2020), and hypotheses should be tested by geodynamic modeling.

2.3.2. *Mantle fabrics imaged by seismic anisotropy*

Only azimuthal anisotropy will be considered here. A medium is azimuthally anisotropic when its seismic velocity depends on the wave propagation (or polarization) direction. Azimuthal anisotropy in the mantle is mostly attributed to strain-induced lattice-preferred orientation of olivine crystals (Nicolas and Christensen 1987). It is therefore a proxy of mantle deformation and/or mantle flow. As the crust has a more heterogeneous mineralogical composition than the mantle, crustal anisotropy may be caused by preferentially aligned joints or microcracks in fault gouges for example, by thin layering in sediments, or by highly foliated metamorphic rocks. The most often used marker of seismic anisotropy is shear-wave splitting (SWS) of core-refracted phases, mostly SKS and/or SKKS from teleseismic earthquakes (Silver and Chan 1991). The crustal contribution to the SWS observations from SK(K)S phases is a question of debate as the measurement integrates the effect of all sources of anisotropy along the raypath. This question is quickly resolved in most publications by considering that the delay time between the two split waves, namely the one polarized in the fast direction and the one polarized in the slow direction, which is often larger than 1 s, is too large for anisotropy to be confined to the crust. Fry et al. (2010) and Schippkus et al. (2020) are the only publications to date that infer azimuthal anisotropy in the crust, for Switzerland and the Vienna basin, respectively. Both publications derive azimuthal anisotropy of Rayleigh waves from ambient noise correlations at periods that are sensitive to crustal structure. They discuss fast-velocity directions (FVD) with respect to regional structure and stress field rather than the strength of anisotropy. However, both publications document strong variations in direction of fast velocity with depth (or period), suggesting a weak contribution of the crust to total azimuthal anisotropy in the studied regions. As most studies suggest that the mantle is isotropic below 400–600 km (see review in Savage 1999 and references therein), SWS is considered as a proxy of strain in the upper mantle.

The numerous observations of SWS from core phases in the greater Alpine region are shown in Figure 2.7. Most fast polarization directions (FPD) tend to be parallel to the trend of the belt while anisotropy is stronger in the external than in the internal parts of the belt (e.g. Barruol et al. 2004; Barruol et al. 2011; Bokelmann et al. 2013; Qorbani et al. 2015; Salimbeni et al. 2018). These observations have led Barruol et al. (2004, 2011) to propose an asthenospheric source of anisotropy in the Western Alps, where present-day or recent mantle flow induced by the retreat of the Apenninic slab would be deflected by the European slab. This hypothesis has been taken up and extended by Salimbeni et al. (2018) who provided additional measurements in the Western Alps and the Po plain. Their new observations contradict the proposal by Vignaroli et al. (2008) of a counterclockwise toroidal flow across the gap between the Alpine and Apenninic slabs. They instead propose that fossil fabrics may be still preserved within the Alpine and Apenninic slabs to explain the weak SWS delay times and the null anisotropy measurements in the Po plain. In the external Western Alps, Salimbeni et al. (2018) follow Barruol et al. (2004) in their proposal that arc-parallel FPD are induced by a suction effect at the scale of the supraslab mantle in response to the progressive rollback of the Apenninic slab. The absence of a counterclockwise toroidal flow at the northern edge of the Apenninic slab could enhance the suction effect in the mantle of the Western Alps.

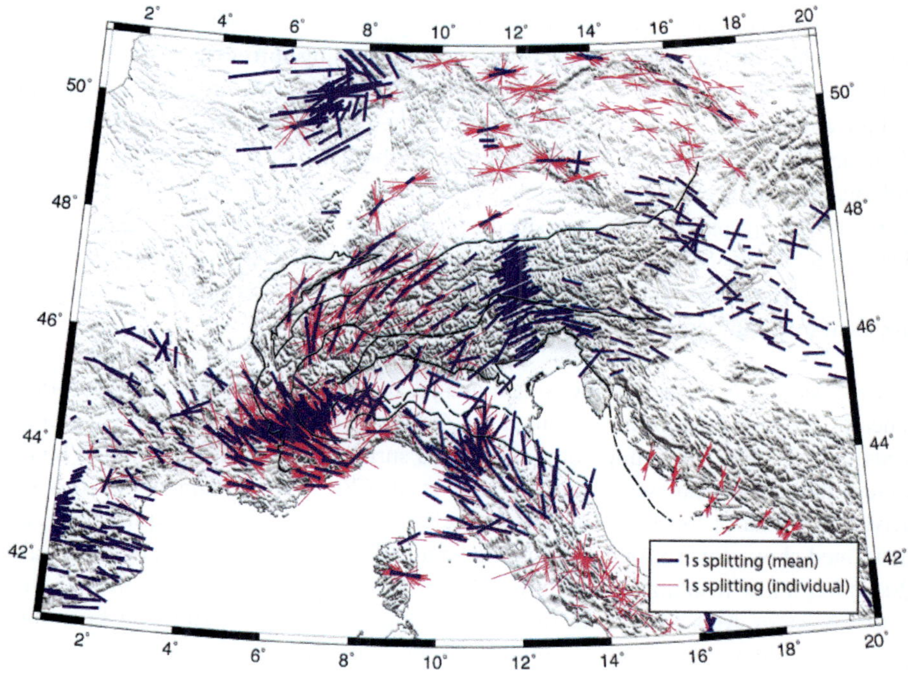

Figure 2.7. *Map of shear-wave splitting measurements in and around the Alps. Station-averaged measurements are shown as thick blue vectors, while individual measurements (from a single earthquake) are shown as thin pink vectors. Vector orientation indicates the fast-polarization (or fast-velocity) direction. Its length is proportional to the time delay between the fast and the slow split phases. Data extracted from the SplitLab SWS database (Barruol et al. 2009; Wüstefeld et al. 2009) complemented with data from Salimbeni et al. (2018)*

In the Central and Eastern Alps, Bokelmann et al. (2013) and Qorbani et al. (2015) observed a similar simple pattern but with a 45° abrupt change at longitude ~12°E from SW–NE azimuths of the FPD in the west to NW–SE azimuths in the east. This change spatially coincides with the change in the dip of the fast-velocity slab in the upper mantle discussed in section 2.2.1.3, from southward west of 12°E, to vertical-northeastward east of 12°E. To explain the dependence of the splitting parameters with the back-azimuth of the incident wave, Qorbani et al. (2015) proposed a two-layer source of anisotropy east of the transition zone. The deeper layer would have a similar FPD as the Central Alps (NE–SW), in relation with the eastward-to-vertically dipping high-velocity body interpreted as a detached European slab, while asthenospheric flow above the detached slab would explain the NW–SE FPD of the upper layer.

The quasi-vertical incidence angle of core phases (SKS, SKKS) used in SWS measurements hinders their ability to provide compelling information on the depth of the source of anisotropy. In a recent paper, Löberich and Bokelmann (2020) apply to the Central Alps a new procedure based on the use of non-vertical SKS arrivals. Their detailed analysis of SWS observations in the northern Central Alps supports with quantitative arguments the hypothesis proposed by Barruol et al. (2011) that the source of the arc-parallel FPD is asthenospheric flow around the Alpine keels. The characteristics of SWS measurements in the southern Central Alps are more complicated and may contain effects of frozen anisotropy in the subduction slabs, as proposed by Salimbeni et al. (2018).

Last but not least, three-dimensional images of azimuthal (and radial) anisotropy beneath the Alps have been computed by Zhu et al. (2015) using FWI of regional earthquake records (including both long-period surface waves and short-period body waves) on a European scale, and by Hua et al. (2017) using joint inversions of arrival time data of local earthquakes and teleseismic events on the Alpine belt. As mentioned above for isotropic velocity, the azimuthal anisotropy model of Zhu et al. (2015) has a low resolution and shows a large-scale variation of FVD from NW–SE in the Western Alps to N–S in the Eastern Alps at depths of 100–200 km. The spatial resolution is better in the model of Hua et al. (2017) that displays a similar arc-parallel trend of FVD as the SWS measurements. The difference in anisotropy amplitude between the outer and the inner parts of the range is however much less clear in their model than in the SWS results in Figure 2.7. The greatest variation in anisotropic parameters with depth is observed around 100 km, and these parameters vary little in the upper mantle. Consequently, the 3D model of Hua et al. (2017) does not provide convincing evidence on the source of azimuthal anisotropy. The origin of this problem is perhaps to be found in the assumption of horizontal azimuthal anisotropy, which is wrong in strongly dipping subduction slabs.

2.4. Conclusion

Compared to the Himalaya–Tibet system or the Andes, the Alps are a small mountain range. Yet they have a special place in all textbooks on continental collision because the amount of geological data available is unmatched for any other orogen, to paraphrase Kissling (1993) but replacing "geophysical" by "geological". Since the early 2000s, several geophysical experiments have been conducted in an attempt to bridge the huge gap between the level of geological and geophysical knowledge. Indeed, the Alps, as part of the intricate subduction–collision system of the Mediterranean region, have a complex structure that is difficult to probe. New imaging methods, such as ANT or FWI, have improved the resolution and reliability of images of the crust and upper mantle. New seismic experiments with close station spacing have brought the resolution of tomography closer to the scale of geological observations, but such experiments are generally confined to a profile and tomography is limited to a 2D section. In combination with permanent networks, the AlpArray seismic experiment (2016–2020) has provided uniform coverage of the greater Alpine region and an unprecedented wealth of data for one of the most emblematic collision belts. The analysis of this unique data set has just started and it will continue for several years ahead. For studies of the crust and mantle structure of the Alps, the data revolution has only just begun.

2.5. Acknowledgments

This chapter was written while I was P.I. of the AlpArray-FR project, the French component of the AlpArray initiative, which was funded by Agence Nationale de la Recherche (contract ANR-15-CE31-0015) and Labex OSUG@2020 (Investissement d'Avenir, ANR-10-LABX-56). The CIFALPS project, which was also key in my recent work on the Western Alps, is supported by the China Academy of Sciences (CAS) and the National Natural Science Foundation of China (NSFC). I warmly thank my colleagues of the CIFALPS team who made it possible for our field and research work on the Western Alps to be carried out in such a pleasant and stimulating atmosphere. I also thank Kästle for his useful plots that compare the results of the numerous tomographies of the Alpine mantle. Reviews by Bokelmann and an anonymous reviewer were of great help to improve this chapter.

2.6. References

Aki, K., Christoffersson, A., Husebye, E.S. (1976). Three-dimensional seismic structure of the lithosphere under Montana Lasa. *Bull. Seis. Soc. Am.*, 66(2), 501–524.

AlpArray Seismic Network (2014). Eastern Alpine Seismic Investigation (EASI) – AlpArray Complimentary Experiment. AlpArray Working Group. doi: 10.12686/alparray/xt_2014.

AlpArray Seismic Network (2015). AlpArray Seismic Network (AASN) temporary component. AlpArray Working Group. Other/seismic network. doi:10.12686/alparray/z3_2015.

Babuška, V., Plomerova, J., Granet, M. (1990). The deep lithosphere in the Alps: A model inferred from P residuals. *Tectonophysics*, 176, 137–165.

Barruol, G., Deschamps, A., Coutant, O. (2004). Mapping upper mantle anisotropy beneath SE France by SKS splitting indicates Neogene asthenospheric flow induced by Apenninic slab roll-back and deflected by the deep Alpine roots. *Tectonophysics*, 394, 125–138.

Barruol, G., Wüstefeld, A., Bokelmann, G. (2009). SKS splitting database. Université de Montpellier, Laboratoire Géosciences. doi:10.18715/sks_splitting_database.

Barruol, G., Bonnin, M., Pedersen, H., Bokelmann, G.H.R., Tiberi, C. (2011). Belt-parallel mantle flow beneath a halted continental collision: The Western Alps. *Earth and Planetary Science Letters*, 302, 429–438.

Bayer, R., Carozzo, M.T., Lanza, R., Miletto, M., Rey, D. (1989). Gravity modelling along the ECORS-CROP vertical seismic profile through the Western Alps. *Tectonophysics*, 162, 203–218.

Behm, M., Brueckl, E., Chwatal, W., Thybo, H. (2007). Application of stacking and inversion techniques to three-dimensional wide-angle reflection and refraction seismic data of the Eastern Alps. *Geophys. J. Int.*, 170, 275–298.

Beller S., Monteiller, V., Operto, S., Nolet, G., Paul, A., Zhao L. (2018). Lithospheric architecture of the Southwestern Alps revealed by multi-parameter teleseismic full-waveform inversion. *Geophys. J. Int.*, 212, 1369–1388.

Bokelmann, G., Qorbani, E., Bianchi, I. (2013). Seismic anisotropy and large-scale deformation of the Eastern Alps. *Earth and Planetary Science Letters*, 383, 1–6.

Bonvalot, S., Balmino, G., Briais, A., Kuhn, M., Peyrefitte, A., Vales, N., Biancale, R., Gabalda, G., Moreaux, G., Reinquin, F. et al. (2012). World Gravity Map. Bureau Gravimetrique International (BGI), CGMW-BGI-CNES-IRD (eds)., Paris.

Blundell, D., Freeman, R., Mueller, S., Button, S. (eds) (1992). *A Continent Revealed – The European Geotraverse*. Cambridge University Press, Cambridge.

Campillo, M. and Paul, A. (2003). Long range correlations in the diffuse seismic coda. *Science*, 299, 547–549.

Cattaneo, M. and Eva, C. (1990). Propagation anomalies in Northwestern Italy by inversion of teleseismic residuals. *Terra Nova*, 2, 577–584.

Chopin, C. (1984). Coesite and pure pyrope in high-grade blueschists of the Western Alps: A first record and some consequences. *Contributions to Mineralogy and Petrology*, 86, 107–118.

Closs, H. and Labrouste, Y. (1963). Recherches séismologiques dans les Alpes Occidentales au moyen des grandes explosions en 1956, 1958 et 1960. In *Année Géophysique Internationale*, *XII*, *2*. CNRS, Paris.

Di Stefano, R., Kissling, E., Chiarabba, C., Amato, A., Giardini, D. (2009). Shallow subduction beneath Italy: Three-dimensional images of the Adriatic–European–Tyrrhenian lithosphere system based on high-quality P-wave arrival times. *J. Geophys. Res.*, 114, B05305.

Diehl, T., Husen, S., Kissling, E., Deichmann, N. (2009). High-resolution 3-D P-wave model of the Alpine crust. *Geophys. J. Int.*, 179, 1133–1147.

Dueker, K.G. and Sheehan, A.F. (1997). Mantle discontinuity structure from midpoint stacks of converted P-to-S waves across the Yellowstone hotspot track. *Journal of Geophysical Research*, 102, 8313–8327.

Dumont T., Schwartz, S., Guillot, S., Simon-Labric, T., Tricart, P., Jourdan, S. (2012). Structural and sedimentary records of the Oligocene revolution in the Western Alpine arc. *Journal of Geodynamics*, 56–57, 18–38.

Fichtner, A., van Herwaarden, D.-P., Afanasiev, M., Simutė, S., Krischer, L., Çubuk-Sabuncu, Y., Taymaz, T., Colli, L., Saygin, E., Villaseñor, A. et al. (2018). The collaborative seismic earth model: Generation 1. *Geophysical Research Letters*, 45, 4007–4016.

Fox, M., Herman, F., Kissling, E., Willett, S.D. (2015). Rapid exhumation in the Western Alps driven by slab detachment and glacial erosion. *Geology*, 43(5), 379–382.

Frei, W., Heitzmann, P., Lehner, P. (1990). Swiss NFP-20 research program of the deep structure of the Alps. *Mém. Soc. Géol. Fr.*, 156, 29–46.

Fry, B., Deschamps, F., Kissling, E., Stehly, L., Giardini, D. (2010). Layered azimuthal anisotropy of Rayleigh wave phase velocities in the European Alpine lithosphere inferred from ambient noise. *Earth Planet. Sci. Lett.*, 297, 95–102.

Giacomuzzi, G., Chiarabba, C., De Gori, P. (2011). Linking the Alps and Apennines subduction systems: New constraints revealed by high-resolution teleseismic tomography. *Earth and Planetary Science Letters*, 301, 531–543.

Guillot, S., Hattori, K., Agard, P., Schwartz, S., Vidal, O. (2009). Exhumation processes in oceanic and continental subduction contexts: A review. In *Subduction Zone Geodynamics*, Lallemand, S., Funiciello, F. (eds). Springer-Verlag, Berlin/Heidelberg.

Heit, B., Weber, M., Tilmann, F., Haberland, C., Jia, Y., Carraro, C., Walcher, G., Franceschini, A., Pesaresi, D. (2017). The Swath-D seismic network in Italy and Austria. *GFZ Data Services*. doi: 10.14470/MF7562601148.

Heit, B., Weber, M., Haberland, C., Tilmann, F., Hemmleb, S., Schwarz, S., Handy, M., Jia, Y., Pesaresi, D. (2018). The AlpArray SWATH-D experiment. *Geophysical Research Abstracts Volume 20 – 2018 EGU General Assembly*, EGU2018-11509.

Hetényi, G., Molinari, I., Clinton, J., Bokelmann, G., Bondar, I., Crawford, W.C., Dessa, J-X., Doubre, C., Friederich, W., Fuchs, F. et al. (2018a). The AlpArray seismic network: A large-scale European experiment to image the Alpine orogeny. *Surv. Geophys.*, 39, 1009–1033.

Hetényi, G., Plomerová, J., Bianchi, I., Exnerová, H.K., Bokelmann, G., Handy, M.R., Babuška, V., AlpArray-EASI Working Group (2018b). From mountain summits to roots: Crustal structure of the Eastern Alps and Bohemian Massif along longitude 13.3°E. *Tectonophysics*, 744, 239–255.

Hua, Y., Zhao, D., Xu, Y. (2017). P wave anisotropic tomography of the Alps. *J. Geophys. Res. Solid Earth*, 122, 4509–4528.

Jolivet, L. and Faccenna, C. (2000). Mediterranean extension and the Africa–Eurasia collision. *Tectonics*, 19, 1095–1106.

Kästle, E.D., El-Sharkawy, A., Boschi, L., Meier, T., Rosenberg, C.L., Bellahsen, N., Cristiano, L., Weidle, C. (2018). Surface wave tomography of the Alps using ambient-noise and earthquake phase velocity measurements. *J. Geophy. Res.: Solid Earth*, 123, 1770–1792.

Kästle E.D., Rosenberg, C.L., Boschi, L., Bellahsen, N., Meier, T., El-Sharkawy, A. (2020). Slab break-offs in the Alpine subduction zone. *International Journal of Earth Sciences*, 109, 587–603.

Kissling, E. (1993). Deep structure of the Alps – What do we really know? *Physics of the Earth and Planetary Interiors*, 79, 87–112.

Koulakov, I., Kaban, M.K., Tesauro, M., Cloetingh, S. (2009). P- and S-velocity anomalies in the upper mantle beneath Europe from tomographic inversion of ISC data. *Geophys. J. Int.*, 179, 345–366.

Kummerow, J., Kind, R., Oncken, O., Giese, P., Ryberg, T., Wylegalla, K., Scherbaum, F., TRANSALP Working Group (2004). A natural and controlled source seismic profile through the Eastern Alps: TRANSALP. *Earth and Planetary Science Letters*, 225, 115–129.

Lardeaux, J.M., Schwartz, S., Tricart, P., Paul, A., Guillot, S., Béthoux, N., Masson, F. (2006). A crustal-scale cross-section of the south-western Alps combining geophysical and geological imagery. *Terra Nova*, 18, 412–422.

Lippitsch, R., Kissling, E., Ansorge, J. (2003). Upper mantle structure beneath the Alpine orogen from high resolution teleseismic tomography. *J. Geophy. Res.*, 108, B82376.

Löberich, E. and Bokelmann, G. (2020). Mantle flow under the Central Alps: Constraints from non-vertical SKS shear-wave splitting. *Solid Earth Discussions* [preprint], 1–4.

Lombardi, D., Braunmiller, J., Kissling, E., Giardini, D. (2008). Moho depth and Poisson's ratio in the Western–Central Alps from receiver functions. *Geophys. J. Int.*, 173, 249–264.

Lu, Y., Stehly, L., Paul, A., AlpArray Working Group (2018). High-resolution surface wave tomography of the European crust and uppermost mantle from ambient seismic noise. *Geophys. J. Int.*, 214, 1136–1150.

Lu, Y., Stehly, L., Brossier, R., Paul, A., AlpArray Working Group (2020). Imaging Alpine crust using ambient noise wave-equation tomography. *Geophys. J. Int.*, 222, 69–85.

Lüschen E., Borrini, D., Gebrande, H., Lammerer, B., Millahn, K., Neubauer, F., Nicolich, R., TRANSALP Working Group (2006). TRANSALP – Deep crustal Vibroseis and explosive seismic profiling in the Eastern Alps. *Tectonophysics*, 414, 9–38.

Lyu, C., Pedersen, H.A., Paul, A., Zhao, L., Solarino, S., CIFALPS Working Group (2017). Shear-wave velocities in the upper mantle of the western Alps: New constraints using array analysis of seismic surface waves. *Geophys. J. Int.*, 210, 321–331.

Maffione, M., Speranza, F., Faccenna, C., Cascella, A., Vignaroli, G., Sagnotti, L. (2008). A synchronous Alpine and Corsica–Sardinia rotation. *J. Geophys. Res. Solid Earth*, 113, B03104.

Malusà, M.G., Faccenna, C., Baldwin, S.L., Fitzgerald, P.G., Rossetti, F., Balestrieri, M.L., Danisik, M., Ellero, A., Ottria, G., Piromallo, C. (2015). Contrasting styles of (U)HP rock exhumation along the Cenozoic Adria–Europe plate boundary (Western Alps, Calabria, Corsica). *Geochem. Geophys. Geosyst.*, 16. doi:10.1002/2015GC005767.

Malusà, M.G., Anfinson, O.A., Dafov, L.N., Stockli D.F. (2016). Tracking Adria indentation beneath the Alps by detrital zircon U-Pb geochronology: Implications for the Oligocene–Miocene dynamics of the Adriatic microplate. *Geology*, 44(2), 155–158.

Mitterbauer, U., Behm, M., Brückl, E., Lippitsch, R., Guterch, A., Keller, G.R., Koslovskaya, E., Rumpfhuber, E.-M., Šumanovac, F. (2011). Shape and origin of the East-Alpine slab constrained by the ALPASS teleseismic model. *Tectonophysics*, 510, 195–206.

Molinari, I. and Morelli, A. (2011). EPcrust: A reference crustal model for the European Plate. *Geophys. J. Int.*, 185, 352–364.

Molinari, I., Verbeke, J., Boschi, L., Kissling, E., Morelli, A. (2015a). Italian and Alpine three-dimensional crustal structure imaged by ambient-noise surface-wave dispersion. *Geochem. Geophys. Geosyst.*, 16. doi:10.1002/2015GC006176.

Molinari, I., Argnani, A., Morelli, A., Basini, P. (2015b). Development and testing of a 3-D seismic velocity model of the Po plain sedimentary basin, Italy. *Bull. Seism. Soc. Am.*, 105, 753–764.

Nicolas, A. and Christensen, N.I. (1987). Formation of anisotropy in upper mantle peridotites – A review. In *Composition, Structure and Dynamics of the Lithosphere – Asthenosphere System, Volume 16*, Fuchs, K. and Froidevaux, C. (eds). Am. Geophys. Un., Washington, DC.

Nicolas, A., Hirn, A., Nicolich, R., Polino, R., ECORS-CROP Working Group (1990). Lithospheric wedging in the western Alps inferred from the ECORS-CROP traverse. *Geology*, 18, 587–590.

Nocquet, J.-M., Sue, C., Walpersdorf, A., Tran, T., Lenôtre, N., Vernant, P., Cushing, M., Jouanne, F., Masson, F., Baize, S. et al. (2016). Present-day uplift of the western Alps. *Scientific Reports*, 6, 28404.

Paffrath, M., Friederich, W., AlpArray Working Group (2020). Teleseismic P-wave travel time tomography of the Alpine upper mantle using AlpArray seismic network data. *EGU General Assembly*, EGU2020-13779..

Paul, A., Cattaneo, M., Thouvenot, F., Spallarossa, D., Béthoux, N., Fréchet, J. (2001). A three-dimensional crustal velocity model of the south-western Alps from local earthquake tomography. *J. Geophys. Res.*, 106, 19367–19389.

Piana Agostinetti, N. and Amato, A. (2009). Moho depth and Vp/Vs ratio in peninsular Italy from teleseismic receiver functions. *J. Geophys. Res.*, 114, B06303.

Piromallo, C. and Faccenna, C. (2004). How deep can we find the traces of Alpine subduction? *Geophys. Res. Lett.*, 31, L06605.

Piromallo, C. and Morelli, A. (2003). P-wave tomography of the mantle under the Alpine–Mediterranean area. *J. Geophys. Res.*, 108, B22065.

Potin, B. (2016). Les Alpes occidentales : tomographie, localisation de séismes et topographie du Moho. PhD Thesis, Université Grenoble Alpes.

Qorbani, E., Bianchi, I., Bokelmann, G. (2015). Slab detachment under the Eastern Alps seen by seismic anisotropy. *Earth and Planetary Science Letters*, 409, 96–108.

Reynard, B. (2013). Serpentine in active subduction zones. *Lithos*, 178, 171–185.

Salimbeni, S., Malusà, M.G., Zhao, L., Guillot, S., Pondrelli, S., Margheriti, L., Paul, A., Solarino, S., Aubert, C., Dumont, T. et al. (2018). Active and fossil mantle flows in the western Alpine region unravelled by seismic anisotropy analysis and high-resolution P-wave tomography. *Tectonophysics*, 731, 35–47.

Savage, M. (1999). Seismic anisotropy and mantle deformation: What have we learned from shear wave splitting? *Rev. Geophys.*, 37, 65–106.

Schippkus, S., Zigone, D., Bokelmann, G., AlpArray Working Group (2020). Azimuthal anisotropy in the wider Vienna basin region: A proxy for the present-day stress field and deformation. *Geophys. J. Int.*, 220, 2056–2067.

Shapiro, N.M. and Campillo, M. (2004). Emergence of broadband Rayleigh waves from correlations of the ambient seismic noise. *Geophys. Res. Lett.*, 31, L07614.

Shapiro, N.M., Campillo, M., Stehly, L., Ritzwoller, M.H. (2005). High-resolution surface-wave tomography from ambient seismic noise. *Science*, 307, 1615–1618.

Silver, P.G. and Chan, W.W. (1991). Shear-wave splitting and subcontinental mantle deformation. *J. Geophys. Res.*, 96, B10, 16429–16454.

Solarino S., Malusà, M.G., Eva, E., Guillot, S., Paul, A., Zhao, L., Aubert, C., Dumont, T., Pondrelli, S., Salimbeni, S. et al. (2018). Mantle wedge exhumation beneath the Dora-Maira (U)HP dome unravelled by local earthquake tomography (Western Alps). *Lithos*, 296–299, 623–636.

Spada, M., Bianchi, I., Kissling, E., Piana Agostinetti, N., Wiemer, S. (2013). Combining controlled-source seismology and receiver function information to derive 3-D Moho topography for Italy. *Geophys. J. Int.*, 194, 1050–1068.

Spakman, W. (1990). Tomographic images of the upper mantle below central Europe and the Mediterranean. *Terra Nova*, 2, 542–553.

Spakman, W. (1991). Delay-time tomography of the upper mantle below Europe, the Mediterranean, and Asia Minor. *Geophys. J. Int.*, 107, 309–332.

Stehly, L., Fry, B., Campillo, M., Shapiro, N.M., Guilbert, J., Boschi, L., Giardini, D. (2009). Tomography of the Alpine region from observations of seismic ambient noise. *Geophys. J. Int.*, 178, 338–335.

Sternai, P., Sue, C., Husson, L., Serpelloni, E., Becker, T.W., Willett, S.D., Faccenna, C., Di Giulio, A., Spada, G., Jolivet, L. et al. (2019). Present-day uplift of the European Alps: Evaluating mechanisms and models of their relative contributions. *Earth–Science Reviews*, 190, 589–604.

Tape, C., Liu, Q., Maggi, A., Tromp, J. (2009). Adjoint tomography of the Southern California crust. *Science*, 325, 988–992.

Thouvenot, F., Paul, A., Fréchet, J., Béthoux, N., Jenatton, L., Guiguet, R. (2007). Are there superposed Mohos in the southwestern Alps? New seismic data from fan-profiling reflections. *Geophys. J. Int.*, 170, 1180–1194.

Thurber, C.H. (1983). Earthquake locations and three-dimensional crustal structure in the Coyote Lake area, central California. *J. Geophys. Res.*, 88, 8226–8236.

Vignaroli, G., Faccenna, C., Jolivet, L., Piromallo, C., Rossetti, F. (2008). Subduction polarity reversal at the junction between the Western Alps and the Northern Apennines, Italy. *Tectonophysics*, 450, 34–50.

Vinnik, L.P. (1977). Detection of waves converted from P to SV in the mantle. *Phys. Earth Planet. Int.*, 15, 39–45.

Wagner, M., Kissling, E., Husen, S. (2012). Combining controlled-source seismology and local earthquake tomography to derive a 3-D crustal model of the western Alpine region. *Geophys. J. Int.*, 191, 789–802.

Waldhauser F., Kissling, E., Ansorge, J., Mueller, S. (1998). Three-dimensional interface modelling with two-dimensional seismic data: The Alpine crust–mantle boundary. *Geophys. J. Int.*, 135, 264–278.

Waldhauser, F., Lippitsch, R., Kissling, E., Ansorge, J. (2002). High-resolution teleseismic tomography of upper-mantle structure using an a priori three-dimensional crustal model. *Geophys. J. Int.*, 150, 403–414

Wüstefeld, A., Bokelmann, G.H.R., Barruol, G., Montagner, J.-P. (2009). Identifying global seismic anisotropy patterns by correlating shear-wave splitting and surface waves data. *Phys. Earth Planet. Int.*, 176(3–4), 198–212.

Zahorec, P., Papčo, J., Pašteka, R., Bielik, M., Bonvalot, S., Braitenberg, C., Ebbing, J., Gabriel, G., Gosar, A., Grand, A. et al. (2021). The first pan-Alpine surface-gravity database, a modern compilation that crosses frontiers. *Earth Syst. Sci. Data*, 13, 2165–2209.

Zhao, L., Paul, A., Guillot, S., Solarino, S., Malusà, M.G., Zheng, T., Aubert, C., Salimbeni, S., Dumont, T., Schwartz, S. et al. (2015). First seismic evidence for continental subduction beneath the Western Alps. *Geology*, 43(9), 815–818.

Zhao, L., Paul, A., Solarino, S., RESIF (2016a). Seismic network YP: CIFALPS temporary experiment (China–Italy–France Alps seismic transect). *RESIF – Réseau Sismologique et géodésique Français*. doi: 10.15778/RESIF.YP2012.

Zhao, L., Paul, A., Malusà, M.G., Xu, X., Zheng, T., Solarino, S., Guillot, S., Schwartz, S., Dumont, T., Salimbeni, S. et al. (2016b). Continuity of the Alpine slab unraveled by high-resolution P-wave tomography. *J. Geophys. Res.: Solid Earth*, 121, 8720–8737.

Zhao, L., Malusà, M.G., Yuan, H., Paul, A., Guillot, S., Lu, Y., Stehly, L., Solarino, S., Eva, E., Lu, G. et al. (2020). Evidence for a serpentinized plate interface favouring continental subduction. *Nature Communications*, 11, 2171.

Zhu, H., Bozdag, E., Tromp, J. (2015). Seismic structure of the European upper mantle based on adjoint tomography. *Geophys. J. Int.*, 201, 18–52.

3

Topography and Landscape Evolution of the Alps

Gerit GRADWOHL[1], Kurt STÜWE[1], Jörg ROBL[2] and Moritz LIEBL[2]
[1] *Earth Sciences, University of Graz, Austria*
[2] *Geography and Geology, University of Salzburg, Austria*

The European Alps are one of the best-known mountain ranges on Earth. However, despite being one of the most studied orogens in the world, surprisingly little is known about the formation of its landscapes or the age of its topography. This paper summarizes the state of knowledge on topography, uplift history and landscape formation of the range. Different approaches and techniques are used to showcase and date these aspects. For example, (i) morphometric analysis shows that the Alps are characterized by decreasing slopes at high elevations with no notable difference between the Eastern and Western Alps. (ii) Sediment fluxes to the surrounding basins started about 35 Ma and have shown a substantial increase since the Pliocene. (iii) Thermochronometric data show evidence for strongly accelerated exhumation in the last 2 Myr, in particular in the Western Alps. (iv) Cosmogenic nuclide concentrations like ^{10}Be and ^{26}Al record Holocene erosion rates that vary dramatically between glaciated and ice-free regions, but are overall indicative of substantial surface uplift and incision during the last 4–5 Myr. Integrating these different datasets into a model for the geological surface uplift history shows that topographic evolution for the last 45 Myr occurred rather slowly until about 5 Ma and accelerated thereafter. Neither tectonics nor climate is responsible as the sole reason for the young surface uplift. For example, for the Western Alps many studies consider climatic driven erosion as a process that is responsible for a substantial portion of the surface uplift. Conversely, in the Eastern Alps, most studies suggest that the majority of the uplift is due to tectonic causes. Overall, it appears to be a combination of both processes that must be invoked to explain the observed topographic pattern.

Geodynamics of the Alps 1,
coordinated by Claudio L. ROSENBERG and Nicolas BELLAHSEN. © ISTE Ltd 2023.

3.1. Introduction

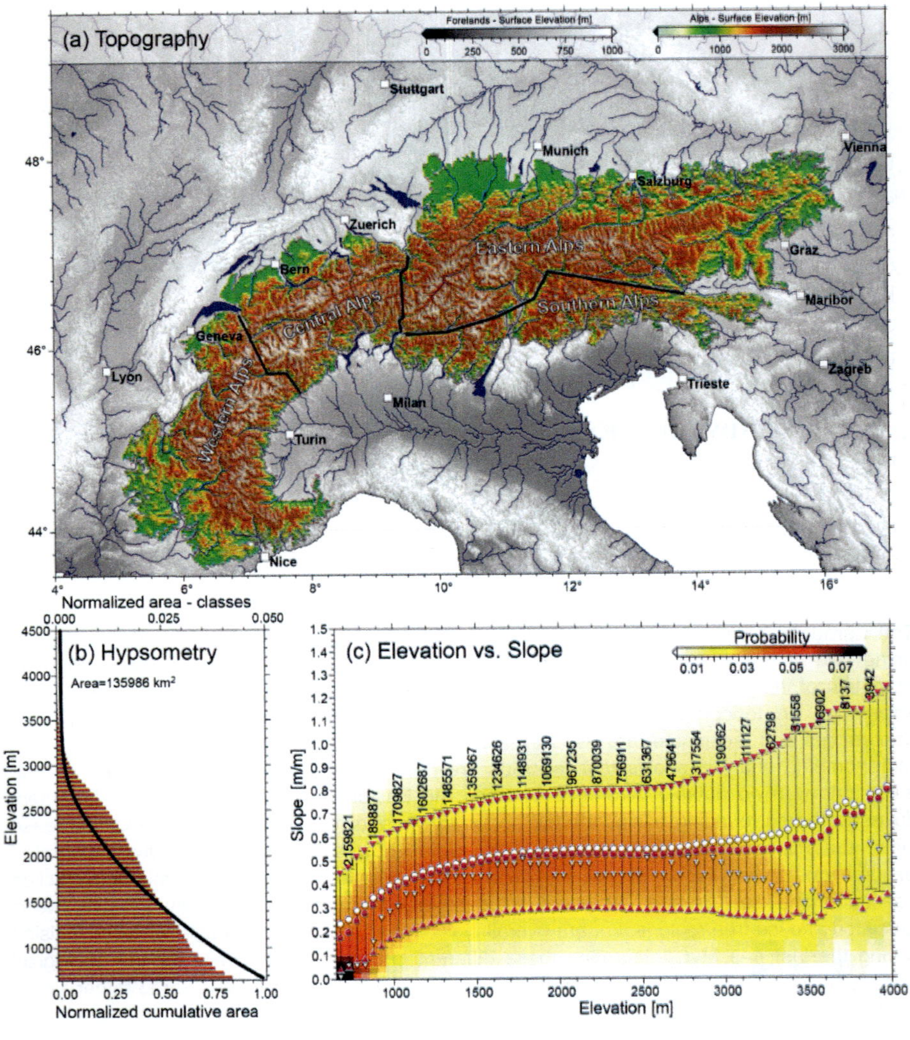

Figure 3.1. *Topographic metrics of the Alps. (a) Topographic map of the Alps above 650 m above sea level with major rivers and lakes. The borders of Western, Central, Eastern and Southern Alps are indicated by black solid lines. (b) Hypsometric curve of the Alps (solid black line) and normalized area of each elevation class (red bars). (c) Slope–elevation plot of the Alps. White circles with black error bars represent the mean slope and one standard deviation for each 50 m elevation slice. Red circles and red triangles represent the median values (50 percentile) with ends at 15.8 and 84.1 percentiles. White triangles represent the modal values for each elevation slice. The numbers above every third elevation slice are given*

The European Alps are one of the most studied mountain ranges in the world. With some 1,000 km in length from Vienna to Marseille, some 250 km in width from Lake Constance to Lake Garda and rising at Mont Blanc up to 4,809 m above sea level, they are also the largest and highest mountain range, located entirely in Europe (Figure 3.1). The topography of the Alps has been a major obstacle for politically motivated movements for millennia (Romans, Turkish invasions, modern refugee crisis) and its nature, origin and evolution has attracted geologists and tourists alike for centuries (e.g. Stüwe and Homberger 2012). Its present state is home to a rich alpine flora and fauna and it acts as an important water supplier for the adjacent countries (Baumgartner et al. 1983). Not least, the Alps have been a region of early settlement since the Paleolithic (Bächler 1940). As such, the topography of the range and its evolution over geological time are of high interest.

This topography is extremely variable. Apart from steep and high glaciated mountain peaks and razor-sharp ridges, the Alps also carry gentle alpine pastures, plateaus and hilly lowlands (Schuster and Stüwe, in print). In places, these widely contrasting environments can be found right next to each other and the fact that different geological processes shaped the landscape is obvious, even to an untrained eye. However, much of the morphology, its evolution and the processes that shaped the mountain range are still subject to modern scientific debates.

Historically, the scientific puzzling about the origin of the Alpine topography has led to the birth of some of the most important modern geological and geomorphological concepts, including that of plate tectonics itself that evolved during the early 20th century (Şengör 2009). Prior to the advent of plate tectonics, mountain ranges such as the Alps were thought to be fixed in geographic position and have formed by vertical motions only – a concept that is still entrenched in the intuitive understanding of mountain building by many non-geologists today. Within this so-called "fixistic" or "permanence" theory of the 19th century, all land masses are thought to remain in a stable position over geological time and topography is simply thought to be the consequence of vertical motion only (see, for example, Şengör 2014). The recognition of similar biota on – for example – both sides of the Atlantic was thought to be caused by wanderings of life forms across land bridges that rise and fall above and below sea level during different periods in geological time. The Central American land bridge between the Americas was thought to be a modern-day analogue. Although such fixistic thinking was still popular in the 1920s when Hans Stille postulated the various stages of mountain building in terms of the "geosynclinal" theory (Stille 1919), it had then already been superseded by the interpretation of the Alps in terms of horizontal motions: the theory of nappe tectonics. Eduard Suess (1883) and Marcel Alexandre Bertrand around the same time were among the first to understand that mountain ranges are formed by horizontal transport that exceeds the vertical motions by several orders of magnitude. This concept was consolidated by Wegener (1912), who formulated the theory of continental drift. Today, we understand the Alps as the product of plate motions within the modern concepts of geodynamics with both vertical and horizontal motions. The widespread occurrence of metamorphic rocks, which are exhumed from many tens of kilometers in depth at the surface of the Alps, testifies of the substantial *vertical* motions of rocks that occur during the process of mountain building. These vertical motions are caused by horizontal transport of both, tectonics and the enormous transport of the major Alpine rivers that fill the basins surrounding the Aps with kilometer-thick sequences of sediments. Overall, the topography of the range is known to be the result of the combined efforts of tectonic and climatic drivers, their dynamic interaction and their feedback processes (e.g. Wittmann et al. 2007; Whipple 2009; Champagnac et al. 2012; Robl et al. 2015).

Nevertheless, some of the first-order morphological features such as the geometry of drainage divides, the course of the major drainages or even the age of topography present puzzling problems on their origin and geomorphological evolution: some rivers run through wide valleys, others cut narrow channels through the rock and in places they seem to alternate. Lake floors like those of the north Italian lakes reach hundreds of meters below sea level (Finckh et al. 1984) and enormous unexplored underground landscapes of caves in the limestone ranges were created by dissolution of these rocks. The paleo-flow patterns, the ages of caves and many other geomorphological problems of the Alps are far from being understood.

In this chapter, we present aspects of the topography of the Alps and its development over the geological past: the landscape evolution. In describing the topography, we make use of some modern metrics that have not been derived for the Alps on the scale of the mountain range as a whole and we summarize some important concepts of its morphology. When describing the landscape evolution, we present this for the last 45 Ma and introduce some modern discussions on the drivers that shape it today. We begin by presenting a summary of the present topography.

3.2. The Alps today

Geographically, the range forms a principal European divide for drainage, climate and culture (Figure 3.1(a)) and, hypsometrically, half of their area are above an elevation of 1,500 m above sea level (for all interconnected Alpine regions above 650 m, Figure 3.1(b)). The Alps may be geographically divided into Western, Central, Eastern and Southern Alps. The line separating the Western Alps from the Central Alps is generally considered to run from the Rhône valley across the St. Bernhard pass into the Aosta valley. According to Stüwe and Homberger (2012), the line between Central and Eastern Alps is drawn from Lake Constance in the north to Lake Como in the south, following the Rhine valley. The line between Eastern and Southern Alps follows the east-west striking valleys of Gail, Puster and Adda. This subdivision is a mixture of geologic and cultural divisions and other divisions are also known. For example, Kissling and Schlunegger (2018) place the boundary between Eastern and Central Alps near the Guidicarie line. The Swiss Jura, the Bohemian Massif and the Slovenian Karst are often not considered to be part of the Alps, but the Steiner Alps and the Pohorje massif are. In the south, the Alpine chain merges topographically (but not geologically) into the Italian Appenine and in the east into the Dinarides. Except for the boundary between Southern and Eastern Alps, the boundaries described above do not correspond exactly to the geological boundaries between, for example, the Austroalpine and the Penninic domains.

Morphologically, the Alps are well-known for famous sub-ranges like the Vercors (Mont Aiguille), Bernese Oberland (including the Eiger), the Wallis (including the Matterhorn and Monte Rosa), the High Tauern (Großglockner), the Lower Tauern and many others. These ranges feature a total of almost 3,000 peaks that have a prominence of more than 250 m with almost 30 of those being above 4,000 m in elevation and another 520 of those being at least 3,000 m high (Figure 3.2(a)), with most high peaks being located in the Western Alps. Correspondingly, the geophysical relief, that is, the difference between maximum and minimum elevation for a given length scale, of the range for a 20 km sized sliding window is up to about 2,500 m and is significantly larger in the west than in the east (Figure 3.2(b)).

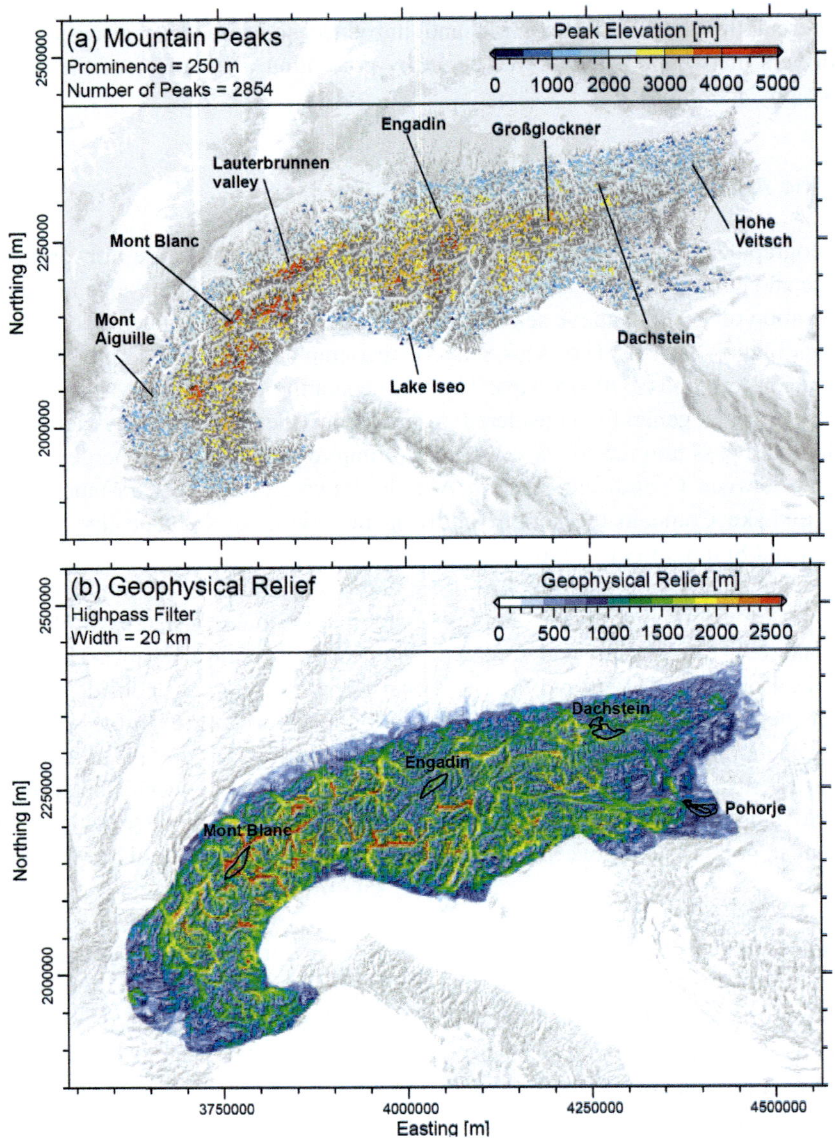

Figure 3.2. *Two maps showing the relief of the Alps. Note that the highest peaks, as well as the highest relief are located in the Western Alps*

COMMENT ON FIGURE 3.2.– *(a) Mountain peaks with prominence of at least 250 m. Locations of photos shown in Figure 3.3 are indicated. (b) Geophysical relief from a sliding window with 20 km width. The marked areas correspond to the regions for which the slope-elevation distribution is discussed in the text.*

Figure 3.3. *Photographs from different landscapes across the Alps taken from www.alpengeologie.org (Stüwe and Homberger 2012) and www.luftbildsteiermark.at (Stüwe and Homberger 2018)*

COMMENT ON FIGURE 3.3.– *(a) Mont Blanc, the highest peak of the Alps. (b) Großglockner, the highest peak of Austria with the retreating Pasterze glacier. (c) Lauterbrunnen valley, a textbook example of a glacially carved U-shaped valley. (d) Lake Iseo with Montisola island, one of the largest islands in a fresh water lake in Europe. (e) Mürz valley, a fluvial valley with gentle slopes that escaped Pleistocene glaciations. (f) Hohe Veitsch in the Mürzsteg Alps, a low relief landscape at high elevation. (g) Dachstein plateau in winter, an example of planation surfaces found in the Northern Calcareous Alps. (h) Mont Aiguille in the Vercors area, a high plateau in the Western Alps.*

The term "Alpine landscape" is defined by the landforms that are encountered when looking at the range. Figure 3.3 illustrates the enormous variability of landscapes found in the range (Stüwe and Homberger 2012). Typically, mountains exceeding 3,000 m are characterized by a glaciated landscape. Ice-filled cirques bordered by steep ridges and nunataks as seen for example on Mont Blanc (Figure 3.3(a)). Even though glaciers are retreating, distinct landforms as legacy of cold climate conditions stay visible after the ice has disappeared (Figures 3.3(b) and (c)). At lower elevations, major rivers of the Alps run in U-shaped valleys, which represent rather glacial than fluvial landforms. This indicates that these valleys were excavated by glaciers during the glaciation periods (Figure 3.3(c)). Alpine lakes with hundreds of meters of depth formed at glacially over-deepened valleys (Figure 3.3(d)). In contrast, areas that escaped major glacial impact often evidence a differently dissected landscape with gullies, ridges and gentle, but V-shaped valleys (Figure 3.3(e)). One peculiar type of landform of the Alps are planation surfaces that occur up to an elevation of 3,000 m. These very gentle landscapes are usually bordered by far steeper flanks toward the lowlands and are found in many parts of the Alps, but particularly in the east (Figures 3(f)–(h)).

3.3. Deriving the history of topography

Landscape is the result of many processes occurring at the same time or subsequently to each other. In general, mountain landscapes reflect the action of constructive and destructive processes driven by tectonics and the climate (e.g. Frisch et al. 1998; Champagnac et al. 2012; Robl et al. 2015). Erosion due to surface processes (climatic driver), but also crustal extension, flexural bending, slab pull or subsidence (tectonic driver), counteract rising topography due to tectonically driven rock uplift. The net elevation of landforms and topography itself is the product of the fine interplay of these processes and is called surface uplift (England and Molnar 1990; Stüwe and Barr 1998).

If constructive and destructive processes are balanced, the topography is said to be in a topographic steady state (Willet and Brandon 2002; Robl et al. 2017a). In the European Alps most landforms are in geomorphic disequilibrium (e.g. Korup and Schlunegger 2007; Valla et al. 2010; Hergarten et al. 2010; Robl et al. 2017b; Winterberg and Willett 2019). As such, rock uplift and erosion are not in balance and it is not trivial to derive the history of elevation and topography. In an effort to do so, Hergarten et al. (2010) used a simple uplift model to infer that only 40% of the time and rock uplift needed to reach the geomorphologic equilibrium are attained in the Alps today. The main driving force for the initial build-up of topography is the continent–continent collision of the European and the Adriatic lithospheric plate (e.g. Tricart 1984; Schmid et al. 1996; Bucher et al. 2003; Handy et al. 2015). Apart from the associated crustal thickening, isostatic rebound from slab break-off processes at depth and rebound from erosional unloading are the principal processes contributing to rock uplift (e.g. Fox et al. 2015). When talking about destruction of topography, glacial and fluvial erosion are the key factors. However, beside the importance of tectonics and the climate, lithology is also a principle control of landscape evolution. Substrate properties affect the pace landscapes adjust to changes in climate or tectonics and eventually control the steady-state geometry of mountain ranges (Robl et al. 2015, 2017a, 2017b; Stutenbecker et al. 2016). Flexural isostatic rebound of the lithosphere due to enhanced erosion and deglaciation since the last glacial maximum might have had (and still have) their fair share when it comes to surface uplift in certain areas across the Alps (Champagnac et al. 2009; Sternai et al. 2012, 2019; Mey et al. 2016). The range of the isostatic signal depends on the flexural wavelength of the lithosphere, which is commonly 100–200 km (Watts 2001).

In order to derive the processes led to the present-day relief of the Alps, it is necessary to consider the different available datasets. These include morphometric analysis (section 3.1), sediment budget (section 3.2), thermochronometric data (section 3.3.) and surface dating methods (section 3.4), but also include the direct dating of elevation via paleoaltimetric methods (e.g. Campani et al. 2012; Mulch 2016).

3.3.1. *Morphometric analysis*

Analysis of the morphology of the surface topography is clearly one of the most direct and obvious tools to infer its history (e.g. Wobus et al. 2006). Among many other metrics, two specific morphological features have proved to be useful for this: Channel profiles and the statistics of slope–elevation data.

3.3.1.1. *Channel profiles*

Rivers and streams do not only showcase that drainage patterns commonly go hand in hand with major fault systems (e.g. Robl et al. 2008, 2017b; Keil and Neubauer 2011; Trost et al. 2020), but their longitudinal profiles can also tell us about the dynamics of active deformation (e.g. Norton et al. 2008; Valla et al. 2010; Kirby and Whipple 2012). In geomorphic equilibrium, where surface uplift rates are balanced by erosion rates, channel profiles are graded, which means that a channel erodes at the same rate everywhere along the stream. The ability of a river to incise into its bedrock increases with channel slope and discharge (where the contributing drainage area is often used as a proxy for discharge). To approach a balance between erosion rates and surface uplift rates along the channel, an increase in contributing drainage area in downstream direction must eventually be matched by a decrease in channel slope. A commonly used relationship between channel slope and contributing drainage area is the so-called Flint's law (Flint 1974), which represents the stream-power law (Howard 1994) for steady state. Flint's law allows modeling steady-state channel profiles, where departures from this equilibrium shape indicate a spatial or temporal variability in climatic conditioning, tectonic forcing or bedrock properties (erodibility). Distinct changes in the channel slope are called knickpoints. The fitting of segments between knickpoints can be interpreted in terms of rock and surface uplift stages (Wobus et al. 2006; Bartosch and Stüwe 2019).

Figure 3.4 shows channel profiles for the four main drainages of the Alps. Rhône and Po flow into the Mediterranean Sea, while Rhine and Danube have the North Sea and the Black Sea as a base level, respectively. On the scale of Figure 3.4, the Po shows three distinct knickpoints in its channel profile. The first one occurs ~35 km downstream of its source and marks the transition from the Dora-Maira Massif to the Po Plain with a sudden decrease in channel slope. The second knickpoint is located around Turin and shows steeper slopes downstream, which correlates well with post-Miocene activity of thrust faults (Perrone et al. 2013). The last one lies in the area of Piacenza. The channel slopes located downstream of the knickpoint are increasing as well. This area is known for recent transpressive tectonic activities, which might explain the knickpoint (Boccaletti et al. 2011). The channel profile of the Rhône as far as Lyon is strongly influenced by the Pleistocene glacial cycles, which does not allow interpretations in terms of uplift (Cyr et al. 2014) even though the area around the upper Rhône recently experienced differential uplift (Kahle et al. 1997; Schlatter et al. 2005). Most deviations from an equilibrium profile fall together with changes in erodibility (Stutenbecker et al. 2016) or glacial geomorphology. Indeed, the significant drop 40 km downstream of Geneva is caused by human interference with the natural river flow, that is, the 100 m high Génissiat dam. Interestingly, the lower portion of the channel profile of the Rhône undercuts the profile of the Rhine. This may be attributed to an upwards migrating knickpoint due to aberrations in the slope distribution and associated changes in erosion rate as a result of the Mediterranean major base level drop during the Messinian (Gargani 2004). However, it also may be interpreted as the consequence of a shorter flow path (Schlunegger and Mosar 2011; Winterberg and Willett 2019).

Despite also sharing a glacial history, the Rhine shows a graded channel profile from its origin to Lake Constance. Downstream of Lake Constance, the Rhine Graben represents a different base level (Uehlinger et al. 2009). The Danube, starting in Donaueschingen, does not have its origin in the Alpine Chain. Nevertheless, it drains the entire Eastern Alps. When looking at its channel profile, two major knickpoints are visible. The first one lies approximately 50 km downstream of Donaueschingen and corresponds well with the Danube Sink, where the Danube loses water to sinkholes (Knop 1878; Hötzl 1996; Robl et al. 2008a). The next downstream segment appears to be somewhat equilibrated, with a dam system downstream of Ingolstadt creating a local step, until a few kilometers upstream of Passau. There, slopes slightly increase with the transition from the Molasse Basin to the erosion-resistant rocks of the Bohemian Massif (Robl et al. 2008a).

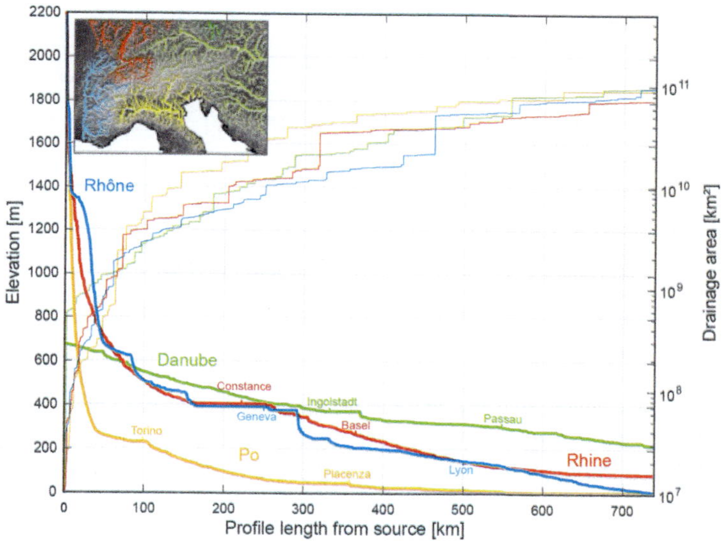

Figure 3.4. *Channel profiles of the four largest drainages of the Alps: Po (yellow), Rhône (blue), Rhine (red) and Danube (green) and their associated drainage areas. The inset shows the tributary network of these four rivers in their according colors*

3.3.1.2. *Slope–elevation statistics*

A good tool to place numbers on topography is the relationship between slope and elevation. Hergarten et al. (2010) produced the first modern study which recognized that the mean slope of the Alps decreases with elevation (see also Kühni and Pfiffner 2001), by using an approach which normalizes slope for catchment size. They suggested three possible interpretations for this relationship: (i) higher erosion rates at high surface elevations, maintaining equilibrium by high erosion on less steep slopes, (ii) absence of surface uplift at high elevations, maintaining equilibrium by low erosion rates and (iii) geomorphic disequilibrium due to immature topography. They suggested that immature topography is the most likely scenario, because the highest parts of the Alps show low erodibility and high threshold slopes and recent surface uplift rate measurements show no decrease of surface uplift rates in high areas.

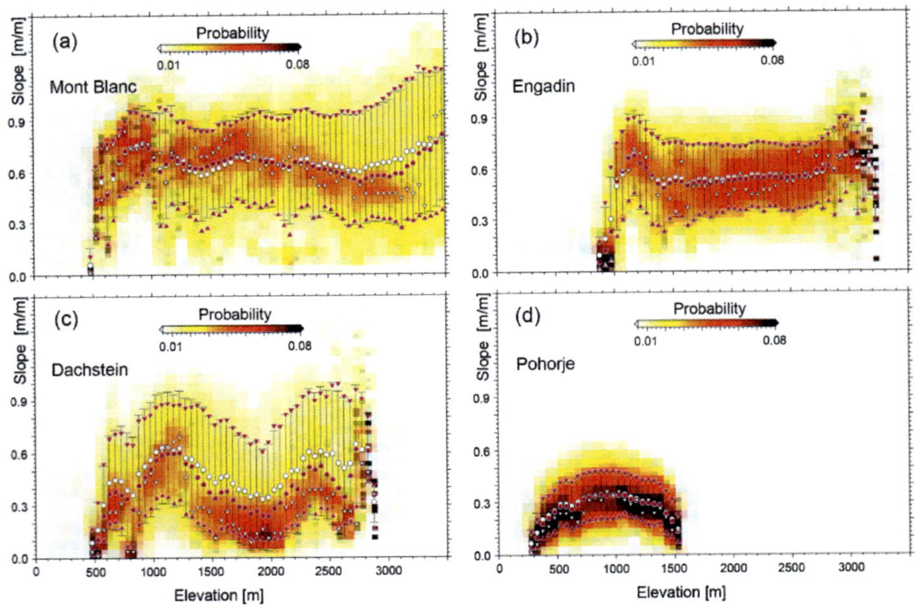

Figure 3.5. *Slope–elevation plots of four areas across the Alps from Robl et al. (2015). a) Mont Blanc region. b) Engadin region. c) Dachstein. d) Pohorje. The locations of these regions are shown in Figure 3.2(b). For further information, see text*

Robl et al. (2015) recognized a similar peculiarity in slope, even without consideration of catchment size. When performing this analysis for the entire European Alps (Figure 3.1(c)), an increase in slope up to about 1,800 m is visible. Above this elevation up to about 2,900 m, the mean and median slopes remain constant at about 27°. For elevations higher than 2,900 m, an increase in slope up to 39° at summit regions is possible, but the mode values decrease up until 3,700 m. This is comparable to the findings of Kühni and Pfiffner (2001), who found 25° between 1,700 m and 2,700 m to be the threshold slope for the Swiss Alps. In Figure 3.5, four selected examples of Alpine topography are depicted in terms of slope and elevation that differ with respect to tectonics, lithology and glacial history.

The Mont Blanc Massif (Figures 3.3(a) and 3.5(a)) represents an Alpine landscape with strong glacial imprint and is characterized by trunk valleys, cirques and high peak topography. At lower elevations a high number of both high and low slopes are observed. This is typical for trunk valleys with their flat valley floors and steep flanks. The highest modal values for slope are found at elevations between 1,500 m and 2,000 m, which roughly coincides with the proposed equilibrium line altitude (ELA) of the last glacial maximum (LGM) (Ivy-Ochs et al. 2006, 2008). Above this altitude where spacious cirques start to occur, topographic gradients decrease. The slope–elevation distribution is therefore consistent with a landscape formed by glaciers.

The Engadin region (Figure 3.5(b)) was glaciated several times during the Pleistocene as well, but it does not show the same abundance of glacial landforms as the Mont Blanc region anymore. Instead, numerous gullies and channels dissect the surface indicating that the turn over from glacial to fluvial landforms is almost completed. Remaining glacial landforms are only locally observed. The slope–elevation distribution starts with very low topographic gradients at the valley floor and steepens along glacially formed flanks. A maximum in slope is reached at about 1,100 m followed by relict planation surfaces on valley shoulders at about 1,300 m. Above this altitude slopes increase up to the summit regions. Since the Engadin region lacks a wide range of topographic gradients due to missing glacial landforms, it shows generally lower slopes compared to the Mont Blanc Massif. Comparing these two areas, the persistence (and therefore the decay rate) of non-equilibrium landforms shows a strong lithology dependence with higher erodible/less stable rocks in the Engadin region compared to the Mont Blanc region.

The Dachstein Massif (Figures 3.3(g) and 3.5(c)) represents an Alpine landscape with glaciation history but low surface process rates. Due to the abundance of karstification, most run-off occurs underground, which leads to low surface erosion rates. The low topographic gradients at 500 m and 700 m represent the valley floors north and south of the Dachstein Massif, respectively. Between 900 m and 1,400 m, the slopes show bimodal distribution, bearing the highest mean and modal values. Planation surfaces and steep near-vertical walls mark this elevation range. Between 1,400 m and 2,100 m, very low slopes occur which correlates with a fluvial gravel-bearing paleo-surface of inferred Oligocene age (Frisch et al. 2001). In the Northern Calcareous Alps, many elevated planation surfaces are known to contain these fluvial gravels. The relief steepens up to 2,500 m but slopes decrease again at the glaciated regions near the summit. The reduction of surface run-off through karstification nicely shows the lithology dependence of landform persistence.

The Pohorje Massif (Figure 3.5(d)) at the southeastern corner of the Alps is representative for a region which was never glaciated during the Pleistocene glaciation cycles. It consists of steep, incised channels at low altitudes with knickpoints separating them from the higher, gentle headwaters where a spacious paleo-surface occurs (Robl et al. 2008a; Legrain et al. 2015). Between 300 m and 500 m, a strong increase in the slope occurs, followed by a significant drop in mean and modal values between 500 m and 700 m. This low topographic gradient correlates with the occurrence of a dry and wide Drava river parallel valley. Then slopes increase again until 1,000 m. In this elevation range, deeply incised valleys occur. Above 1,000 m slopes decrease significantly toward the peaks where paleo-surfaces dominate. Note the low standard deviation of this never glaciated region when compared to the slope–elevation distribution of the heavily glaciated Mont Blanc Massif.

In summary, it may be said that there is no clearly apparent difference in morphometric trends between the Eastern and Western Alps. However, when looking at slope-elevation plots of specific areas with varying glacial history and landform persistence (lithology), different trends may be discerned. In the Alps, different aspects of quantifiable geomorphic criteria have been interpreted in terms of a variety of different surface uplift or erosion histories (e.g. Kühni and Pfiffner 2001; Anders et al. 2010; van der Beek et al. 2010; Norton et al. 2010a; Valla et al. 2010; Herman et al. 2011; Sternai et al. 2011, 2012, 2013; Schlunegger and Norton 2013), but the interpretative tool of slope-elevation statistics (Kühni and Pfiffner 2001; Hergarten et al. 2010; Robl et al. 2015) appears to indicate relict surfaces for the never-glaciated Pohorje Massif, glacial conditioning of an earlier paleo-landscape for the Dachstein Massif, fluvial overprint in soft lithologies for the once-glaciated Engadin and glacial shaping for the Mont Blanc Massif.

3.3.2. *Sediment budget*

As soon as topography rises, erosion will begin to compensate for the state of disequilibrium through increased erosion rates until a new state of equilibrium is established (e.g. Ahnert 1970; Adams 1985; Whipple 2001). Higher erosion rates result in increased sediment fluxes from the mountain range toward the adjacent basins. Thus, sediment budgets from surrounding sedimentary basins may be interpreted in terms of the history of topography. The Alps are surrounded by a series of sedimentary basins, which accommodate the sediments that have been eroded off the range since onset of surface topography formation in the Oligocene (Frisch 1998). These basins include the northern and southern foreland basins, the Pannonian Basin, the Rhône Delta in the Tyrrhenian Sea, the Danube Delta in the Black Sea and the Rhine Graben (e.g. Bernet et al. 2004). England (1981) already recognized that the sediment volumes of these basins can be used to constrain exhumation processes and aspects of topography development.

When looking at the sediment budget of Alpine catchments in more detail, past erosion rates can be determined with high temporal but low spatial resolution (Kuhlemann et al. 2002). Changes in erosion rates can have multiple controlling factors. Apart from the relief in the geologic past, changes in base level (eustatic sea-level change), climate change and differences in erodibility can take important roles for the increase and decrease of sediment discharge. If these factors show minor significance, the sediment budget will be strongly controlled by the relief of the orogen (Einsele 1992). To find a fitting interpretation for the observed sediment budget of a certain time frame, a comparison with tectonic events, climate and sea level is needed (Kuhlemann 2007; Willett 2010).

Figure 3.6 shows the sediment yield from the European Alps since the Oligocene, split into Eastern and Western Alps (Kuhlemann et al. 2001; Willett 2010). Although these data and their interpretation is highly debated (see below) and associated with large uncertainties (see the original data in Kuhlemann 2000), it is a much used dataset and is described here. Since the oldest sediments in the Molasse basin show late Eocene ages, this age appears to be a good starting point to evaluate the sedimentary fluxes of the Alps. The first significant change occurred around 30–28 Ma with a doubling of sediment accumulation rates. This was correlated with collision and slab break-off of the Penninic oceanic lithosphere below the Periadriatic line at depth (von Blanckenburg and Davies 1995) and the resulting surface uplift (Kuhlemann 2007). Between 24 Ma and 21 Ma, the sediment yield increased again. This trend is clearly visible for the Western Alps but only slightly visible in the Eastern Alps and is interpreted to correspond to the exhumation of the Aar massif (Schlunegger et al. 1997). The sediment budget of the entire Alps decreased strongly during the Miocene, interrupted by a short increase between 18 Ma and 17 Ma. This is interpreted to be the result of extension with a short phase of decreased lateral extrusion leading to surface uplift (Frisch et al. 2000). Between 12 Ma and 10 Ma, the Eastern Alps show a rise in sediment discharge rates by 35% not represented in the Western Alps and are at least partly explained by the deceleration of extension (Kuhlemann et al. 2001). At around 5.5 Ma, a sharp rise in sediment yield is visible for the Western Alps. This signal seems to be somewhat delayed for the Eastern Alps, showing a less pronounced increase at 4 Ma and a sudden rise in sediment load at around 2 Ma. The Western Alps show generally higher cumulative sediment discharge rates than the Eastern Alps but depict similar trends nonetheless. Willett (2010) split the sediment budget of the Southern Alps originally summarized by Kuhlemann et al. (2002) into a western and eastern part and added the latter to the Eastern Alps (Figure 3.6). By doing so, the strong increase in sediment yield at ~5.5 Ma becomes also apparent in the Eastern Alps. An increase in sediment discharge at mid to high latitudes for the last 5 million years is also visible across the globe (Zhang et al. 2001; Molnar 2004).

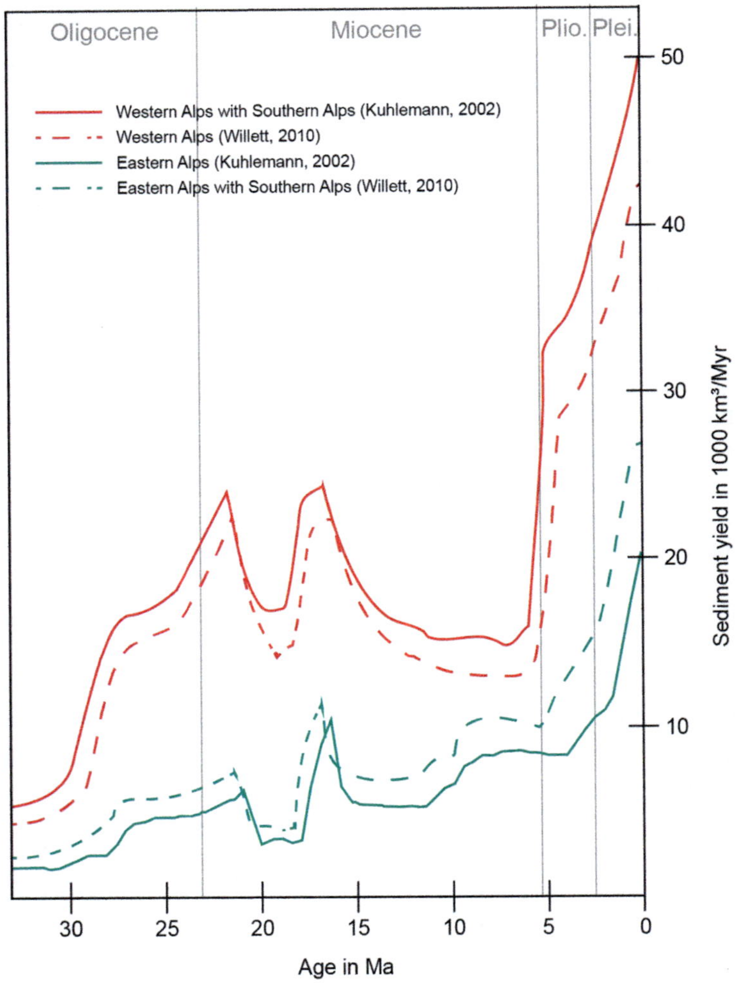

Figure 3.6. *Sediment budget of the Eastern (blue) and Western (red) Alps over the last 33 Ma from Kuhlemann et al. (2002)*

COMMENT ON FIGURE 3.6.– *The solid lines include the whole of the Southern Alps into the Western Alps (Kuhlemann et al. 2002), while the dashed lines count the sediments derived from the eastern Southern Alps to the Eastern Alps (Willett 2010).*

The data and their interpretation presented above are subject to debate. For example, hiatuses in sedimentation are problematic, since they become harder to identify with increasing time into the past. Also, younger sediments did not undergo the same compaction as older sediments. Therefore, sedimentation rates in more recent times seem higher if mentioned factors are ignored (Sadler 1999). In fact, many authors challenge the idea of a strong increase in sedimentation rates (and erosion rates) during the latest Cenozoic and argue that the young sediment record is being overestimated (e.g. Schumer and Jerolmack 2009; Willenbring and von Blanckenburg 2010). Nonetheless, this topic remains debated and the data set is much used (Herman and Champagnac 2016; Willenbring and Jerolmack 2016; Norton and Schlunegger 2017).

3.3.3. *Thermochronometric data*

Thermochronometric systems date temperatures along the cooling path of certain minerals and are therefore subject to a series of assumptions on the relationship between cooling and exhumation. Due to diverse closure temperatures of different systems, individual cooling steps can be dated and the lower the closure temperature, the better is the use of such data for the interpretation of surface processes like uplift. The most frequently used low-temperature thermochronometric systems and their respective closing temperatures for a common orogenic cooling rate around 15°C/Myr are: (U-Th)/He in apatite, which closes at ~60°C (Farley 2000), apatite fission tracks at ~110°C (Ketcham et al. 1999), (U-Th)/He in zircon at ~180°C (Reiners et al. 2004; effectively covers a wide range of temperatures depending on the accumulated radiation damage, Guenthner et al. 2013) and zircon fission tracks, which records ~230°C (Brandon et al. 1998). To obtain the depths of the closure temperatures in the Earth's crust, assumptions on the thermal structure in the past have to be made (Malusà and Fitzgerald 2019) and for very slow cooling rates the minerals residence time in the partial annealing zone has to be considered (Green et al. 1989). The temporal differences between the dated closure depths give exhumation rates over time (provided that multiple thermochronometric systems are analyzed, e.g. Vernon et al. 2008). Erosion rates and incision rates on a million year timescale can also be determined (e.g. Valla et al. 2011; Legrain et al. 2015). Furthermore, by looking at cooling histories of not only one spot in the field but also the surrounding areas, exhumation mechanisms can be inferred (e.g. Hejl 1997).

Many studies have contributed to the pool of low-temperature thermochronometric ages from the Alps and are summarized in Figure 3.7(a). An overview of the modeled exhumation rates (Fox et al. 2014) over the last 32 million years derived from this dataset of cooling ages was presented by Fox et al. (2016) (Figure 3.7(b)). This study shows that the first areas to show fast exhumation rates of up to 0.8 km/Myr at 30 Ma lie in the internal arc of the Western Alps. However, it should be noted that significantly older near surface exhumation is known from isolated "cold spots", that is, rocks that cooled to below 100°C in the Early Eocene and before, in the Eastern Alps (Hejl 1997). These ancient rapid exhumation rates decrease significantly to 0.3 km/Myr at around 20 Ma. Simultaneously, exhumation rates across the External Massifs of the Western Alps show an increase to 0.6 km/Myr (Glotzbach et al. 2011). The Tauern Window and the Lepontine Dome yield a similar advent of high exhumation rates at 20 Ma (e.g. Fügenschuh et al. 1997; Campani et al. 2010; Boston et al. 2017). In the Tauern Window, differences between the east and west are visible. The eastern (Wölfler et al. 2011, 2012) and the western Tauern Window (Fügenschuh et al. 1997) show unroofing times of 18 and 15 Ma, respectively. The Glarus region starts to display enhanced exhumation rates at 4 Ma (e.g. Vernon et al. 2008). Looking at the latest evolution, a significant increase in exhumation rates since 2 Ma can be seen in the Western Alps, which is not visible in the Eastern Alps (Herman et al. 2013; Figure 3.7(b)), but we note that this interpretation was challenged by Schildgen et al. (2018). This concerns in particular rapid exhumation in the Mont Blanc and Belledonne External Massifs (Seward and Mancktelow 1994; Leloup et al. 2005; Glotzbach et al. 2011; Fox et al. 2015). Unfortunately, large areas of the Alps do not bear suitable lithologies for thermochronometric analyses (e.g. Northern Calcareous Alps).

Apart from studying in situ rocks, there is also the possibility to examine thermochronometric ages of sedimentary rocks and sediments of the pro- and retro-side basins of the Alpine wedge (e.g. Hurford et al. 1984; Garver and Brandon 1994; Spiegel et al. 2000; Bernet et al. 2004, 2009). The age difference of the cooling age of the analyzed thermochronometric system and the deposition age gives the so-called lag time (Garver et al. 1999), that is, the time that the measured grain needs to travel from the closure depth to the final point of deposition. If the transport time is insignificant compared to the error of the applied method, the lag time reflects erosion rate (Bernet et al. 2004, 2009). Detrital thermochronometry can also be used for provenance analysis by comparing detrital grain ages with ages of bedrock samples (e.g. Hurford et al. 1984; Hurford and Carter 1991; Carter 1999). For a more detailed insight into this topic, see Chapter 4 by Bernet et al. (2023) in Volume 3.

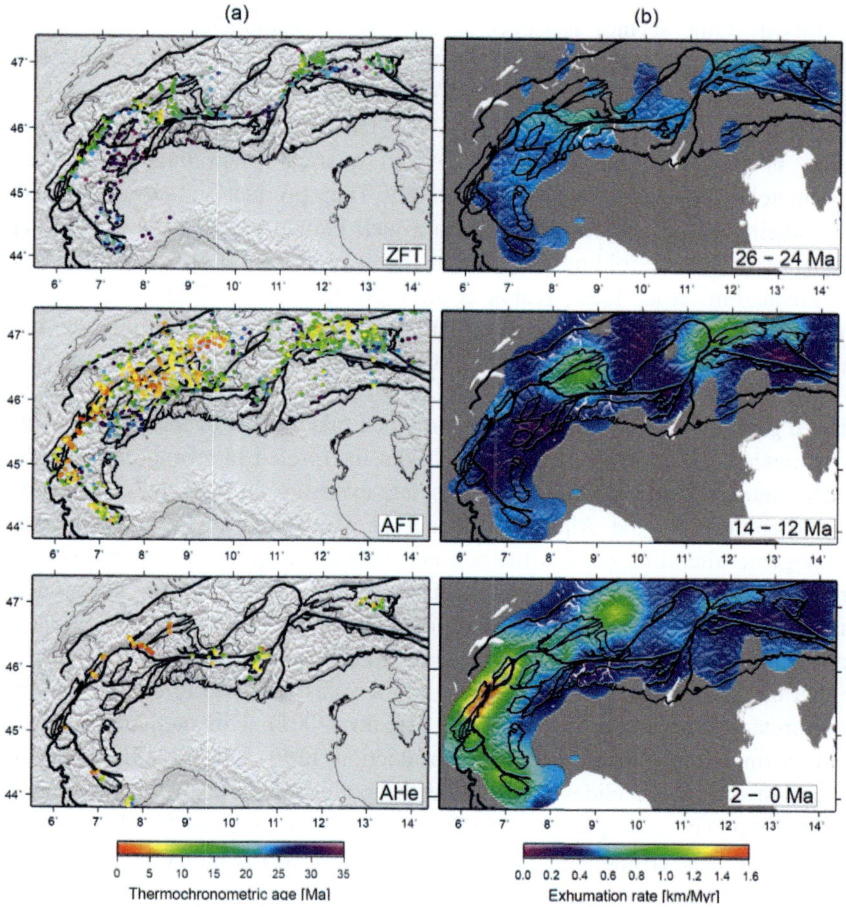

Figure 3.7. *Low-temperature thermochronometric data from the Alps*

COMMENT ON FIGURE 3.7.– *(a) Zircon fission track (ZFT), apatite fission track (AFT) and (U-Th)/He in apatite (AHe) data points with their respective ages. (b) Exhumation rates for the time steps 26–24 Ma, 14–12 Ma and 2–0 Ma from a linear inversion model. Adapted from Fox et al. (2016).*

3.3.4. *Direct surface dating*

Direct dating of the surface and its elevation history can be done with optically stimulated luminescence, paleoaltimetric, paleontological, sedimentological and a series of other methods. However, one of the most commonly used methods to date landforms near the surface has become the use of cosmogenic nuclide methods. Cosmogenically generated isotopes with geological applications include the radioactive nuclides ^{10}Be, ^{14}C, ^{26}Al and ^{36}Cl, as well as the stable nuclides ^{3}He and ^{21}Ne (Dunai 2010; Ivy-Ochs and Kober 2013). The most widely used in situ formed nuclides in the Alps are ^{10}Be and ^{26}Al in quartz, which have half-lives of approximately 0.7 Ma and 1.4 Ma, respectively. Their production rates as well as the ratio of their production rates in quartz are well known and show a simple depth dependence with an e-folding penetration depth of some 60 cm (Lal and Arnold 1985; Nishiizumi et al. 1989; Kubik et al. 1998; Gosse and Phillips 2001). Apart from telling us the age of the observed surface, in situ cosmogenic nuclide dating can also help us to reconstruct the rate of its destruction.

Three general ideas are being exploited when making use of these two isotopes: (i) The concentration of ^{10}Be and ^{26}Al may be interpreted in terms of an exposure age. This interpretation assumes instantaneous exposure through the last meters of the rock column. In the Alps, the impact of snow cover, vegetation, soil and anthropogenic alteration on this method are difficult to quantify on time scales of interest (i.e. 10^4 years). (ii) The concentration of ^{10}Be and ^{26}Al may be interpreted in terms of continuous accumulation during erosion-driven exhumation of the last few meters, that is, in terms of erosion rate (von Blanckenburg 2005). (iii) Finally, the time that has passed since the measured grain was last exposed to cosmic rays may be interpreted as burial age (Granger and Muzikar 2001). This method exploits the known cosmogenic $^{26}Al/^{10}Be$ surface production ratio and the different decay constants of the two nuclides to compute the period of burial from the measured $^{26}Al/^{10}Be$ ratio. The error on these ages is usually in the range of 5%. However, the decay rates of these cosmogenic radionuclides limit these dating methods to the time scale of some 5 Ma.

Erosion rate measurements on a 10^4 years time scale using ^{10}Be have been performed for a series of regions in the Alps (e.g. Wittmann et al. 2007, 2016; Norton et al. 2008, 2010b; Delunel et al. 2010; Glotzbach et al. 2013; Figure 3.8). They vary widely between 50 and 2,000 mm/kyr with a distinct trend of higher erosion rates in glaciated regions and lower rates (typically below 100 mm/kyr) in regions not affected by Pleistocene glaciations (Dixon et al. 2016). However, aside from glacial conditioning (e.g. Norton et al. 2010b; Dixon et al. 2016), there are more factors controlling erosion rates such as lithology, precipitation, temperature, mean basin slope and changes in base level. However, these factors mostly represent local erosional environments and their weighting might differ completely when comparing different areas. Thus, correlation of orogen-scale erosion rates remains challenging.

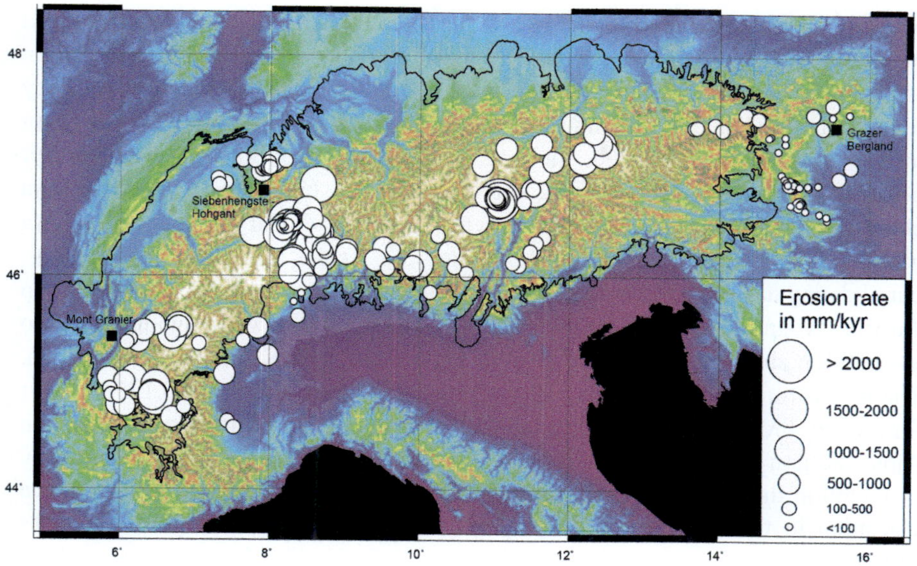

Figure 3.8. *Cosmogenic nuclide-derived erosion rates on a 10^4 year time scale across the Alps. The black line marks the boundary of the last glacial maximum*

COMMENT ON FIGURE 3.8.– *Note that the unglaciated eastern end of the Alps shows significantly lower erosion rates than the rest of the range. Produced from the data compilation of Dixon et al. (2016).*

Burial age dating efforts in the Alps are very limited, but due to the abundance of karst caves at various topographic levels, this method is likely to advance our understanding of surface uplift history in the future. Many karst caves in the Alps record "keyhole" passages, which are a result of a cave originally forming more or less horizontally below the groundwater table (at phreatic conditions) and subsequently falling dry, cutting a canyon at vadose conditions through the original tube. Such evidence for phreatic–vadose transitions may be interpreted in terms of increasing vertical distance to the local base level (i.e. surface uplift of peaks and plateaus next to antecedent drainages, or base level lowering). Such caves can bear siliceous sediments, which were deposited at the latest when the cave was in flooding range of the draining river for the last time. Hence, $^{26}Al/^{10}Be$ ratios of these sediments can be used to constrain the minimum age of cave-formation (and therefore date the former base level) as a proxy for surface uplift of the Alps. For example, the formation of the Siebenhengste-Hohgant cave system in Switzerland, with cave levels ranging from 558 m a.s.l. up to 1,800 m a.s.l., has been constrained by dating siliceous sediments using ^{10}Be and ^{26}Al (Häuselmann 2007; Häuselmann et al. 2007; Figure 3.8). For the cave level at 1,800 m, the oldest burial age they found was 4.4 Ma. By sampling several speleogenetic levels, the valley incision history could be reconstructed, showing rather low incision rates of 0.12 km/Myr before 0.8 Ma and an increase to 1.2 km/Myr thereafter. Interestingly, in a cave at an elevation of 1,705 m a.s.l. at Mont Granier in France, a similar burial age of 4.3 Ma was determined (Hobléa et al. 2011; Figure 3.8). At the eastern end of the Alps, burial dating has been performed in caves of the never glaciated Grazer Bergland in Austria (Wagner et al. 2010; Figure 3.8). They sampled several cave levels up to over 500 m above the local base level and found the oldest burial age to be 4 Ma. The reconstructed incision history shows low mean incision rates of 0.125 km/Myr for the last 4 Ma. When looking at the data of Wagner et al. (2010) from the Eastern Alps, incision rates show a decrease since 2.5 Ma, contrasting the latest valley evolution in the Swiss Alps (Häuselmann 2007). This may be due to differences in glaciation in the two areas. The incision rates at the eastern fringe of the Alps compare quite well with the incision rates of the Swiss Alps before 0.8 Ma. These incision rates in the Grazer Bergland region may also be directly interpreted in terms of surface uplift rates of the surrounding highlands because it is observed that the principal drainages of the region are antecedent streams, which transect the Alpine orogen–Pannonian basin transition without a knickpoint. Wagner et al. (2011) also correlated planation surfaces of the area with sampled cave levels, constraining their age as well. Such a correlation is only possible if erosion rates have been low enough to ensure that the elevation of cave levels does not differ dramatically from the corresponding planation surfaces. By backward modeling the number of nuclides prior to burial, paleo-erosion rates can be gauged (assuming the local production rates of ^{10}Be and ^{26}Al). This approach resulted in low pre-burial erosion rates of ~20 m/Myr for sediments in the sampled caves from Wagner et al. (2011).

3.4. Landscape evolution since the Eocene

The methods described in the last section allow us to reconstruct aspects of the landscape evolution of the Alps over the last 45 Myr (Figure 3.9; Schuster and Stüwe, in print). Quite generally, it may be said that the formation of the present-day landscape and topography started with the termination of subduction of the Penninic Ocean. Even though the distribution of land and sea then differed strongly from today, rocks that testify of these different environments from this time can still be found in the Alps.

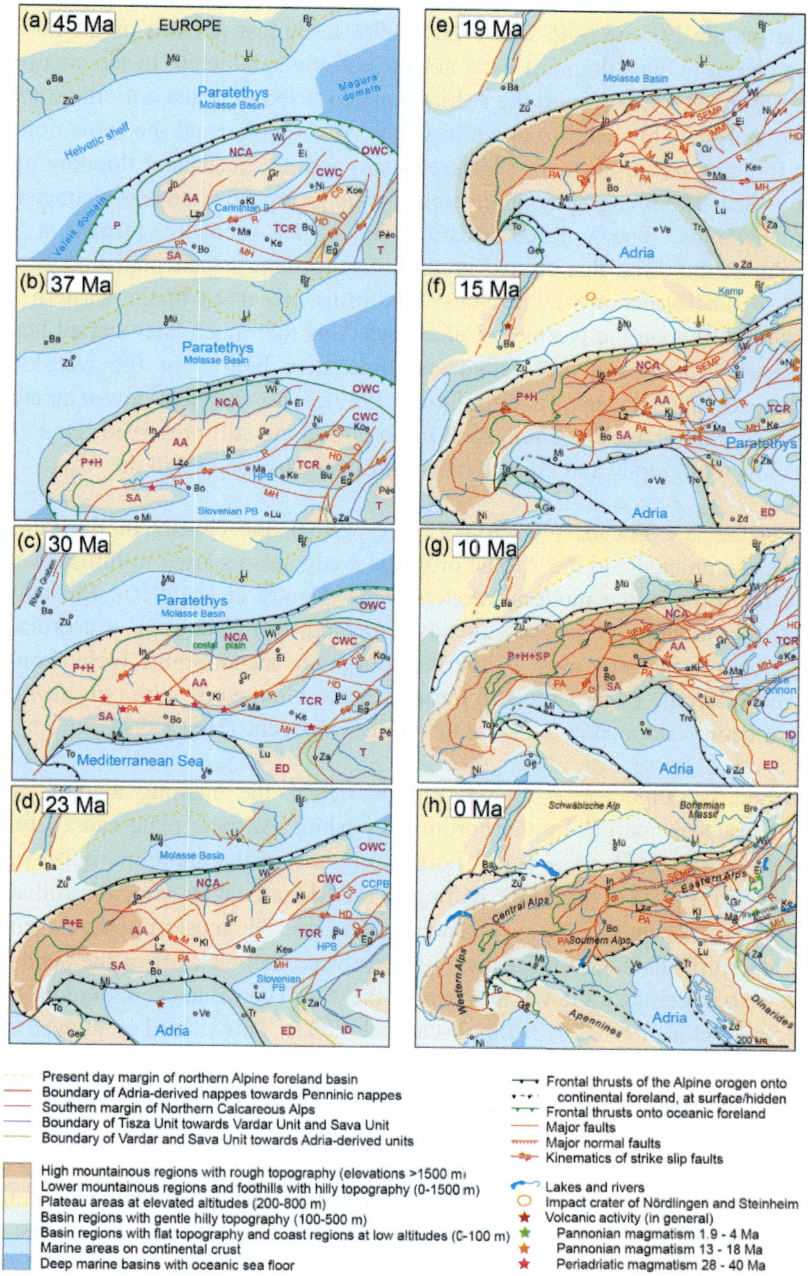

Figure 3.9. *Tectonic and topographic evolution of the Alps since 45 Ma. After Schuster and Stüwe (in print) but with an extension to the west. For details, see text*

In the middle Eocene (~45 Ma, Figure 3.9(a)), the last remains of the northern Penninic Ocean (Valais domain) were not yet entirely subducted. In the area of the modern Eastern Alps, however, the Adriatic plate started to thrust onto the southern European margin. This led to rock uplift of the Austroalpine and the formation of a flexural foreland basin (northern Molasse Basin) and its subsequent flooding by the Paratethys Sea. The topography on the non-marine European plate was characterized by flat lowlands. The uplifted Austroalpine rocks probably formed a hilly topography at this time that started eroding and forming a peneplanation surface. This led to the formation of the oldest landforms preserved in the Eastern Alps today: the plateaus on the Northern Calcareous Alps that much later started hosting the Augenstein-landscape (see Oligocene time slice below) or the well-known Nockfläche in the Niedere Tauern (Exner 1949). The fact that low-temperature thermochronometric ages reaching back up to 50 Ma are still preserved (Hejl 1997) suggests low topography and erosion rates for Eocene landscapes in the Alps.

During the late Eocene (~37 Ma, Figure 3.9(b)), the closure of the Penninic Ocean was completed and the Alpine orogenic wedge also started to thrust over the Helvetic platform in the modern Western Alps (Handy et al. 2010). Most of the areas of the modern Alps had already emerged from the sea. According to thermochronometric data, most of the medium-grade metamorphic rocks from the Alpine metamorphic event were still buried several kilometers, with only few areas in the east being exhumed in the early to middle Paleogene (Hejl 1997).

In the early Oligocene (~30 Ma, Figure 3.9(c)), the ongoing loading of the southern European margin by the northward pushing Adriatic plate caused further subsidence of the shelf areas and a northward shift of the shoreline north of the Paratethys Sea (Frisch et al. 1998). Rifting of the Rhine Graben, which is still active today, commenced and caused subsidence in that area. The most important event for the surface uplift history of the Alps, however, was the break-off of the subducted Penninic slab underneath the western part of the Alps (von Blanckenburg and Davis 1995; Schmid et al. 1996, 2013; Lippitsch et al. 2003). This caused an uneven uplift and exhumation regime across the range: while the Western Alps experienced accelerated uplift and exhumation, the Eastern Alps east of the Inntal fault, were not or only insignificantly affected. Together with the Western Carpathians, they formed hilly lowlands terminating in coastal plains bordering the Paratethys Sea (Frisch et al. 1998). On the plateaus of the Northern Calcareous Alps, remnants of these coastal plains are still preserved as conglomerates and (redeposited) gravels, also called the Augenstein Formation (Frisch et al. 2001). The peneplanation of these surfaces on Triassic rocks indicates that these planes formed after overlying Jurassic and younger rocks had eroded. All major rivers of the Eastern Alps drained to the north into the northern Molasse Basin at this time (Frisch et al. 1998). The Periadriatic Fault changed its shear sense from sinistral to dextral right after the intrusion of the main Periadriatic plutons and associated volcanism (Mancktelow et al. 2001).

From late Oligocene to early Miocene (~23 Ma, Figure 3.9(d)), the removal of the Penninic slab that occurred at 30 Ma caused further surface uplift of the Western Alps with the development of the first mountainous regions west of the Inn valley (Frisch et al. 1998, 2001). According to pollen data, the Western Alps reached elevations of 1,800 m already in the Oligocene (Fauquette et al. 2015). This increased topography gave rise to enhanced erosion and thus exhumation (Kuhlemann et al. 2001). The increased sediment flux caused the transition from basin underfill to overfill in the Molasse basin (Sinclair et al. 1991; Sinclair and Allen 1992). This caused the area north of the Western Alps to fall dry. In the east, the last remnant of Penninic oceanic lithosphere subducted westwards underneath the present-day Carpathians.

In the early Miocene (~19 Ma, Figure 3.9(e)), the subducting plate underneath the Carpathian Arc began rapid rolling-back to the east (Royden et al. 1983; Ren et al. 2012). The opened space was filled by the adjacent continental units, including the Alpine orogenic wedge. This led to lateral extrusion of the Alps, causing the formation of the major orogen-parallel strike-slip faults, which strongly affect the present-day drainage pattern of the Eastern Alps (Ratschbacher et al. 1991; Frisch et al. 1998; Robl et al. 2008b; Wölfler et al. 2011; Bartosch et al. 2017). Contemporaneous to crustal thinning in the easternmost part of the Alps and the Pannonian region, north-south shortening of the Adriatic Indenter led to crustal thickening in the Tauern Window region (Ratschbacher et al. 1989). Enhanced erosion due to higher topography and extensional detachments perpendicular to the Alpine orogen initiated accelerated exhumation of the Lepontine Dome (Simplon (Mancktelow 1992; Boston et al. 2017)) and the Tauern Window (Brenner, Katschberg; Fügenschuh et al. 1997; Wölfler et al. 2011, 2012). North of the Alps, subsidence occurred due to the propagating Alpine orogenic wedge and/or internal slab loads underneath the foreland (Kissling and Schlunegger 2018; Garefalakis and Schlunegger 2019). The northern Molasse Basin was subsequently re-flooded by the redevelopment of the sea arm connecting the Paratethys and the Mediterranean Sea. The orogen adjacent basins were substantially filled with sediment due to enhanced erosion (Kuhlemann et al. 2001).

During the middle Miocene (~15 Ma, Figure 3.9(f)), indentation led to crustal thickening and hence further rock uplift of the Alps, leading to enhanced deformation in the Southern Alps (Castellarin and Cantelli 2000). The Lepontine dome, the Aar and Gotthard Massifs continued the rapid exhumation that had its maximum already at 20 Ma (Boston et al. 2017; Herwegh et al. 2017) and the Tauern Window now shows increased exhumation rates (Wölfler et al. 2012). In contrast, the western External Massifs stay at lower rates, indicating foreland propagation at a later time for the westernmost Alps (Fox et al. 2016). In addition to the change in tectonic regime, the climate got colder and caused a global sea level drop due to major growth of the East Antarctic ice sheet (Flower and Kennett 1994). This caused the Molasse basin to successively fall dry from east to west. The still ongoing eastward extension led to the formation of several intramontane basins by fault-controlled pull-apart processes along the east-west oriented strike-slip faults defining the drainage network of the Eastern Alps (Bartosch et al. 2017). At the same time, these basins were filled by local clastic sediments, the Paratethys transgressed south of the Alps to the west, flooding the easternmost basins and forming a narrow connection to the Adriatic Sea. The fact that some of these intramontane basins bear marine sediments shows that the mean elevation of the Eastern Alps was still not substantial. In the Western Alps, orogen-parallel extension has been attributed to lateral extrusion due to the indentation and rotation of the Apulian microplate and the opening of the Ligurian Sea (Champagnac et al. 2006).

In late Miocene times (~10 Ma, Figure 3.9(g)), the east–west extension had declined in the east due to termination of the subduction underneath the Carpathian Arc, causing onset of basin inversion and rock and surface uplift (Bada et al. 2007). The brackish remnants of the isolated western Paratethys, known as Lake Pannon, formed the eastern boarder of the modern Alps. The central and eastern Northern Calcareous Alps became substantially uplifted and started to erode (Frisch et al. 1998). In the Western Alps, Mont Blanc, Aiguille Rouge and Belledonne External Massifs were still buried at kilometers depth but now show accelerated exhumation (Fox et al. 2016). However, this interpretation of Fox et al. (2016) is questionable, as all the external massifs (at least Gotthard, Aar, Belledonne) in the Western Alps have already cooled below the cooling temperatures of most low-temperature geochronological systems before that time (e.g. Bernet et al. 2001; Glotzbach et al. 2010; and references therein).

In the latest Miocene, the most important event affecting the topography of the Alps occurred in the Messinian (~6 Ma). The Mediterranean Sea became isolated from the ocean due to closure of the waterways to the Atlantic Ocean (e.g. Krijgsman et al. 1999). This led to the desiccation of the Mediterranean Sea and the formation of evaporites from highly saline waters. The Messinian Salinity Crisis was responsible for the biggest area of dry land below sea level known worldwide. The drop in base level also significantly increased incision of rivers draining into the relics of the Mediterranean Sea, accompanied by a drastic increase in erosional fluxes into the surrounding sedimentary basins (Willet et al. 2006; Figure 3.6). This created deep canyons that are submerged underneath the Po Plain today. However, relicts of these incisions may be seen in the large Italian lakes at the southern border of the Alps, although these were later superimposed by glacial carving (Bini et al. 1978; Finckh 1978). This complicated evolution of the valleys since the Messinian may partly be evidenced by channel profiles for rivers like the Adige that bear very heterogeneous and geomorphologically unequilibrated channel profiles (Robl et al. 2008a, 2017b). Similarly, the influence of a lowered base level during the Messinian can also be seen in the Rhône valley in reconstructions of the original valley floor underneath the sediments from seismic and drilling data (Gargani 2004). The Messinian event terminated with the re-flooding of the Mediterranean Sea around 5.3 Ma.

From the Pliocene until today (5.3 Ma–0 Ma, Figure 3.9(h)), the Alps underwent a serious build-up of topography. Erosion rates may have significantly increased since 5 Ma (Kuhlemann 2007), but the increase in apparent sediment flux may also be a reflection of an increased eroding region and sediment recycling as a consequence of foreland basin inversion (e.g. Cederbom et al. 2003). Nevertheless, both processes are reflection of a rejuvenation of rock uplift. At the eastern border of the Alps, surface uplift of at least 500 m since the Pliocene is proposed, implying that the rock uplift accelerated since this time (Wagner et al. 2010; Legrain et al. 2014, 2015). Around the same time, the highest still recognizable drainage networks from Mount Granier and Siebenhengste formed at the western and northwestern border of the Alps, respectively (Häuselmann 2007; Häuselmann et al. 2007; Hobléa et al. 2011). With onset of major glaciation in the Alps during the Pleistocene, the carving of U-shaped valleys and cirques with steep faces dominated the processes sculpting the landscape. In combination with surface uplift also visible in the non-glaciated parts of the Alps, these young glacial cycles are responsible for most of our Alpine landforms seen today. But the Alps are far from being done growing. Vertical displacements measured with GPS/GNSS stations show growth rates exceeding 2 mm/yr (Nocquet et al. 2016; Sternai et al. 2019; Figure 3.10). However, it is noted that GPS-determined uplift rates are gathered from short-term measurements and therefore not directly comparable to uplift rates on a geological timescale.

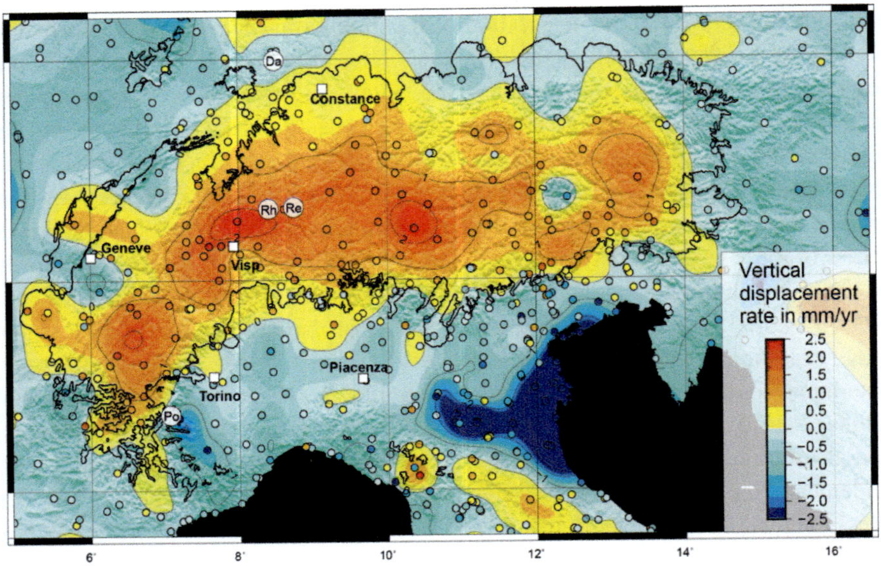

Figure 3.10. *Present-day vertical displacement rates measured with GPS/GNSS across the Alps, modified from Sternai et al. (2019). Circles indicate the location of the measuring GPS/GNSS stations*

COMMENT ON FIGURE 3.10.– *Note that the interpretation of vertical motion presented here is partly inconsistent with geological evidence, in particular for the eastern end of the Alps.*

3.5. Drivers of topography and landscape evolution

In view of the surface uplift history described in the previous sections, the question about the underlying drivers arises quickly. Clearly, the evolution of a landscape mirrors the interplay of rock uplift and erosion to produce surface uplift. Even in one of the best studied mountain ranges in the world, assessing the drivers of the evolution of our observed topography is still fuelling active debates (e.g. Baran et al. 2014; Dixon et al. 2016; Sternai et al. 2019). The two opposing stands argue about whether the young uplift and increase in denudation rates is mainly due to climatic or tectonic control. These two drivers have been discussed for time slices as old as the Oligocene (Schlunegger and Norton 2015), but we focus here on the last substantial uplift pulse at post-Miocene times.

The tectonic control may best be understood by summarizing today's deformation regime. The current rotation pole of the Adriatic microplate with respect to the European plate is generally placed within the Western Alps and the western Po Basin (e.g. Calais et al. 2002; Weber et al. 2010) with a counterclockwise rotational motion. This rotational motion, which has been active since at least the late Oligocene (Vialon et al. 1989; Schmid and Kissling 2000; Collombet et al. 2002), results in different deformation regimes across the Alps: transtension or minor compression in the Western Alps, dextral shear in the Central Alps and compression in the Eastern Alps (e.g. Sue et al. 2007; Walpersdorf et al. 2015; Serpelloni et al. 2016). This is consistent with our knowledge that deformation in the Central Alps due to north–south convergence stopped at least since the Pliocene (Schmid et al. 1996) and the locus of uplift and exhumation migrated into the core of the orogen (Fox et al. 2016; Figure 3.7). At depth, teleseismic tomography implies a southward dipping lithospheric slab underneath the Central Alps, a detached European slab underneath the Western Alps and a reversal in subduction polarity underneath the Eastern Alps with a northeastward-dipping Adriatic slab (Lippitsch et al. 2003), or even potential slab detachment underneath the Eastern Alps as well (Qorbani et al. 2015). However, a high-resolution tomography model from Zhao et al. (2016) shows a clear lateral connection of the European slab from the Western to the Central Alps, ruling out slab break-off as a cause for Cenozoic topography development.

The climatic control is governed by a continuous cooling of the atmosphere since the mid Miocene thermal maximum (Zachos et al. 2001). This cooling accelerated during the Messinian and ultimately resulted in glaciation of the Alps in the Pleistocene. The Miocene–Pliocene transition was also characterized by a change to wetter conditions and the desiccation of the Mediterranean from 5.8 to 5.3 Ma, as well as a dramatic increase in erosional fluxes into the sedimentary basins surrounding the Alps (Willett et al. 2006; Figure 3.6). Increased erosional fluxes around this time are observed globally (Molnar 2004).

In view of the global increase of sediment fluxes at 5 Ma (Molnar 2004), it is tempting to interpret the climatic cooling since the Messinian as the main driver for rock and likely also surface uplift, in particular since no global tectonic phenomena are known for that time (Molnar and England 1990). For the Alps, this is in fact the conclusion of Willett et al. (2006) and Champagnac et al. (2007) who argue that erosion-induced uplift due to a wetter climate since the Messinian (augmented by the Mediterranean base level drop) accounts for some 50% of the surface uplift. Such an interpretation implies that this surface uplift is predominantly seen in the peaks, with most erosion occurring in the valleys and the mean elevation of the range actually decreasing. Champagnac et al. (2008) showed how this erosion driven isostatic response is also responsible for the tilting of some foreland regions. Such tilting and erosion of the foreland basin was also recognized by Cederbom et al. (2004) who extrapolated this tilting into the orogen to argue that some 6.5 km were eroded off the Western Alps since the Pliocene. However, Mey et al. (2016) argue that 90% of the present-day surface uplift is due to deglaciation-induced isostatic rebound, given that much of the material eroded from the Alps was deposited within Alpine valleys. Nevertheless, in general it is agreed that considerably less than 100% of the surface uplift in the central axis of the Alpine orogen can be attributed to these climatic drivers. As deformation in the Central and Western Alps is known to have ceased at the time of these processes, deep seated drivers in the lithosphere are likely to be partly responsible for these vertical motions in the central orogen, showing that both, tectonic and climatic drivers contribute to uplift.

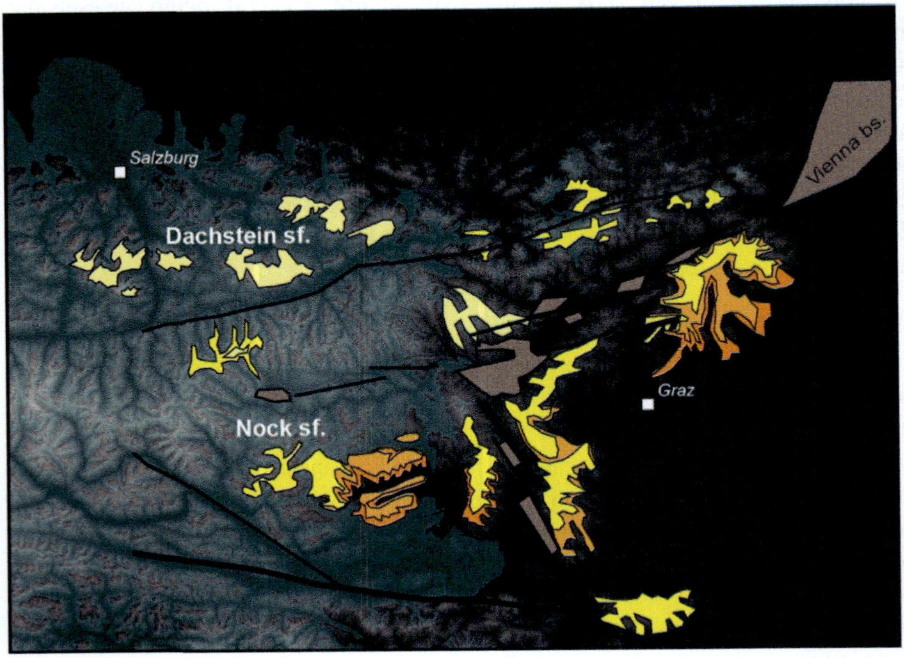

Figure 3.11. *Simplified schematic map of the planation surfaces in the Eastern Alps (from studies of Winkler-Hermaden 1957; Hejl 1997; Frisch et al. 2001; Wagner et al. 2011; Legrain et al. 2015; Dertnig et al. 2017; Bartosch and Stüwe 2019)*

COMMENT ON FIGURE 3.11.– *Wagner et al. (2011) provided some correlation of some of these surfaces. According to them, the shown surfaces correspond roughly and schematized to: light yellow = the Dachstein-surface at about 2,500 m a.s.l.; bright yellow = the Nock- or Kor-surface; light orange = the Wolscheneck or*

Hubenhalt levels; and dark orange = the lower levels that are known as the Trahütten, Hochstraden, Stadelberg/Zahrerberg levels below about 600 m above base level. Some major faults and pull-apart basins along them are shown by black solid lines and brown areas, respectively. Ice extent of the LGM is indicated in blue (Ehlers et al. 2011).

In the Eastern Alps, deep seated tectonic drivers as the principal cause of a long wavelength post-Miocene uplift and exhumation event have been argued for some time (e.g. Lippitsch et al. 2003; Genser et al. 2007; Wagner et al. 2010; Robl et al. 2015). Stepwise arrangement of planated fluvial relict landscapes is abundant, in particular in areas that have escaped the Pleistocene glaciations and present evidence for substantial Pliocene surface uplift (Figure 3.11). Correlations of these planation surfaces with associated knickpoints in fluvial channels and with burial ages from different cave levels record post-Miocene surface uplift of some 500 m since 4 Ma (Robl et al. 2008a; Wagner et al. 2010, 2011; Legrain et al. 2014; Dertnig et al. 2017). Because of the documented accelerated uplift before the Pleistocene glaciations commenced and outside of their geographic extent, deep seated drivers are inferred (Legrain et al. 2014). However, too little surfaces across the Alps were dated this way to enable interpretations covering the whole Alpine range. Interestingly, elevated low relief surfaces in the Eastern Alps can be found in areas that were glaciated, as well as unglaciated during the last glacial maximum, for example the Augenstein landscape of the Northern Calcareous Alps or the Nockfläche in the Niedere Tauern region (Figure 3.11). To entangle the influence of glaciers and a possible young tectonic surface uplift pulse on these landscapes, further efforts regarding the coverage of mapped and dated planation surfaces, combined with numerical landscape evolution modeling, have to be made.

When looking at recent erosion rates, differences between Western and Eastern Alps are small, except for the unglaciated easternmost part, where erosion rates are significantly lower (Dixon et al. 2016; Figure 3.8). Dixon et al. (2016) showed that erosion rates in the Eastern Alps are controlled by the prevailing erosional regime, which is primarily governed by the climate. Glacial erosion was long thought to be more effective than fluvial erosion and was often used to explain an increase in erosion rates, for example, by lowering the equilibrium-line altitude (e.g. Häuselmann et al. 2007). Apart from lowering the equilibrium-line altitude by generally decreasing the mean annual temperature, increased glaciation can also be achieved by uplifting lower areas above the existing ELA. In this case, tectonics has a direct impact on the glacial extent and could initiate the feedback loop of erosion and uplift. However, over 10^4 years and longer time scales fluvial and glacial erosion seem to erode the landscape at a similar pace (Koppes and Montgomery 2009). When encountering combinations of fluvial and glacial erosion, for example a U-shaped valley with a V-shape at the bottom, the fluvial incision is usually considered to have occurred after the glaciation event. However, since implausibly high erosion rates would be needed to explain the incision after glaciation, some gorges are believed to have formed contemporary or even before glacial advance (Montgomery and Korup 2011).

3.6. Conclusion

We presented a number of state-of-the-art research methods and their applications in modern neotectonic studies that have been used to derive the age and evolution of landscapes and surface elevation of the European Alps. Combining all the gathered information, we come to the following conclusions:

– The landscape evolution of the Alps started about 35 Ma, but some landforms can be traced back to about 45 Ma. However, the surface uplift in the Alps did not occur linearly since this time. A substantial part of this uplift is as young as the Pliocene. Deep lithospheric processes appear to be the best explanation for this in the Eastern Alps, but in the Western Alps a substantial portion of this has been attributed to climatic drivers.

– Even if recent sediment fluxes were being overestimated due to sampling effects, a sudden rise in sediment fluxes to the sedimentary basins surrounding the Alps since the Pliocene seems reasonable. It correlates well with the young incision and surface uplift observed in both the Eastern and Western Alps.

– The existence of planation surfaces, particularly in the Eastern Alps, is interpreted to be the reflection of an immature geomorphological state and the young surface uplift since the Pliocene. The occurrence of these surfaces in regions with differing glacial history remains puzzling and deserves further investigation but it is possible that such surfaces in glaciated regions are also remains of pre-glaciation planation events.

– Since the onset of Pleistocene glaciations, the Alps were reshaped with an abundance of glacial features. Differences in ^{10}Be-derived erosion rates between glacially reshaped and never glaciated regions vary from less than 500 mm/kyr to more than 1,000 mm/kyr.

– Despite the difference in geophysical relief, the slope–elevation distribution and erosion rates do not show a clear trend differentiating the Eastern Alps from the Western Alps. A bimodal distribution of low slopes at low and high elevations is unusual in comparison with many other mountain belts. Minor differences may depend on the varying glacial histories.

3.7. Acknowledgments

We thank M. Fox, R. Schuster and P. Sternai for kindly sharing their working files from their publications to produce Figures 3.7, 3.9 and 3.10, respectively. We also thank T. Bartosch for helping us to create the channel profiles and R. Watson for handing over a well sorted pool of literature on the topic. We also thank M. Bernet, F. Schlunegger and an anonymous reviewer for their constructive comments, as well as C.L. Rosenberg and N. Bellahsen for editorial handling. This work was financially supported by the Austrian Science Fund (FWF-project P31609-NBL).

3.8. References

Adams, J. (1985). Large-scale tectonic geomorphology of the Southern Alps, New Zealand. In *Tectonic Geomorphology*, Morisawa, M. and Hack, J.T. (eds). Allen and Unwin, Winchester.

Ahnert, F. (1970). Functional relationships between denudation, relief, and uplift in large mid-latitude drainage basins. *Am. J. Sci.*, 268, 243–263.

Anders, A.M., Mitchell, S.G., Tomkin, J.H. (2010). Cirques, peaks, and precipitation patterns in the Swiss Alps: Connections among climate, glacial erosion, and topography. *Geology*, 38(3), 239–242.

Bächler, E. (1940). *Das alpine Paläolithikum der Schweiz im Wildkirchli, Drachenloch und Wildenmannlisloch: die ältesten Menschlichen Niederlassungen aus der Altsteinzeit des Schweizerlandes*. Birkhäuser & Cie, Basel.

Bada, G., Horváth, F., Dövényi, P., Szafián, P., Windhoffer, G., Cloetingh, S. (2007). Present-day stress field and tectonic inversion in the Pannonian basin. *Glob. Planet. Change*, 58, 165–180.

Baran, R., Friedrich, A.M., Schlunegger, F. (2014). The late Miocene to Holocene erosion pattern of the Alpine foreland basin reflects Eurasian slab unloading beneath the western Alps rather than global climate change. *Lithosphere*, 6, 124–131.

Bartosch, T. and Stüwe, K. (2019). Evidence for pre-Pleistocene landforms in the Eastern Alps: Geomorphological constraints from the Gurktal Alps. *Austrian J. Earth Sci.*, 112, 84–102.

Bartosch, T., Stüwe, K., Robl, J. (2017). Topographic evolution of the Eastern Alps: The influence of strike-slip faulting activity. *Lithosphere*, 9, 384–398.

Baumgartner, A., Reichelt, E., Weber, G. (1983). *Der Wasserhaushalt der Alpen. Niederschlag, Verdunstung, Abfluß und Gletscherspende im Gesamtgebiet der Alpen im Jahresdurchschnitt für die Normalperiode 1931-1969*. Oldenbourg, Munich, Vienna.

van der Beek, P.A., Valla, P.G., Herman, F., Braun, J., Persano, C., Dobson, K.J., Labrin, E. (2010). Inversion of thermochronological age–elevation profiles to extract independent estimates of denudation and relief history II: Application to the French Western Alps. *Earth Planet. Sci. Lett.*, 296, 9–22.

Bernet, M., Zattin, M., Garver, J.I., Brandon, M.T., Vance, J.A. (2001). Steady-state exhumation of the European Alps. *Geology*, 29, 35–38.

Bernet, M., Brandon, M.T., Garver, J.I., Molitor, B. (2004). Downstream changes of Alpine zircon fission-track ages in the Rhone and Rhine rivers. *J. Sediment. Res.*, 74, 82–94.

Bernet, M., Brandon, M., Garver, J., Balestieri, M.L., Ventura, B., Zattin, M. (2009). Exhuming the Alps through time: Clues from detrital zircon fission-track thermochronology. *Basin Res.*, 21, 781–798.

Bernet, M., van der Beek, P., Valla, P.G., Glotzbach, C. (2023). The detrital thermochronological record of exhumation of the European Alps since the early Oligocene. In *Geodynamics of the Alps 3*, Rosenberg, C.L., Bellahsen, N. (eds). ISTE Ltd, London, and John Wiley & Sons, New York.

Bini, A., Cita, M.B., Gaetani, M. (1978). Southern Alpine lakes – Hypothesis of an erosional origin related to the Messinian entrenchment. *Mar. Geol.*, 27, 271–288.

von Blanckenburg, F. (2005). The control mechanisms of erosion and weathering at basin scale from cosmogenic nuclides in river sediment. *Earth Planet. Sci. Lett.*, 237, 462–479.

von Blanckenburg, F. and Davies, J.H. (1995). Slab breakoff: A model for syncollisional magmatism and tectonics in the Alps. *Tectonics*, 14, 120–131.

Boccaletti, M., Corti, G., Martelli, L. (2011). Recent and active tectonics of the external zone of the Northern Apennines (Italy). *Int. J. Earth Sci.*, 100, 1331–1348.

Brandon, M.T., Roden-Tice, M.K., Carver, J.I. (1998). Late Cenozoic exhumation of the Cascadia accretionary wedge in the Olympic Mountains, northwest Washington State. *Bull. Geol. Soc. Am.*, 110, 985–1009.

Calais, E., Nocquet, J.M., Jouanne, F., Tardy, M. (2002). Current strain regime in the Western Alps from continuous Global Positioning System measurements, 1996-2001. *Geology*, 30, 651–654.

Campani, M., Mancktelow, N., Seward, D., Rolland, Y., Müller, W., Guerra, I. (2010). Geochronological evidence for continuous exhumation through the ductile-brittle transition along a crustal-scale low-angle normal fault: Simplon Fault Zone, central Alps. *Tectonics*, 29, TC3002.

Campani, M., Mulch, A., Kempf, O., Schlunegger, F., Mancktelow, N. (2012). Miocene paleotopography of the Central Alps. *Earth Planet. Sci. Lett.*, 337, 174–185.

Castellarin, A. and Cantelli, L. (2000). Neo-Alpine evolution of the Southern Eastern Alps. *J. Geodyn.*, 30, 251–274.

Cederbom, C.E., Sinclair, H.D., Schlunegger, F., Rahn, M.K. (2004). Climate-induced rebound and exhumation of the European Alps. *Geology*, 32, 709–712.

Champagnac, J.D., Molnar, P., Anderson, R.S., Sue, C., Delacou, B. (2007). Quaternary erosion-induced isostatic rebound in the western Alps. *Geology*, 35, 195–198.

Champagnac, J.D., van der Beek, P., Diraison, G., Dauphin, S. (2008). Flexural isostatic response of the Alps to increased Quaternary erosion recorded by foreland basin remnants, SE France. *Terra Nov.*, 20, 213–220.

Champagnac, J.D., Schlunegger, F., Norton, K., von Blanckenburg, F., Abbühl, L.M., Schwab, M. (2009). Erosion-driven uplift of the modern Central Alps. *Tectonophysics*, 474, 236–249.

Champagnac, J.D., Molnar, P., Sue, C., Herman, F. (2012). Tectonics, climate, and mountain topography. *J. Geophys. Res. Solid Earth*, 117. doi: 10.1029/2011JB008348.

Collombet, M., Thomas, J.C., Chauvin, A., Tricart, P., Bouillin, J.P., Gratier, J.P. (2002). Counterclockwise rotation of the western Alps since the Oligocene: New insights from paleomagnetic data. *Tectonics*, 21, 1–15.

Cyr, A.J., Granger, D.E., Olivetti, V., Molin, P. (2014). Distinguishing between tectonic and lithologic controls on bedrock channel longitudinal profiles using cosmogenic 10Be erosion rates and channel steepness index. *Geomorphology*, 209, 27–38.

Delunel, R.. van der Beek, P.A., Carcaillet, J., Bourlès, D.L., Valla, P.G. (2010). Frost-cracking control on catchment denudation rates: Insights from in situ produced ^{10}Be concentrations in stream sediments (Ecrins–Pelvoux massif, French Western Alps). *Earth Planet. Sci. Lett.*, 293, 72–83.

Dertnig, F., Stüwe, K., Woodhead J., Stuart F.M., Spötl, C. (2017). Constraints on the Miocene landscape evolution of the Eastern Alps from the Kalkspitze region, Niedere Tauern (Austria). *Geomorphology*, 299, 24–38.

Dixon, J.L., von Blanckenburg, F., Stüwe, K., Christl, M. (2016). Glaciation's topographic control on Holocene erosion at the eastern edge of the Alps. *Earth Surf. Dyn.*, 4, 895–909.

Dunai, T.J. (2010). *Cosmogenic Nuclides*. Cambridge University Press, Cambridge.

Ehlers, J., Gibbard, P.L., Hughes, P.D. (eds) (2011). *Quaternary Glaciations – Extent and Chronology: A Closer Look*. Elsevier, Amsterdam.

Einsele, G. (1992). *Sedimentary Basins: Evolution, Facies and Sediment Budget*. Springer-Verlag, Berlin.

England, P. (1981). Metamorphic pressure estimates and sediment volumes for the Alpine orogeny: An independent control on geobarometers? *Earth Planet. Sci. Lett.*, 56, 387–397.

England, P. and Molnar, P. (1990). Surface uplift, uplift of rocks, and exhumation of rocks. *Geology*, 18, 1173–1177.

Exner, C. (1949). Beitrag zur Kenntnis der jungen Hebung der östlichen Hohen Tauern. *Mitt. Geogr. Ges. Wien*, 91, 186–196.

Farley, K.A. (2000). Helium diffusion from apatite: General behavior as illustrated by Durango fluorapatite. *J. Geophys. Res. Solid Earth*, 105, 2903–2914.

Fauquette, S., Bernet, M., Suc, J.P., Grosjean, A.S., Guillot, S., van der Beek, P., Jourdan, S., Popescu, S.M., Jiménez-Moreno, G., Bertini, A. et al. (2015). Quantifying the Eocene to Pleistocene topographic evolution of the southwestern Alps, France and Italy. *Earth Planet. Sci. Lett.*, 412, 220–234.

Finckh, P. (1978). Are southern Alpine lakes former Messinian canyons? – Geophysical evidence for preglacial erosion in the southern Alpine lakes. *Mar. Geol.*, 27, 289–302.

Finckh, P., Kelts, K., Lambert, A. (1984). Seismic stratigraphy and bedrock forms in perialpine lakes. *Bull. Geol. Soc. Am.*, 95, 1118–1128.

Flint, J. (1974). Stream gradient as a function of order, magnitude, and discharge. *Water Resour. Res.*, 10, 969–973.

Flower, B.P. and Kennet, J.P. (1994). The middle Miocene climatic transition: East Antarctic ice sheet development, deep ocean circulation and global carbon cycling. *Palaeogeogr. Palaeoclimatol. Palaeoecol.*, 108, 537–555.

Fox, M., Herman, F., Willet, S.D., May, D.A. (2014). A linear inversion method to infer exhumation rates in space and time from thermochronometric data. *Earth Surf. Dyn.*, 2, 47–65.

Fox, M., Herman, F., Kissling, E., Willett, S.D. (2015). Rapid exhumation in the Western Alps driven by slab detachment and glacial erosion. *Geology*, 43, 379–382.

Fox, M., Herman, F., Willett, S.D., Schmid, S.M. (2016). The exhumation history of the European alps inferred from linear inversion of thermochronometric data. *Am. J. Sci.*, 316, 505–541.

Frisch, W., Kuhlemann, J., Dunkl, I., Brügel, A. (1998). Palinspastic reconstruction and topographic evolution of the Eastern Alps during late tertiary tectonic extrusion. *Tectonophysics*, 297, 1–15.

Frisch, W., Dunkl, I., Kuhlemann, J. (2000). Post-collisional orogen-parallel large-scale extension in the Eastern Alps. *Tectonophysics*, 327, 239–265.

Frisch, W., Kuhlemann, J., Dunkl, I., Székely, B. (2001). The Dachstein paleosurface and the Augenstein Formation in the Northern Calcareous Alps – A mosaic stone in the geomorphological evolution of the Eastern Alps. *Int. J. Earth Sci.*, 90, 500–518.

Fügenschuh, B., Seward, D., Mancktelow, N. (1997). Exhumation in a convergent orogen: The western Tauern Window. *Terra Nov.*, 9, 213–217.

Garefalakis, P. and Schlunegger, F. (2019). Tectonic processes, variations in sediment flux, and eustatic sea level recorded by the 20 Myr old Burdigalian transgression in the Swiss Molasse basin. *Solid Earth*, 10, 2045–2072.

Gargani, J. (2004). Modelling of the erosion in the Rhone valley during the Messinian crisis (France). *Quat. Int.*, 121, 13–22.

Garver, J.I. and Brandon, M.T. (1994). Fission-track ages of detrital zircons from Cretaceous strata, southern British Columbia: Implications for the Baja BC hypothesis. *Tectonics*, 13, 401–420.

Garver, J.I., Brandon, M.T., Roden-Tice, M., Kamp, P.J.J. (1999). Exhumation history of orogenic highlands determined by detrital fission-track thermochronology. *Geol. Soc. London, Spec. Publ.*, 154, 283–304.

Genser, J., Cloetingh, S.A.P.L., Neubauer, F. (2007). Late orogenic rebound and oblique Alpine convergence: New constraints from subsidence analysis of the Austrian Molasse basin. *Glob. Planet. Change*, 58, 214–223.

Glotzbach, C., Reinecker, J., Danišík, M., Rahn, M., Frisch, W., Spiegel, C. (2010). Thermal history of the central Gotthard and Aar massifs, European Alps: Evidence for steady state, long-term exhumation. *J. Geophys. Res. Earth Surf.*, 115, F03017.

Glotzbach, C., van der Beek, P.A., Spiegel, C. (2011). Episodic exhumation and relief growth in the Mont Blanc massif, Western Alps from numerical modelling of thermochronology data. *Earth Planet. Sci. Lett.*, 304, 417–430.

Glotzbach, C., van der Beek, P., Carcaillet, J., Delunel, R. (2013). Deciphering the driving forces of erosion rates on millennial to million-year timescales in glacially impacted landscapes: An example from the Western Alps. *J. Geophys. Res. Earth Surf.*, 118, 1491–1515.

Gosse, J.C. and Phillips, F.M. (2001). Terrestrial in situ cosmogenic nuclides: Theory and application. *Quat. Sci. Rev.*, 20, 1475–1560.

Granger, D.E. and Muzikar, P.F. (2001). Dating sediment burial with in situ-produced cosmogenic nuclides: Theory, techniques, and limitations. *Earth Planet. Sci. Lett.*, 188, 269–281.

Green, P.F., Duddy, I.R., Laslett, G.M., Hegarty, K.A., Gleadow, A.J.W., Lovering, J.F. (1989). Thermal annealing of fission tracks in apatite 4. Quantitative modelling techniques and extension to geological timescales. *Chem. Geol. Isot. Geosci. Sect.*, 79, 155–182.

Guenthner, W.R., Reiners, P.W., Ketcham, R.A., Nasdala, L., Giester, G. (2013). Helium diffusion in natural zircon: Radiation damage, anisotropy, and the interpretation of zircon (U-TH)/He thermochronology. *Am. J. Sci.*, 313, 145–198.

Handy, M.R., Schmid, S.M., Bousquet, R., Kissling, E., Bernoulli, D. (2010). Reconciling plate-tectonic reconstructions of Alpine Tethys with the geological-geophysical record of spreading and subduction in the Alps. *Earth-Sci. Rev.*, 102, 121–158.

Handy, M.R., Ustaszewski, K., Kissling, E. (2015). Reconstructing the Alps–Carpathians–Dinarides as a key to understanding switches in subduction polarity, slab gaps and surface motion. *Int. J. Earth Sci.*, 104, 1–26.

Häuselmann, P. (2007). How to date nothing with cosmogenic nuclides. *Acta Carsologica*, 36, 93–100.

Häuselmann, P., Granger, D.E., Jeannin, P.Y., Lauritzen, S.E. (2007). Abrupt glacial valley incision at 0.8 Ma dated from cave deposits in Switzerland. *Geology*, 35, 143–146.

Hejl, E. (1997). "Cold spots" during the Cenozoic evolution of the Eastern Alps: Thermochronological interpretation of apatite fission-track data. *Tectonophysics*, 272, 159–173.

Hergarten, S., Wagner, T., Stüwe, K. (2010). Age and prematurity of the Alps derived from topography. *Earth Planet. Sci. Lett.*, 297, 453–460.

Herman, F. and Champagnac, J.D. (2016). Plio-Pleistocene increase of erosion rates in mountain belts in response to climate change. *Terra Nov.*, 28, 2–10.

Herman, F., Beaud, F., Champagnac, J.D., Lemieux, J.M., Sternai, P. (2011). Glacial hydrology and erosion patterns: A mechanism for carving glacial valleys. *Earth Planet. Sci. Lett.*, 310, 498–508.

Herman, F., Seward, D., Valla, P.G., Carter, A., Kohn, B., Willett, S.D., Ehlers, T.A. (2013). Worldwide acceleration of mountain erosion under a cooling climate. *Nature*, 504, 423–426.

Herwegh, M., Berger, A., Baumberger, R., Wehrens, P., Kissling, E. (2017). Large-scale crustal-block-extrusion during late alpine collision. *Sci. Rep.*, 7, 1–10.

Hobléa, F., Häuselmann, P., Kubik, P. (2011). Cosmogenic nuclide dating of cave deposits of Mount Granier (Hauts de Chartreuse Nature Reserve, France): Morphogenic and palaeogeographical implications. *Géomorphologie Reli. Process. Environ.*, 17, 395–406.

Hötzl, H. (1996). Origin of the Danube-Aach system. *Environmental Geology*, 27, 87–96.

Howard, A.D. (1994). A detachment-limited model of drainage-basin evolution. *Water Resour. Res.*, 30, 2261–2285.

Hurford, A.J. and Carter, A. (1991). The role of fission track dating in discriminitaion of provenance. In *Developments in Sedimentary Provenance Studies*, Morton, A.C., Todd, S.P., Haughton, P.D.W. (eds). Geol. Soc. London, Spec. Pub., London.

Hurford, A.J., Fitch, F.J., Clarke, A. (1984). Resolution of the age structure of the detrital zircon populations of two Lower Cretaceous sandstones from the Weald of England by fission track dating. *Geol. Mag.*, 121, 269–277.

Ivy-Ochs, S. and Kober, F. (2013). Exposure geochronology. In *Encyclopedia of Quaternary Science: Second Edition*, Elias, S.A., Mock, C.J. (eds). Elsevier B.V., Amsterdam.

Ivy-Ochs, S., Kerschner, H., Reuther, A., Maisch, M., Sailer, R., Schaefer, J., Kubik, P.W., Synal, H.-A., Schlüchter, C. (2006). The timing of glacier advances in the northern European Alps based on surface exposure dating with cosmogenic ^{10}Be, ^{26}Al, ^{36}Cl, and ^{21}Ne. In *In Situ-Produced Cosmogenic Nuclides and Quantification of Geological Processes*, Siame, L.L., Bourlès, D.L., Brown, E.T. (eds). Geological Society of America, Boulder.

Ivy-Ochs, S., Kerschner, H., Reuther, A., Preusser, F., Heine, K., Maisch, M., Kubik, P.W., Schlüchter, C. (2008). Chronology of the last glacial cycle in the European Alps. *J. Quat. Sci.*, 23, 559–573.

Keil, M. and Neubauer, F. (2011). Neotectonics, drainage pattern and geomorphology of the orogen-parallel Upper Enns Valley (Eastern Alps). *Geol. Carpathica*, 62, 279–295.

Ketcham, R.A., Donelick, R.A., Carlson, W.D. (1999). Variability of apatite fission-track annealing kinetics III: Extrapolation to geological time scales. *Am. Mineral.*, 84, 1235–1255.

Kirby, E. and Whipple, K.X. (2012). Expression of active tectonics in erosional landscapes. *J. Struct. Geol.*, 44, 54–75.

Kissling, E. and Schlunegger, F. (2018). Rollback Orogeny model for the evolution of the Swiss Alps. *Tectonics*, 37, 1097–1115.

Knop, A. (1878). Über die hydrographischen Beziehungen zwischen der Donau und der Aachquelle im badischen Oberlande. *Neues Jahrbuch für Mineralogie, Geologie und Paläontologie*, Jahrgang, 350–360.

Koppes, M.N. and Montgomery, D.R. (2009). The relative efficacy of fluvial and glacial erosion over modern to orogenic timescales. *Nat. Geosci.*, 2, 644–647.

Korup, O. and Schlunegger, F. (2007). Bedrock landsliding, river incision, and transience of geomorphic hillslope-channel coupling: Evidence from inner gorges in the Swiss Alps. *J. Geophys. Res. Earth Surf.*, 112, F03027.

Krijgsman, W., Hilgen, F.J., Raffi, I., Sierro, F.J., Wilson, D.S. (1999). Chronology, causes and progression of the Messinian salinity crisis. *Nature*, 400, 652–655.

Kubik, P.W., Ivy-Ochs, S., Masarik, J., Frank, M., Schlüchter, C. (1998). ^{10}Be and ^{26}Al production rates deduced from an instantaneous event within the dendro-calibration curve, the landslide of Köfels, Ötz Valley, Austria. *Earth Planet. Sci. Lett.*, 161, 231–241.

Kuhlemann, J. (2000). Post-collisional sediment budget of circum-Alpine basins (Central Europe). *Mem. Sci. Geol. Padova*, 52, 1–91.

Kuhlemann, J. (2007). Paleogeographic and paleotopographic evolution of the Swiss and Eastern Alps since the Oligocene. *Glob. Planet. Change*, 58, 224–236.

Kuhlemann, J., Frisch, W., Dunkl, I., Székely, B. (2001). Quantifying tectonic versus erosive denudation by the sediment budget: The miocene core complexes of the Alps. *Tectonophysics*, 330, 1–23.

Kuhlemann, J., Frisch, W., Székely, B., Dunkl, I., Kázmér, M. (2002). Post-collisional sediment budget history of the Alps: Tectonic versus climatic control. *Int. J. Earth Sci.*, 91, 818–837.

Kühni, A. and Pfiffner, O.A. (2001). The relief of the Swiss Alps and adjacent areas and its relation to lithology and structure: Topographic analysis from a 250 m DEM. *Geomorphology*, 41, 285–307.

Lal, D. and Arnold, J.R. (1985). Tracing quartz through the environment. *J. Earth Syst. Sci.*, 94, 1–5.

Legrain, N., Stüwe, K., Wölfler, A. (2014). Incised relict landscapes in the eastern Alps. *Geomorphology*, 221, 124–138.

Legrain, N., Dixon, J., Stüwe, K., von Blanckenburg, F.V., Kubik, P. (2015). Post-Miocene landscape rejuvenation at the eastern end of the Alps. *Lithosphere*, 7, 3–13.

Leloup, P.H., Arnaud, N., Sobel, E.R., Lacassin, R. (2005). Alpine thermal and structural evolution of the highest external crystalline massif: The Mont Blanc. *Tectonics*, 24, TC4002.

Lippitsch, R., Kissling, E., Ansorge, J. (2003). Upper mantle structure beneath the Alpine orogen from high-resolution teleseismic tomography. *J. Geophys. Res.*, 108, 2376.

Malusà, M.G. and Fitzgerald, P.G. (eds) (2019). *Fission-track Thermochronology and Its Application to Geology*. Springer International Publishing, Cham.

Mancktelow, N.S. (1992). Neogene lateral extension during convergence in the Central Alps: Evidence from interrelated faulting and backfolding around the Simplonpass (Switzerland). *Tectonophysics*, 215, 295–317.

Mancktelow, N.S., Stöckli, D.F., Grollimund, B., Müller, W., Fügenschuh, B., Viola, G., Seward, D., Villa, I.M. (2001). The DAV and pediatriatic fault systems in the Eastern Alps South of the Tauern window. *Int. J. Earth Sci.*, 90, 593–622.

Mey, J., Scherler, D., Wickert, A.D., Egholm, D.L., Tesauro, M., Schildgen, T.F., Strecker, M.R. (2016). Glacial isostatic uplift of the European Alps. *Nat. Com.*, 7, 13382.

Molnar, P. (2004). Late cenozoic increase in accumulation rates of terrestrial sediment: How might climate change have affected erosion rates? *Annu. Rev. Earth Planet. Sci.*, 32, 67–89.

Molnar, P. and England, P. (1990). Late Cenozoic uplift of mountain ranges and global climate change: Chicken or egg? *Nature*, 346, 29–34.

Montgomery, D.R. and Korup, O. (2011). Preservation of inner gorges through repeated Alpine glaciations. *Nat. Geosci.*, 4, 62–67.

Mulch, A. (2016). Stable isotope paleoaltimetry and the evolution of landscapes and life. *Earth Planet. Sci. Lett.*, 433, 180–191.

Nishiizumi, K., Winterer, E.L., Kohl, C.P., Klein, J., Middleton, R., Lal, D., Arnold, J.R. (1989). Cosmic ray production rates of 10^{Be} and 26^{Al} in quartz from glacially polished rocks. *J. Geophys. Res.*, 94, 17907.

Nocquet, J.M., Sue, C., Walpersdorf, A., Tran, T., Lenôtre, N., Vernant, P., Cushing, M., Jouanne, F., Masson, F., Baize, S. et al. (2016). Present-day uplift of the western Alps. *Sci. Rep.*, 6, 1–6.

Norton, K.P. and Schlunegger, F. (2017). Lack of a weathering signal with increased Cenozoic erosion? *Terra Nov.*, 29, 265–272.

Norton, K.P., von Blanckenburg, F., Schlunegger, F., Schwab, M., Kubik, P.W. (2008). Cosmogenic nuclide-based investigation of spatial erosion and hillslope channel coupling in the transient foreland of the Swiss Alps. *Geomorphology*, 95, 474–486.

Norton, K.P., Abbühl, L.M., Schlunegger, F. (2010a). Glacial conditioning as an erosional driving force in the Central Alps. *Geology*, 38, 655–658.

Norton, K.P., von Blanckenburg, F., Kubik, P.W. (2010b). Cosmogenic nuclide-derived rates of diffusive and episodic erosion in the glacially sculpted upper rhone valley, Swiss Alps. *Earth Surf. Process. Landforms*, 35, 651–662.

Peizhen, Z., Molnar, P., Downs, W.R. (2001). Increased sedimentation rates and grain sizes 2±4Myr ago due to the in¯uence of climate change on erosion rates. *Nature*, 410, 891–897.

Perrone, G., Morelli, M., Piana, F., Fioraso, G., Nicolò, G., Mallen, L., Cadoppi, P., Balestro, G., Tallone, S. (2013). Current tectonic activity and differential uplift along the Cottian Alps/Po Plain boundary (NW Italy) as derived by PS-InSAR data. *J. Geodyn.*, 66, 65–78.

Qorbani, E., Bianchi, I., Bokelmann, G. (2015). Slab detachment under the Eastern Alps seen by seismic anisotropy. *Earth Planet. Sci. Lett.*, 409, 96–108.

Ratschbacher, L., Frisch, W., Neubauer, F., Schmid, S.M., Neugebauer, J. (1989). Extension in compressional orogenic belts: The eastern Alps. *Geology*, 17, 404–407.

Ratschbacher, L., Merle, O., Davy, P., Cobbold, P. (1991). Lateral extrusion in the eastern Alps, Part 1: Boundary conditions and experiments scaled for gravity. *Tectonics*, 10, 245–256.

Reiners, P.W., Spell, T.L., Nicolescu, S., Zanetti, K.A. (2004). Zircon (U-Th)/He thermochronometry: He diffusion and comparisons with $^{40}Ar/^{39}Ar$ dating. *Geochim. Cosmochim. Acta*, 68, 1857–1887.

Ren, Y., Stuart, G.W., Houseman, G.A., Dando, B., Ionescu, C., Hegedüs, E., Radovanović, S., Shen, Y. (2012). Upper mantle structures beneath the Carpathian-Pannonian region: Implications for the geodynamics of continental collision. *Earth Planet. Sci. Lett.*, 349–350, 139–152.

Robl, J., Hergarten, S., Stüwe, K. (2008a). Morphological analysis of the drainage system in the Eastern Alps. *Tectonophysics*, 460, 263–277.

Robl, J., Stüwe, K., Hergarten, S., Evans, L. (2008b). Extension during continental convergence in the Eastern Alps: The influence of orogen-scale strike-slip faults. *Geology*, 36, 963–966.

Robl, J., Prasicek, G., Hergarten, S., Stüwe, K. (2015). Alpine topography in the light of tectonic uplift and glaciation. *Glob. Planet. Change*, 127, 34–49.

Robl, J., Hergarten, S., Prasicek, G. (2017a). The topographic state of fluvially conditioned mountain ranges. *Earth-Sci. Rev.*, 168, 190–217.

Robl, J., Heberer, B., Prasicek, G., Neubauer, F., Hergarten, S. (2017b). The topography of a continental indenter: The interplay between crustal deformation, erosion, and base level changes in the eastern southern Alps. *J. Geophys. Res. Earth Surf.*, 122, 310–334.

Royden, L., Horváth, F., Rumpler, J. (1983). Evolution of the Pannonian Basin System 1: Tectonics. *Tectonics*, 2, 63–90.

Sadler, P.M. (1999). The influence of hiatuses on sediment accumulation rates. *GeoResearch Forum*, 5, 15–40.

Schildgen, T.F., van der Beek, P.A., Sinclair, H.D., Thiede, R.C. (2018). Spatial correlation bias in late-Cenozoic erosion histories derived from thermochronology. *Nature*, 559, 89–93.

Schlatter, A., Schneider, D., Geiger, A., Kahle, H.G. (2005). Recent vertical movements from precise levelling in the vicinity of the city of Basel, Switzerland. *Int. J. Earth Sci.*, 94, 507–514.

Schlunegger, F. and Mosar, J. (2011). The last erosional stage of the Molasse Basin and the Alps. *Int. J. Earth Sci.*, 100, 1147–1162.

Schlunegger, F. and Norton, K.P. (2013). Water versus ice: The competing roles of modern climate and Pleistocene glacial erosion in the Central Alps of Switzerland. *Tectonophysics*, 602, 370–381.

Schlunegger, F. and Norton, K.P. (2015). Climate vs. tectonics: The competing roles of late oligocene warming and alpine orogenesis in constructing alluvial megafan sequences in the north alpine foreland basin. *Basin Res.*, 27, 230–245.

Schlunegger, F., Matter, A., Burbank, D.W., Klaper, E.M. (1997). Magnetostratigraphic constraints on relationships between evolution of the central Swiss Molasse basin and Alpine orogenic events. *Bull. Geol. Soc. Am.*, 109, 225–241.

Schmid, S.M. and Kissling, E. (2000). The arc of the western Alps in the light of geophysical data on deep crustal structure. *Tectonics*, 19, 62–85.

Schmid, S.M., Pfiffner, O.A., Froitzheim, N., Schönborn, G., Kissling, E. (1996). Geophysical-geological transect and tectonic evolution of the Swiss–Italian Alps. *Tectonics*, 15, 1036–1064.

Schmid, S.M., Scharf, A., Handy, M.R., Rosenberg, C.L. (2013). The Tauern Window (Eastern Alps, Austria): A new tectonic map, with cross-sections and a tectonometamorphic synthesis. *Swiss J. Geosci.*, 106, 1–32.

Schumer, R. and Jerolmack, D.J. (2009). Real and apparent changes in sediment deposition rates through time. *J. Geophys. Res. Solid Earth*, 114, 1–12.

Schuster, R. and Stüwe, K. (n.d.). Geological and tectonic setting of Austria. In *Landscapes and Landforms of Austria*, Embleton-Hamann, C. (ed.). Springer, New York.

Şengör, A.M. (ed.) (2009). The large-wavelength deformations of the lithosphere: Materials for a history of the evolution of thought from the earliest times to plate tectonics. In *The Large-Wavelength Deformations of the Lithosphere: Materials for a History of the Evolution of Thought from the Earliest Times to Plate Tectonics*. Geological Society of America, Boulder.

Şengör, A.M. (2014). Pangäische Spiele 2012. *Mitt. Naturwiss. Ver. Steiermark*, 143, 7–48.

Serpelloni, E., Vannucci, G., Anderlini, L., Bennett, R.A. (2016). Kinematics, seismotectonics and seismic potential of the eastern sector of the European Alps from GPS and seismic deformation data. *Tectonophysics*, 688, 157–181.

Seward, D. and Mancktelow, N.S. (1994). Neogene kinematics of the central and western Alps: Evidence from fission-track dating. *Geology*, 22, 803–806.

Sinclair, H.D. and Allen, P.A. (1992). Vertical versus horizontal motions in the Alpine orogenic wedge: Stratigraphic response in the foreland basin. *Basin Res.*, 4, 215–232.

Sinclair, H.D., Coakley, B.J., Allen, P.A., Watts, A.B. (1991). Simulation of Foreland Basin Stratigraphy using a diffusion model of mountain belt uplift and erosion: An example from the central Alps, Switzerland. *Tectonics*, 10, 599–620.

Spiegel, C., Kuhlemann, J., Dunkl, I., Frisch, W., Von Eynatten, H., Balogh, K. (2000). The erosion history of the Central Alps: Evidence from zircon fission track data of the foreland basin sediments. *Terra Nov.*, 12, 163–170.

Sternai, P., Herman, F., Fox, M.R., Castelltort, S. (2011). Hypsometric analysis to identify spatially variable glacial erosion. *J. Geophys. Res. Earth Surf.*, 116, F03001.

Sternai, P., Herman, P., Champagnac, J.D., Fox, M., Salcher, B., Willet, S.D. (2012). Pre-glacial topography of the European Alps. *Geology*, 40, 1067–1070.

Sternai, P., Herman, F., Valla, P.G., Champagnac, J.D. (2013). Spatial and temporal variations of glacial erosion in the Rhône valley (Swiss Alps): Insights from numerical modeling. *Earth Planet. Sci. Lett.*, 368, 119–131.

Sternai, P., Sue, C., Husson, L., Serpelloni, E., Becker, T.W., Willett, S.D., Faccenna, C., Di Giulio, A., Spada, G., Jolivet, L. et al. (2019). Present-day uplift of the European Alps: Evaluating mechanisms and models of their relative contributions. *Earth-Science Rev.*, 190, 589–604.

Stille, H. (1919). Die Begriffe Orogenese und Epirogenese. *Zeitschrift der Dtsch. Geol. Gesellschaft*, 71, 164–208.

Stutenbecker, L., Costa, A., Schlunegger, F. (2016). Lithological control on the landscape form of the upper Rhône Basin, Central Swiss Alps. *Earth Surf. Dyn.*, 4, 253–272.

Stüwe, K. and Barr, T.D. (2000). On the relationship between surface uplift and gravitational extension. *Tectonics*, 19, 1056–1064.

Stüwe, K. and Homberger, R. (2012). *High Above th Alps: A Bird's Eye View of Geology*. Weishaupt Publishing, Gnas.

Stüwe, K. and Homberger, R. (2018). *Steiermark aus der Luft*. Weishaupt Publishing, Gnas.

Sue, C., Delacou, B., Champagnac, J.D., Allanic, C., Tricart, P., Burkhard, M. (2007). Extensional neotectonics around the bend of the Western/Central Alps: An overview. *Int. J. Earth Sci.*, 96, 1101–1129.

Suess, E. (1883). *Das Antlitz der Erde. Bd. Ia (Erste Abtheilung)*. F. Tempsky, Prag und G. Freytag, Leipzig.

Trost, G., Robl, J., Hergarten, S., Neubauer, F. (2020). The destiny of orogen-parallel streams in the Eastern Alps: The Salzach-Enns drainage system. *Earth Surf. Dyn.*, 8, 69–85.

Uehlinger, U., Wantzen, K.M., Leuven, R.S.E.W., Arndt, H. (2009). The Rhine river basin. In *Rivers of Europe*, Tockner, K., Uehlinger, U., Robinson, C.T. (eds). Academic Press, London.

Valla, P.G., van der Beek, P.A., Lague, D. (2010). Fluvial incision into bedrock: Insights from morphometric analysis and numerical modeling of gorges incising glacial hanging valleys (Western Alps, France). *J. Geophys. Res. Earth Surf.*, 115, F02010.

Valla, P.G., Shuster, D.L., van der Beek, P.A. (2011). Significant increase in relief of the European Alps during mid-Pleistocene glaciations. *Nat. Geosci.*, 4, 688–692.

Vernon, A.J., van der Beek, P.A., Sinclair, H.D., Rahn, M.K. (2008). Increase in late Neogene denudation of the European Alps confirmed by analysis of a fission-track thermochronology database. *Earth Planet. Sci. Lett.*, 270, 316–329.

Vialon, P., Rochette, P., Ménard, G. (1989). Indentation and rotation in the western Alpine arc. In *Alpine Tectonics*, Coward, M.P., Dietrich, D., Park, R.G. (eds). Geol. Soc. London, Spec. Pub., London.

Wagner, T., Fabel, D., Fiebig, M., Häuselmann, P., Sahy, D., Xu, S., Stüwe, K. (2010). Young uplift in the non-glaciated parts of the Eastern Alps. *Earth Planet. Sci. Lett.*, 295, 159–169.

Wagner, T., Fritz, H., Stüwe, K., Nestroy, O., Rodnight, H., Hellstrom, J., Benischke, R. (2011). Correlations of cave levels, stream terraces and planation surfaces along the River Mur-Timing of landscape evolution along the eastern margin of the Alps. *Geomorphology*, 134, 62–78.

Walpersdorf, A., Sue, C., Baize, S., Cotte, N., Bascou, P., Beauval, C., Collard, P., Daniel, G., Dyer, H., Grasso, J.R. et al. (2015). Coherence between geodetic and seismic deformation in a context of slow tectonic activity (SW Alps, France). *J. Geodyn.*, 85, 58–65.

Watts, A.B. (2001). *Isostasy and Flexure of the Lithosphere*. Cambridge University Press, Cambridge.

Weber, J., Vrabec, M., Pavlovčič-Prešeren, P., Dixon, T., Jiang, Y., Stopar, B. (2010). GPS-derived motion of the Adriatic microplate from Istria Peninsula and Po Plain sites, and geodynamic implications. *Tectonophysics*, 483, 214–222.

Wegener, A. (1912). Die Entstehung der Kontinente. *Geol. Rundschau*, 3, 276–292.

Whipple, K.X. (2001). Fluvial landscape response time: How plausible is steady-state denudation? *Am. J. Sci.*, 301, 313–325.

Whipple, K.X. (2009). The influence of climate on the tectonic evolution of mountain belts. *Nat. Geosci.*, 2, 97–104.

Willenbring, J.K. and von Blanckenburg, F. (2010). Long-term stability of global erosion rates and weathering during late-Cenozoic cooling. *Nature*, 465, 211–214.

Willenbring, J.K. and Jerolmack, D.J. (2016). The null hypothesis: Globally steady rates of erosion, weathering fluxes and shelf sediment accumulation during Late Cenozoic mountain uplift and glaciation. *Terra Nov.*, 28, 11–18.

Willett, S.D. (2010). Late neogene erosion of the Alps: A climate driver? *Annu. Rev. Earth Planet. Sci.*, 38, 411–437.

Willett, S.D. and Brandon, M.T. (2002). On steady states in mountain belts. *Geology*, 30, 175–178.

Willett, S.D., Schlunegger, F., Picotti, V. (2006). Messinian climate change and erosional destruction of the central European Alps. *Geology*, 34, 613–616.

Winkler-Hermaden, A. (1957). *Geologisches Kräftespiel und Landformung*. Springer Verlag, Vienna.

Winterberg, S. and Willett, S.D. (2019). Greater Alpine river network evolution, interpretations based on novel drainage analysis. *Swiss J. Geosci.*, 112, 3–22.

Wittmann, H., von Blanckenburg, F., Kruesmann, T., Norton, K.P., Kubik, P.W. (2007). Relation between rock uplift and denudation from cosmogenic nuclides in river sediment in the Central Alps of Switzerland. *J. Geophys. Res. Earth Surf.*, 112, F04010.

Wittmann, H., Malusà, M.G., Resentini, A., Garzanti, E., Niedermann, S. (2016). The cosmogenic record of mountain erosion transmitted across a foreland basin: Source-to-sink analysis of in situ ^{10}Be, ^{26}Al and ^{21}Ne in sediment of the Po river catchment. *Earth Planet. Sci. Lett.*, 452, 258–271.

Wobus, C., Whipple, K.X., Kirby, E., Snyder, N., Johnson, J., Spyropolou, K., Crosby, B., Sheehan, D. (2006). Tectonics from topography: Procedures, promise, and pitfalls. In *Tectonics, Climate, and Landscape Evolution*, Willett, S.D., Hovius, N., Brandon, M.T., Fisher, D.M. (eds). Geological Society of America, Boulder.

Wölfler, A., Kurz, W., Fritz, H., Stüwe, K. (2011). Lateral extrusion in the Eastern Alps revisited: Refining the model by thermochronological, sedimentary, and seismic data. *Tectonics*, 30, TC4006 1–15.

Wölfler, A., Stüwe, K., Danišík, M., Evans, N.J. (2012). Low temperature thermochronology in the Eastern Alps: Implications for structural and topographic evolution. *Tectonophysics*, 541–543, 1–18.

Zachos, J., Pagani, H., Sloan, L., Thomas, E., Billups, K. (2001). Trends, rhythms, and aberrations in global climate 65 Ma to present. *Science*, 292, 686–693.

4

Late-Orogenic Extension and Current Tectonics in the Western and Central Alps: A Review

Christian SUE[1,2]
[1] Université de Franche-Comté, Besançon, France
[2] Université Grenoble Alpes, USMB, CNRS, IRD, IFSTTAR, ISTerre, Grenoble, France

The synthesis of observations integrated from structural geology, geochronology, seismotectonics and geodesy allows us to constrain the late-orogenic tectonics around the bend of the Western and Central Alps. Orogen-parallel brittle extension, at the presently exposed levels, developed during Miocene all along the inner zones of the belt, synchronously with North and Westward propagation of shortening in the most external zones. This phase is associated with a complex system of combined normal and strike-slip faults, which carves the internal zones and provides a remarkably stable paleostress field depicted by extensional directions rotating with the backbone of the belt all over the internal Alps. A secondary signal arises from brittle deformation analysis mainly in the Western alpine Arc. It corresponds to orogen-perpendicular extensional tectonics, which could have initiated during the Pliocene. This second extensional phase correlates with the present-day tectonics observed using both seismotectonics and geodesy. The current tectonic scheme is characterized by widespread extension in the highest parts of the belt, mainly the internal zones, and transcurrent to transpressive tectonics at the periphery of the belt (external massifs and foreland), both through seismotectonic and geodetic (horizontal strain) approaches. However, the main geodetic signal corresponds to a preeminent widespread uplift correlated to the topography in the Central and Northwestern Alps. Indeed, a partial correlation is also observed between uplift and extension, except in the Southern part of the alpine arc. Such a correlation and the absence of shortening across the orogen point out the role of isostasy-related models in the current geodynamics of the Alps and raise the matter of intrinsic buoyancy forces versus extrinsic plate tectonics, as the

Geodynamics of the Alps 1,
coordinated by Claudio L. ROSENBERG and Nicolas BELLAHSEN. © ISTE Ltd 2023.

far-field transcurrent context is compatible with Apulia's counterclockwise rotation. The overall isostasy of the belt may be impacted by Glacial Isostatic Adjustments (GIA) and erosional unloading, as well as by mantle convection and slab dynamics. Notwithstanding the relative role of each phenomena, interactions between plate tectonics and buoyancy, rather than a single process, control the ongoing geodynamics of the Alps.

4.1. Introduction

The tectonic history of the Alpine belt, from the Ligurian-Tethyan Ocean opening, then subduction, to the final collisional stages, is well described in the literature (Termier 1903; Argand 1916, 1924; Dewey et al. 1973; Coward and Dietrich 1989; Schmid and Kissling 2000; Schmid et al. 2004, 2017; Handy et al. 2010; Sue and Schmid, 2017) and presented in the present volume. Most of the studies have been dedicated to nappe stacking and collisional processes, resulting in the development of subsequent sedimentary, crustal, then lithospheric collisional prims (Tricart 1984; Nicolas et al. 1990; Stampfli et al. 1998), coevally with the arc-shaped geometry of the Western part of the chain at the edge of the Apulo-Adriatic indenter (Pavoni 1961; Tapponnier 1977). The reader is asked to refer to the relevant chapters of this volume for an accurate description of the overall alpine structure. More recently, there has been an increasing interest toward extensional processes acting at the end of the alpine orogeny.

Extensional tectonics is a classical phenomenon observed in many orogenic contexts (e.g. Fleitout and Froidevaux 1982; Platt 1986; Molnar and Lyon-Caen 1988; Blanckenburg and Davies 1995; Selverstone 2004). In the Alps, it has first been described as a major feature in the Eastern Alps (Behrmann 1988; Selverstone 1988; Ratschbacher et al. 1991a, 1991b; Rosenberg et al. 2015; Bertrand et al. 2015) and in the Central Alps along the Simplon Fault (Amstutz 1954; Steck 1984; Mancktelow 1985, 1987, 1990; Mancktelow and Pavlis 1994). On the contrary, late extension in the Western alpine arc has primarily been described very locally both concerning fault analysis (e.g. Virlouvet et al. 1996; Bistacchi et al. 2000; Malusa et al. 2005) and seismotectonics (e.g. Fréchet 1978; Fréchet and Pavoni 1979; see Sue 1998 and references herein). However, systematic regional studies, then large-scale synthesis, have now shown the prevalence of brittle extension along the Western arc of the Alps that could indeed be connected to the one observed to the North and East (e.g. Sue et al. 2007; Bertrand and Sue 2017). These observations raise the matter of late orogenic extension all along the Alpine Orogeny, and related geodynamic processes, which currently still remain a matter of debate.

The aim of this chapter is to provide an updated overview on the late-orogenic extension around the bend of the Western and Central Alps, which has been described and discussed since the end of the 1990s (Figure 4.1). In continuity of the late alpine (Neogene) brittle extension, the second part of this chapter focuses on the current tectonic activity of the Western Alps, as viewed by seismotectonics and geodesy. The last part is dedicated to an overall discussion, including geodynamic modeling, which could explain the processes acting during these late alpine tectonic phases, and raise some of the remaining open questions. The brittle deformation (2) and active tectonics (3) parts comprise of a brief description of the main methodologies, then a synthesis of the available data, together with synoptic views of the related stress/strain fields and kinematic maps. The general geodynamic interpretations, including some attempts of modeling, focused on the actual large-scale processes, are regrouped in section 4.4.

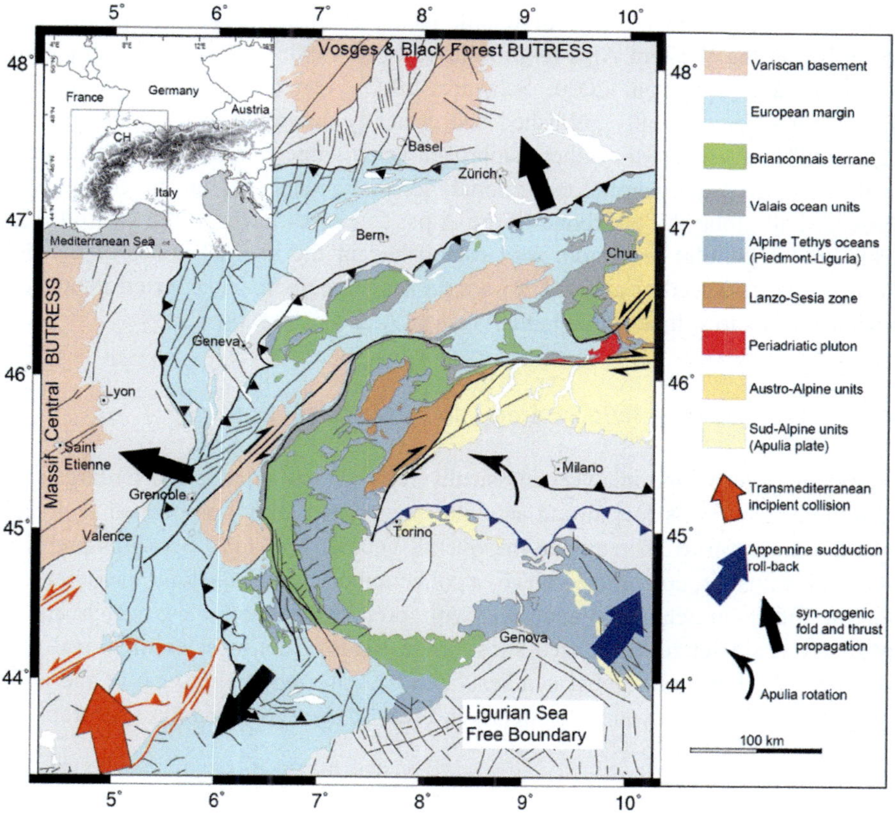

Figure 4.1. *Tectonic framework of the study area (modified from Sue (1998); De Graciansky et al. (2011))*

4.2. Neogene brittle deformation: data analysis

This section presents an up-to-date synthesis of the available brittle deformation-related dataset, which aimed to constrain the local state of stress/strain by the use of dynamic analysis of a collection of meso-scale faults' measurements. Indeed, discussion and characterization of individual structure are out of the scope of this chapter, which focuses on an alpine-scale set of data. Moreover, if we except the role of the Penninc Frontal Thrust reactivated in extension (see below), extensional tectonics in the Western Alps is mainly distributed as a pretty homogeneous deformation in the volume of the Internal Zones, rather than localized on major structures. On the contrary, it appears much more localized in the Eastern Alps, from the Simplon Fault eastward. It is worth noting that most of the faults used in this synthesis represent a late alpine deformation and did not suffer noticeable tilting. This synthesis focuses on Neogene and eventually Pliocene brittle extension and does not discuss older brittle or ductile-brittle extension that have locally been described in the Western Alps (e.g. Ballèvre et al. 1990; Ballèvre and Merle 1993; Bistacchi and Massironi 2000; Schwartz 2002). Overall synthesis of Neogene faulting has been recently published and critically described by Bertrand and Sue (2017). This quantitative synthetic approach bears witness to a complex tectonism, with a major orogen-parallel extensional Neogene signal that can be continuously observed all along the backbone of the Alps, and specifically in the Central Alps. This orogen-parallel extension, well established in the internal zones of the belt, seems to be followed or evolved toward an orogen perpendicular extensional field, which appears to still be active in the Western Alps (see section 4.3).

4.2.1. *Methodology, field strategy and data collection*

Brittle deformation analyses classically rely on two approaches: (i) large-scale morphotectonic tools (aerial and satellite images and DEM) associated with field work, which aim to characterize the overall geometry of fault networks and their related kinematics, combined with (ii) a collection of meso-scale fault measurements on the field, subsequently analyzed with proper methodologies, which aim to precisely characterize the related stress/strain fields at the local and regional scale (Figure 4.2). Later on, global synthesis can be achieved at the scale of the entire orogen (or other geodynamic system), in order to feed geodynamic interpretations and discussions (Figure 4.2).

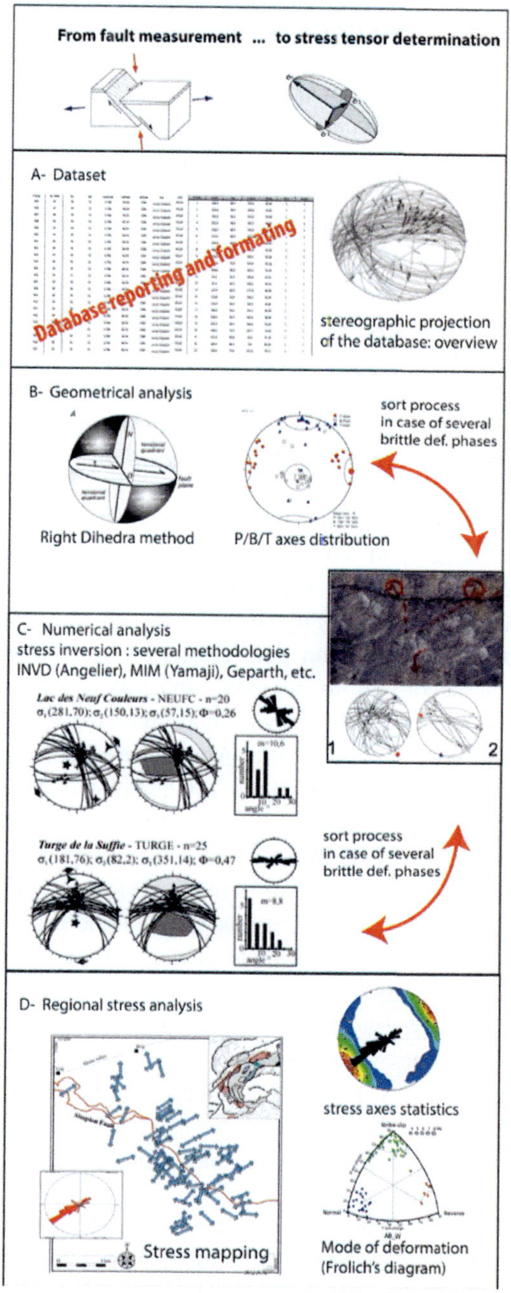

Figure 4.2. *Synoptic sketch of (paleo)stress determination from fault analysis*

COMMENT ON FIGURE 4.2.– *This figure includes (A) database reporting and a first stereographic global outlook of each dataset (each site); (B) geometrical stress analysis using the right-dihedra method and P/B/T axes distribution in 3D, to get an idea of the quality and self-compatibility of the dataset, which allow the determination of multi-phases brittle deformations and the data to be sorted; (C) numerical inversion using various methodologies (e.g. INVD, MIM, Gepharth method, etc.), providing quality indicators, such as histograms of differential angles, that also allow us to investigate and sort eventual multi-phases. The given examples correspond to two extensional tensors in the same area (Brianonnais-Ubaye; top: orogen-perpendicular extension; bottom: orogen-parallel-extension); (D) stress mapping, together with a mode of deformation characterization (Frolich's diagram), and statistical distribution of the stress axes at the local/regional scale (example map from the Simplon area). As an overall Alpine synthesis, this chapter only describes the (D) step. (A), (B) and (C) are available in the relevant detailed publications. See text for details.*

In the Alps, systematic field work all around the bend of the Western/Central Alps, further East from the Tauern Window and all the way into the Pannonian basin, have been led since the 1990s by several research teams. The classical field strategy consisted of concentrating on field efforts on a given part of the belt for each field campaign, according to regional objectives. All of them provided important datasets of faults, with several thousand single measurements. This chapter focuses on the Western and Central Alps, where numerous regional papers have been published (Sue and Tricart 1999, 2002, 2003; Bistacchi and Massironi 2000; Tricart et al. 2006; Champagnac et al. 2003, 2004, 2006; Grosjean et al. 2004; Ciancaleoni and Marquer 2008; Perrone et al. 2011; Beucher et al. 2017; Allanic 2012; Bauve et al. 2014). They are synthetized hereafter in a single analysis (see also Bertrand and Sue (2017)), which emphasizes the coherency arising from these independent studies. To give an idea of the field work, all in one, the overall dataset used in this paper comprises more than 10,000 single fault measurements, allowing us to compute 675 single stress/strain tensors in the Western and Central Alpine realm, which finally provides an accurate view of the paleo-stress field(s) in the Alps from the Neogene onward. This chapter focuses on the quantitative paleo-stress approach to discuss the late alpine dynamics. Similar methods are also applied to the current stress field constrained because of seismotectonics tools.

From a methodological viewpoint, stress/strain inversion methods from systematic analysis of brittle deformation measured in the field are based on the Wallace (1951) and Bott's (1959) principle, stating that a given fault ruptures parallel to the direction of maximum resolved shear stress within the fault plane. On the basis of this principle, long lasting theoretical and methodological works have been developed since the

1970s (e.g. Carey and Brunier 1974; Angelier and Mechler 1977; Angelier and Goguel 1979; Angelier 1989; Angelier 1990; Gephart 1990; Twiss and Unruh 1998; Yamaji 2000), providing various software solutions. The main methods used in the different works developed along the alpine belt are as follows: right-dihedra method (e.g. Angelier and Mechler 1977; Pfiffner and Burkhard 1987), P-B-T axes statistic method (Turner 1953; Marrett and Allmendinger 1990), numerical standard direct inversion (Angelier 1990; Sperner et al. 1993) and MIM method (Yamaji 2000). These methods are implemented in various software such as the TectonicFP one, which allow efficient separation of an kinematically inhomogeneous set of fault data into homogeneous subsets (Ortner et al. 2002; Ortner 2003). A major (unsolved) issue arising from these approaches concerns the stress versus strain theoretical background. However, this debate is beyond the scope of this chapter. The reader is invited to refer to the updated and precise critical discussion presented by Lacombe (2012) on this topic. Whatever the method used to investigate stress or strain fields, a set of basic assumptions allowing proper inversion are supposed to be verified: (i) rock volume isotrope and homogeneous; (ii) incremental displacements along faults; (iii) independent individual faults slip; (iv) uniform stress/strain field during a given tectonic phase and (v) non-rotational deformation. In the field, the structural geologist observes the direction and sense of slip on the fault using various slickensides, mineral fibers and grooves. The sense of movement is given, among other indicators, by steps, Riedel criteria, fibers and drag-folds (e.g. Petit 1987, Angelier 1989). Figure 4.2 presents a synoptic sketch of a stress inversion process to illustrate the methodological background.

4.2.2. *Observations: some typical structures*

Whatever the specific topics investigated, structural geology always begins by observation of structures and/or deformation markers, either in the field (meso-scale), at larger scale (remote sensing, DEM, etc.), or at smaller scale (thin-section scale, microtectonic) (Ramsay and Hubert 1983). Regarding brittle deformation, once the structures are recognized in the field, fault measurements, then state-of-the-art geometrical and/or numerical investigations allow us to quantitatively analyze the related stress fields, and in fine to feed geodynamic interpretations. To illustrate brittle deformation in the alpine realm, Figure 4.3 presents remarkable faults spanning the belt, specifically along the internal zones. They correspond to late alpine faulting, in the way that they all crosscut alpine compression-related structures: nappes contacts, alpine schistosities, folds, etc. These faults illustrate both extensional and transcurrent tectonics. Note that Figure 4.3 only illustrates the kind of structures that can be observed in the field, but they do not allow quantitative regional analyses in terms of kinematics and dynamics, which is the step forward.

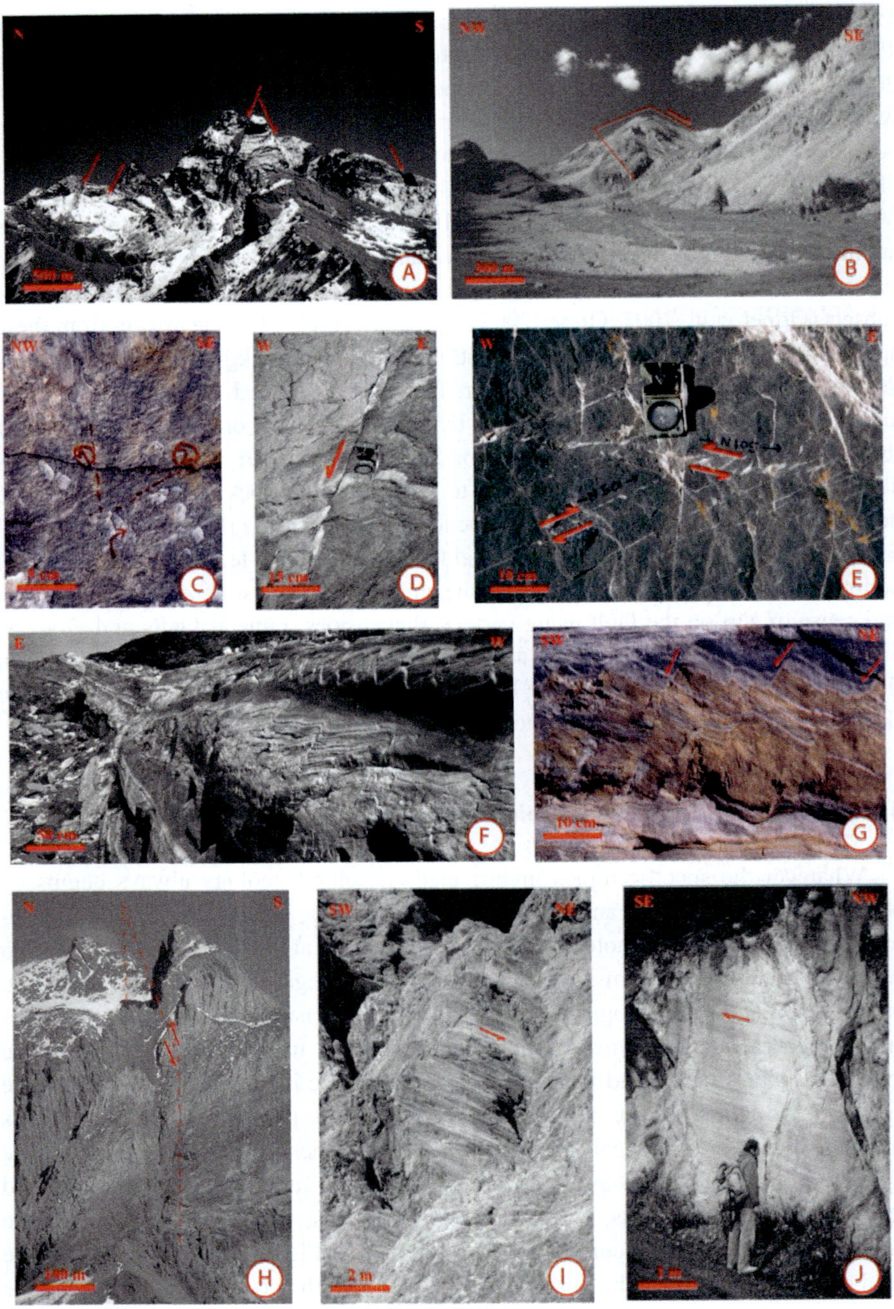
N
S
A
NW
SE
B
NW
SE
C
W
E
D
W
E
E
E
W
F
SW
NE
10 cm
G
N
S
H
SW
NE
2 m
I
SE
NW
J

Figure 4.3. *Field illustrations of some late-alpine faults and related brittle deformation structures around the bend of the Western and Central Alps*

COMMENT ON FIGURE 4.3.– *(A) Conjugated normal faults carving the Viso massif, SW Alps. (B) The Chaberton normal fault, which imprints the present-day morphology, NE of Briançon. (C) Cross-cutting calcite fibers at the Iseran pass, Vanoise massif with (1) extension and (2) left lateral strike-slip. (D) Micro-scale and related drag fold in calcshists, Melezin massif, NE of Briançon. (E) Calcite tension-gashes system resulting from transcurrent brittle deformation, Briançonnais zone. (F) Brittle-ductile transition (limestone/dolomite alternances) in the vicinity of the Simplon pass. (G) Close-up of F showing centimetric-scale tilted blocks. (H) The Font Sancte left-lateral strike-slip fault, Queyras massif. (I) Striated fault mirror along the Serennes fault, Val d'EscreinsL. (J) Striated fault mirror (seen at the top) along the Font Sancte left-lateral strike-slip fault, Houerts Peak, Queyras. (K) Normal fault mirror and fault gauge, Tête d'Oreac area, S of Briançon. (L) Pousterle pass normal fault and associated catacalsite, S of Briançon. (M) Large scale view of K, Tête d'Oreac area, S of Briançon. (N) Syn-schistosity folds cross-cut by normal faults, Vanoise massif. (O) Normal fault system, Aosta valley. (P) Aerial view of the right-lateral Serennes strike-slip fault and its morphological imprint, Queyras.*

4.2.3. Quantitative analysis of brittle deformation and kinematic synthesis

4.2.3.1. *Overall synthesis*

The dynamic analysis of about 10,000 single fault measurements reported in the Western and Central Alps provided a set of 675 single paleo-stress/strain tensors computed around the belt since the 1990s (see detailed references hereafter); they are synthetized in this chapter on the basis of Bertrand and Sue's (2017) critical analysis. As such a large dataset is difficult to present in detail, subsets of tensors have been regrouped per homogeneous areas, which in general correspond to a given independent study. In terms of the mode of deformation observed along the belt, the large majority of tensors correspond to extensional tectonics, which is the first-order tectonic signal that arises from the synthesis (see the ternary diagrams according to Frolich's (1992) method, Figure 4.4). However, this extension is intimately combined with transcurrent tectonics, which appears as the second-order tectonic signal. Moreover, extensional tectonics corresponds to two individualized phases in time. It is worth noting that very few reverse tectonics have been reported, which is paradoxical in a collision belt. Indeed, the following analysis mainly focuses on the $\sigma 3$ orientations.

This synthesis (Figures 4.4 and 4.5, see also Bertrand and Sue (2017)) includes a complete set of paleostress/strain tensors from the Penninic units north of the Argentera massif (Bauve et al. 2014; Beucher et al. 2017), the overall Briançonnais region between the Pelvoux, Viso and Argentera massifs (Sue and Tricart 1999 2002, 2003), the Vanoise area (Champagnac et al. 2006), the Aosta region (Bistacchi and Massironi 2000; Champagnac et al. 2004), the Valais area (Champagnac et al. 2003), the vicinity of the Simplon fault (Grosjean et al. 2004), the Lepontine Dome (Allanic 2012) and the Bergell pluton (Ciancaleoni and Marquer 2008). In addition, a study presenting only the major kinematic directions (without detailed analysis on the tensors) is also integrated in the Cottian Alps (Perrone et al. 2011). Note that these data perfectly fit the regional paleostress fields.

Following the approach of Bertand and Sue (2017), the paleostress fields are reassessed with a homogeneous statistical analysis (Figure 4.4, Table 4.1), including σ1, σ2, σ3 statistical distributions and Frohlich's representations (Frohlich 1992) for each of the 10 sub-domain analyzed, plus one regrouping the Central Alps, one regrouping the Western Alps and one for the overall dataset. The statistical map-view of Figure 4.5 presents the σ3 distributions and main statistical directions for each sub-area. Note that horizontal σ3 may actually correspond to extensional, and/or transtensional, and/or transcurrent tectonics (see caption for further details). The map in Figure 4.5 displays a very steady paleostress signal showing orogen parallel extension, with sub-horizontal σ3 axes running parallel to the main direction of the Alps, all along the backbone of the belt. To the East, in the Tauern Window, the roughly E-W extension characterized by Bertrand et al. (2015) could be associated with the same global scheme. Reiter et al. (2018) reported exactly that extension slightly oblique to orogen strike in the Engadine area. Despite the lack of quantitative paleostress data in the Engadine vicinity, we can propose a structural continuity in terms of paleostress from East to West. Indeed, a similar major signal arises from the synthesis: namely extensional axes displaying a steady orogen parallel direction from the Bergell to the Valais areas (ENE-WSW extension), then turning with the structural trend of the Western Alpine arc and following the arcuate shape of the belt. This main tectonic signal, representative of a global process acting at the scale of the entire Alps, is still difficult to interpret (see discussion below).

CODE (Fig.4)	*reference*	*N-data*	*specific localization*	*average dominant s3 dir. [0-180] (°)*	** secondary s3 dir. [0-180] (°)*
A	Ciancaleoni & Marquer 2008	111	*see ref.*	56	
B	Allanic, 2012	105	*see ref.*	64	
C	Grosjean et al. 2004	60	*see ref.*	58	
L	Bistacchi et Massironi 2000	-	*see ref. D2 phase*	30	
D	Champagnac et al. 2006	90	lat>46°	63	
E	Champagnac et al. 2006	58	45,5°<lat<46	27	116
F	Champagnac et al. 2006	68	lat<45,5°	21	85
K	Peronne et al.	-	*see ref.*	145	95
G	Sue & Tricart 2004	28	lat>44,75°	160	84
H	Sue & Tricart 2004	30	lat<44,75°	115	
J	Bauve et al. 2014	46	*see ref.*	94	
I	Beucher, 2009	79	*see ref.*	101	
Central Alps	A to D	366		60	
Western Alps	E to J	309		**	**
TOTAL	A to J	675		**	**

** significant occurrence of orogen perpendicular secondary extension (see discussion in Sue et al., 2007)*
*** not relevant directional statistics due to the arcuate shape of the Western Alps*

Table 4.1. *Statistical parameters of the mean directions of s3 for each local subset of data, plus the overall Central Alps (modified from Bertrand and Sue (2017))*

From a kinematic viewpoint, horizontal σ3 in the alpine realm are associated with normal faulting tensors in approximately 75% of the sites (vertical σ1), and to transcurrent faulting tensors in approximately 25% of them (vertical σ2), whatever the scale investigated by the data, that is, at the local scale, or at the global (alpine) scale. Orogen-parallel normal faulting and strike-slip faulting tensors are kinematically coherent over the entire region covered by this synthesis, the direction of σ3 axes staying subhorizontal and subparallel in each of the sub-area investigated, whatever the types of tensors (strike-slip or extensional, Sue et al. (2007)).

4.2.3.2. *Focus on the Central Alps*

Focusing on the internal zones of the Central Alps, four different studies investigated the paleostress fields from the Bergell area to the Valais area (Champagnac et al. 2003; Grosjean et al. 2004; Ciancaleoni and Marquer 2008; Alanic 2012). All of them found the extensional stress field as the very dominant signal, with mean directions of σ3 axes comprised between N56° and N64° (statistical distributions A, B, C, D on Figures 4.4 and 4.5). The overall statistics (Figure 4.4) regrouping 366 tensors in the region provides a best-fit axis oriented N58°. Indeed, the late alpine brittle deformation in the entire Central Alps bears witness to an exceptionally stable extensional field with less than 10° of variation, along more than 250 km, parallel to the structural strike of the orogen (Table 4.1). This extensional paleostress field is coherent with the major long-lasting Simplon ductile normal fault (e.g. Mancktelow 1990), which suggests a correlated dynamic between this specific structure and the overall extension observed in the entire Central Alps, including the Lepontine Dome to the East and the South Valais to the West.

4.2.3.3. *Focus on the Western Alpine arc*

The paleostress field appears more complex along the arcuate shape of the Western Alps. Each local subset of data still presents the main orogen-parallel extensional signal, which indeed turns with the axis of the arc, and thus prevents overall stereographic statistics from being provided. Nonetheless, the orogen-parallel extension is directly linked to the one observed in the Central Alps as the representative σ3 axes gently turns from the N58° direction of the Central Alps to the N30° direction in the Aosta region (Champagnac et al. 2004) and Vanoise massif (Champagnac et al. 2006). South of the latitude 45°N, the orogen parallel direction of extension turns to N160° in the East of the Pelvoux massif (Briançon region; Sue and Tricart 2003). Finally, in the southern branch of the belt (Ligurian Alps), the direction of extension is oriented between N90° and N100° (Bauve et al. 2014; Beucher et al. 2017).

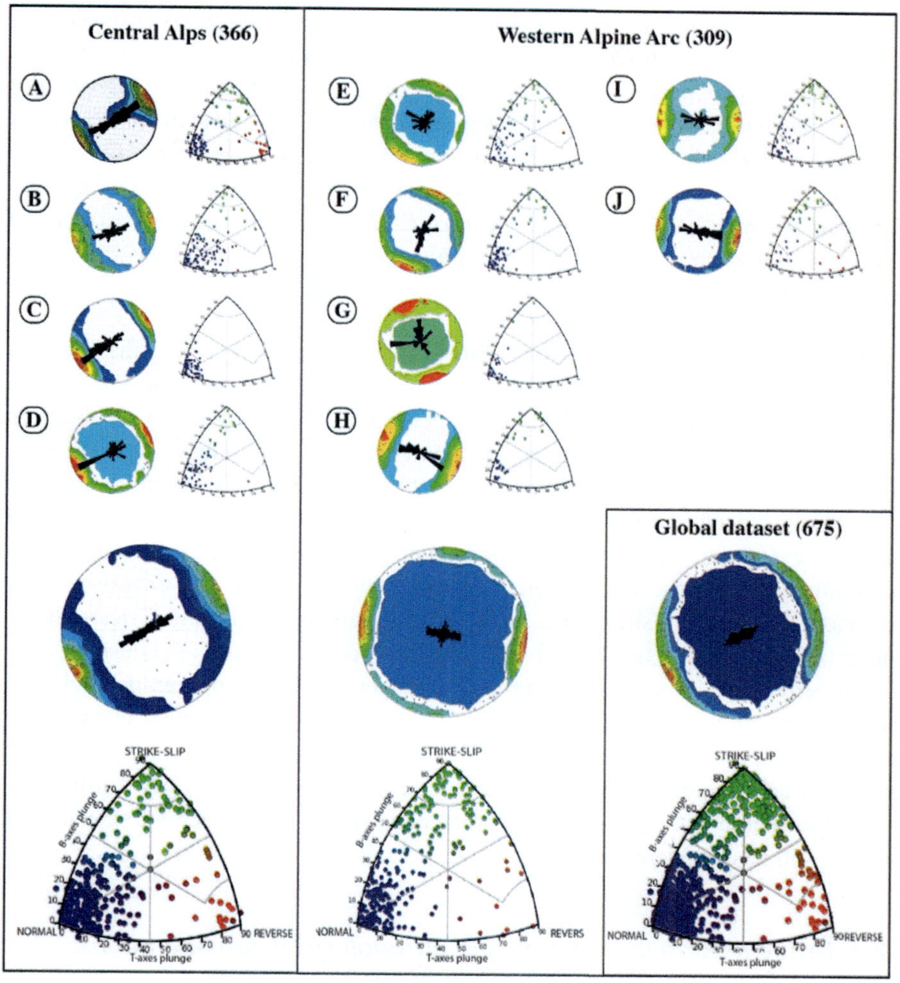

Figure 4.4. *σ3 axes' density diagram distributions and related rose diagrams (stereonets, vectorial distribution) for the 10 sectors analyzed around the bend of the Western and Central Alps (referenced as A to J)*

COMMENT ON FIGURE 4.4.– *See Figure 4.5 for location and references. Numerical values are given in Table 4.1. See Bertrand and Sue (2017) for the description of the statistical methodology for the vectorial distribution. For each subset of data, the Frolich diagram is also drawn to show the related major/minor deformation modes. The global set of data (respectively, Central Alps A to D; Western Alps E to J) statistics and distributions are regrouped in the lower right diagrams (respectively, lower left; lower center). The Frolich diagrams show that the majority of the stress tensors are extensional (blue), whatever the region, with a noticeable proportion of transcurrent tensors (green), and a very small minority of compressional tensors (red, mainly in the Bergell area). The Central Alps exhibits a very stable WNW-ESE σ3 axes' direction, whereas to the West the direction turns with the arc of the belt. Note that the rose diagrams for the sectors F–H display two maxima, associated with the major orogen-parallel Miocene phase of extension, and to the still active orogen-perpendicular one.*

Beyond the complexity arising from the arcuate shape of the belt, a careful analysis shows a second-order extensional signal, less important, with representative σ3 oriented more or less perpendicular to the arc. It can be observed from the Aosta region to the Briançon region (statistical distributions E, F, G, K in Figures 4.4 and 4.5; Sue and Tricart 2002, 2003; Champagnac et al. 2006; Perrone et al. 2011). Indeed, the core of the arc presents a complex brittle system with two phases of extension. Actually, the second-order perpendicular-extension vanishes to the southern tip of the belt, where one can mainly observe the orogen-parallel extension (statistical distributions H, I, J in Figure 4.5; Sue and Tricart 2003; Bauve et al. 2014; Beucher et al. 2017). From crosscut relationships, the orogen-perpendicular extension appears younger than the orogen-parallel one (Sue and Tricart 2003). It is worth noting that the orogen-perpendicular extension is responsible for the reactivation of the crustal penninic front (CPF) as a major normal fault, East and Southeast of the Pelvoux massif (Sue and Tricart 1999). Moreover, this orogen-perpendicular extension corresponds to the seismotectonic signal, and thus can still be considered active.

4.2.4. *Brittle deformation dating and chronological considerations*

A major generic problem arising from brittle deformation analysis concerns the dating of the deformation, as few methods allow us to directly date the occurrence of the rupture on a fault. This question is still a hot topic in the Western and Central Alps. However, some direct dating has been performed on pseudotachylites, clay minerals associated with fault gouges and mineral fibers, focusing on a given fault or fault system. Complementary to these approaches, thermochronology, and more specifically zircon fission track (ZFT), could provide regional information on the brittle deformation ages.

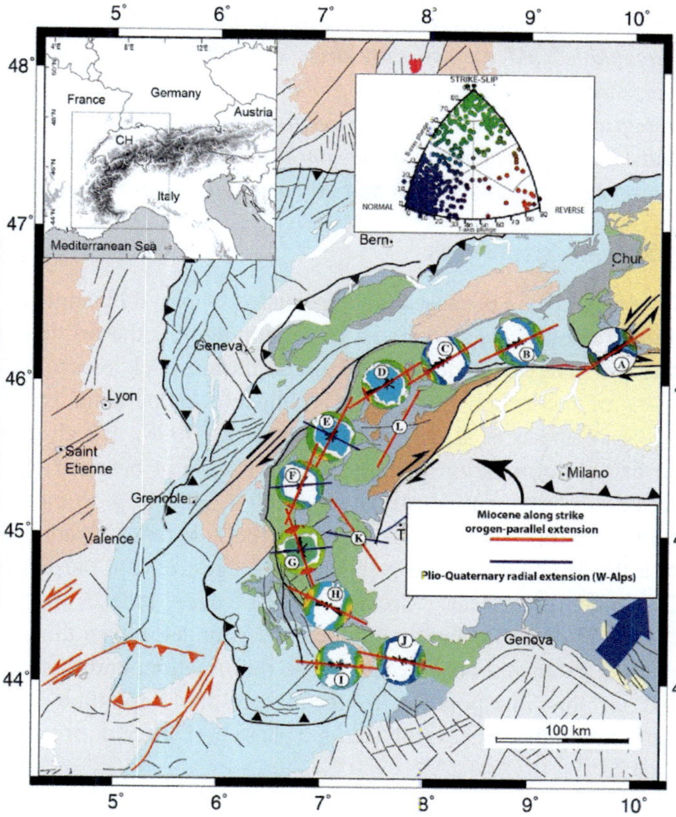

Figure 4.5. *Map view of the data plotted Figure 4.4 (modified from Bertrand and Sue 2017)*

COMMENT ON FIGURE 4.5.– *The stereonet (density diagrams) of the σ3 directions for each sector synthetized in this chapter is represented, together with a red bar underling the main direction of extension (orogen-parallel), and a blue bar for the secondary orogen-perpendicular extension, if it ever arises from the dataset (sites E–G).*

Investigated areas are as follows: the Bergell pluton (A, Ciancaleoni and Marquer 2008), the Lepontine Dome (B, Allanic 2012), the vicinity of the Simplon fault (C, Grosjean et al. 2004), the South-Valais area (D, Champagnac et al. 2003), the Aosta area (E, Champagnac et al. 2004), the Vanoise area (F, Champagnac et al. 2006), the Briançon region between Pelvoux, Viso and Argentera massifs (G and H; Sue and Tricart 1999, 2002, 2003), and the penninic Units, North of the Argentera massif (Bauve et al. 2014; Beucher et al. 2017). Dataset from the Cottian Alps

(K, Perrone et al. 2011) and the Eastern Aosta valley (L, Bistacchi and Massironi 2000) are unavailable, and we can only draw the average direction of extension, without proper statistical analysis. The global Frolich diagram (reported from Figure 4.4) integrates all the single inversion from the sites A to J (675 tensors in total). It establishes the prevalence of extension in terms of brittle deformation at the scale of the internal zones of the orogen and the secondary importance of strike-slip.

4.2.4.1. *Indirect time-constraints*

Bertrand and Sue (2017) proposed an interpolated map of 492 single ZFT ages in the entire Alpine realm in order to give a maximum age for the brittle deformation observed in the belt. They excluded detrital zircon fission-track ages to get rid of the inherited history of the zircon minerals. ZFT closure temperature is estimated at 240 ± 60°C (e.g. Yamada et al. 1995; Brandon et al. 1998), which gives a good proxy for the brittle deformation domain, as the transition between ductile-brittle deformation mode in basement is estimated at 280 ± 30°C (e.g. Stipp et al. 2002). Focusing on the Western and Central Alps, all the subareas considered above in terms of brittle deformation provide Neogene ZFT ages, except the Briançon area, which provides much older (early Cretaceous) ZFT ages. Indeed, these ages only represent a maximum limit for faulting that can be younger. Given the continuity of the orogen-parallel extension observed in brittle deformation, and the younger ZFT ages provided on this interpolation (especially in the Lepontine Dome, about 10 Ma), we can suggest that this overall tectonic phase took place during the Miocene.

This interpretation is strengthened by the analysis of the 32–30 Ma Bergell granitoid (giving an absolute limit for the deformation; Von Blanckenburg 1992), with ZFT ages giving an early Micoene age (23.1 ± 1.4 Ma; Ciancaleoni 2005). In this framework, the late orogenic parallel extension observed in the brittle domain could develop during the Miocene, after the emplacement of the pluton, and its uplift and eastward escape between the right-lateral Tonale fault and the left-lateral Engadine fault during the overall N–S shortening phase (Schmid and Froitzheim 1993). Actually, E-W orogen-parallel extension is not exclusive with N-S shortening processes, and both processes could be coeval or pretty close in time (see discussion below). In the Simplon vicinity, Bolognesi and Bistacchi (2016) reported a E-W normal fault related to orogen-perpendicular phase, which crosscuts and postdates the Simplon Line itself (orogen-parallel extension phase). Another kind of indirect observation is reported in the Briançon region, where the late alpine fault system crosscuts all the shortening structures (Sue and Tricart 1999, 2003) of Oligocene age (such as the CPF; Tricart 1984). Indeed, the brittle extension is also suggested to develop during Miocene in this region. The main problem arises here from the crosscutting of the two extensional signals (orogen-parallel and orogen-perpendicular) that often merge in the field, leading to a so-called "multi-directional extension" system (Sue and Tricart 2002). However, the timing is coherent with the timing proposed northward in the Valais, Simplon, Lepontine and Bergell areas. Moreover, in the Briançon region, striated paleokarst infills of Pliocene age (paleoclimatic indirect considerations, see Mercier (1977); Sue (1998)) probably bear witness to the last phase and allow us to propose an indirect Pliocene age for the perpendicular extension. Furthermore, this last phase still appears to be active in the current seismicity (see below).

To sum up the tectonic context in the core of the alpine arc (Briançon and surroundings), Oligocene compressional phases are followed by Miocene orogen-parallel extension, prior to the development during Plio-quaternary times of the still active orogen-perpendicular extensional tectonics.

4.2.4.2. *Absolute dating of brittle deformation*

Absolute dating of brittle deformation relies on few methods, namely $^{40}Ar/^{39}Ar$ dating on earthquake-related pseudotachylites (e.g. McDougall and Harrison 1999; Van der Pluijm et al. 2001), K–Ar dating on authigenic clay minerals neoformed in fault gauges (e.g. Lyons and Snellenberg 1971; Kralik et al. 1987), or direct dating of slickensides-related mineral fibers such as calcite using the U-Pb method (e.g. Robert and Walker 2016), or hematite using the (U-Th-Sm)/He method (Ault et al. 2015).

In the Alps, most of the effort on absolute dating has been localized in the Central Alps (Simplon, Lepontine Dome, Aar massif). Allanic (2012) provided $^{40}Ar/^{39}Ar$ pseudotachylites ages in the Lepontine Dome, spanning from 7.7 ± 0.1 Ma to 11 ± 2 Ma. These ages are consistent with the ages of Zwingmann and Mancktelow (2004) obtained with K–Ar dating on fault gauges in the vicinity of the Simplon fault (from 6.01 ± 0.18 Ma to 7.17 ± 0.2 Ma) and along the Centovalli fault (8.72 ± 0.3 Ma), and with the dating of synkinematic white mica from the footwall of the Simplon fault (14.6 ± 9.5 and 11.0 ± 0.1 Ma, Campani et al. (2010)), interpreted as the age of the ductile-brittle transition. This set of ages from various methods are consistent with each other, and with the ZFT ages synthetized by Bertrand and Sue (2017). They are related to the orogen parallel extensional phase describe above (Zwingmann and Mancktelow 2004; Allanic 2012). Indeed, they strongly suggest that the climax of the orogen parallel brittle extension took place in the Middle to Upper Miocene (at about 6–12 Ma). In the Aar massif, ages on syn-kinematic phengites sampled in shear-zone at the brittle-ductile transition (from 13.8 ± 0.2 Ma to 12.2 ± 0.2 Ma; Rolland et al. 2009), and ages on fault gouge clay mineral (10.1 ± 0.4 Ma and 7.6 ± 0.4 Ma; Kralik et al. 1992) are coherent with the ages obtained in the Lepontine and Simplon areas, but also with the strike-slip system described in the inner zones in terms of kinematics. Herwegh et al. (2017)

recently obtained comparable ages in the Aar massif for the last deformation phase. Last but not least, a recent and ongoing work (Bilau 2019; Bilau et al. 2021) provided preliminary absolute ages of about 2 Ma as a major signal on hematite fibers sampled on the Pousterle normal fault (see Sue and Tricart (1999) for a structural and tectonic description, and Figure 4.3), which corresponds to the reactivation and inversion of the CPF, in the Southwestern Alps (Briançon region). Notwithstanding the future results of this study and the methodological limits and reproducibility of the ages, which are currently under investigation, this Upper Pliocene age is very important in the framework of the late orogenic extension along the CPF, and in the Briançon region. It confirms the indirect arguments for a relatively young age for the orogen-perpendicular extension phase, and fills the gap between the ongoing seismicity and the brittle deformation observed in the field.

4.3. From Neogene tectonics to current activity

The previous section focused on dynamic analysis of the brittle deformation observed in the field all around the Western and Central Alps, using state of the art structural geology tools. Notwithstanding the improvements and recent developments in fault dating, the overall time-bracketing of the brittle deformation (Neogene to Plio-Quaternary ages) remains a key point. However, the orogen-perpendicular extension, as interpreted by field observations and new ages, seems to correspond to the last tectonic phase. To extend this problem in time, the main arising question concerns the development of such a tectonism in terms of ongoing alpine activity. In that frame, seismotectonics combined with geodesy are two major observables that help to constrain the actual current deformation of the belt. Beyond the geodynamic interpretations they may induce, they also support important implications in terms of natural hazards (e.g. Sue et al. 2007b; Hannouz et al. 2019; Hannouz et al. submitted). This paragraph focuses on active deformation at the orogenic scale. Indeed, it presents a large-scale seismotectonics synthesis *s.l.* and does not aim to discuss individual active faults in the alpine realm (for that purpose, see, for instance, Mathey et al. (2020a); Jomard et al. (2017); Thouvenot et al. (2003)), but to constrain the current stress and strain fields around the belt. The overall seismicity in the Western and Central Alps (see the earthquake density map in Figure 4.6) develops along two main arcs (so-called Brianconnais seismic arc to the West, and Piemontais seismic arc to the East, Rothé (1941); see also Sue et al. (2002)), which concentrate most of the W-alpine seismicity. In the Central Alps, active zones also develop North and South of the Rhone Valley in the Valais area. Beyond these concentrated zones of seismicity, the seismic activity is much more diffuse around the bend of the arc and in its foreland (Figure 4.6), but still provides key information in terms of current deformation of the belt.

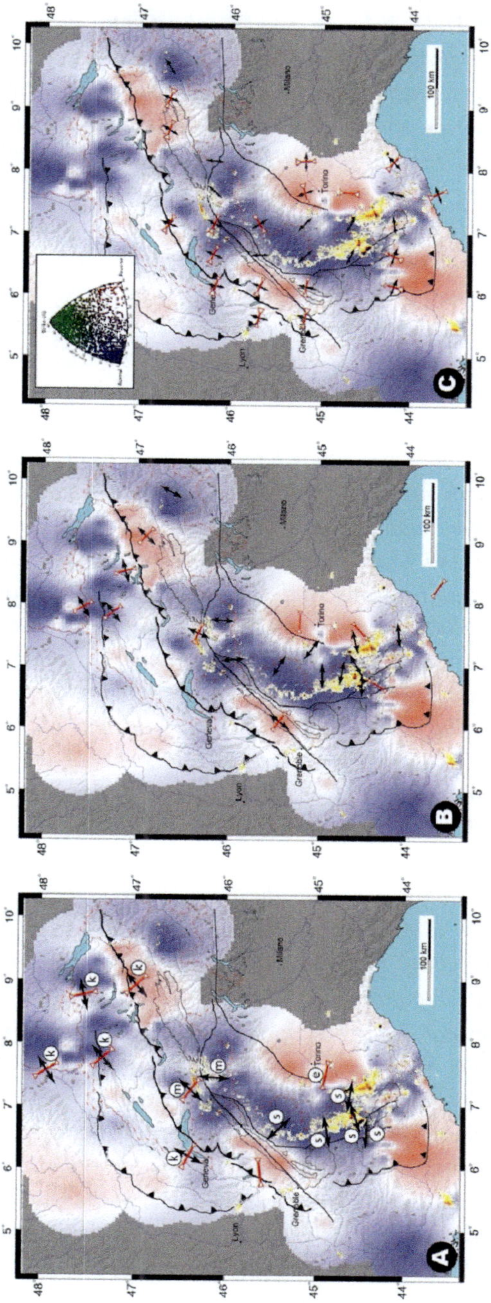

Figure 4.6. *Three snapshots of the current state of stress around the bend of the Alpine belt from inversion of focal mechanisms (see referenced papers for details)*

COMMENT ON FIGURE 4.6.– *Background (blueish to reddish colors) is the map of the mode of deformation from Delacou et al. (2004) (shortening in red; extension in blue; the strike-slip is not properly addressed by this bi-modal approach). The yellow to red then black concentrated dots give the density of the earthquakes in the belt, which mainly spans to the south of the arc along the so-called Briançonnais and Piemontais seismic arc (Sue et al. 2002), and also in the Valais area. Red convergent arrows give s1; black divergent arrows give s3. (A) Stress tensors plotted from Maurer et al. (1997) (m), Eva et al. (1998) (e), Sue et al. (1999) (s) and Kastrup et al. (2004) (k). This composite map corresponds to the synthesis of several regional stress fields computed at the end of the 1990s using various strategies of inversions. It shows (i) a transcurrent state of stress in the North Alpine domain compatible with dextral movements parallel to the orogen's axis, that is, with the counterclockwise rotation of Apulia; (ii) the orogen-perpendicular extension well expressed in the core of the Alpine arc; and (iii) compressive localized areas on both sides of the Western Alps. This map is based on focal mechanism's inversions in boxes chosen by the authors according to the somehow homogeneous type of deformation. (B) Active stress map from Delacou et al. (2004). This map has been built using a synthesis of 389 focal mechanisms in the Alpine realm. Indeed, it integrates and re-assesses the previous studies plotted in map A. It also integrates the work by Baroux et al. (2001) to the south. The database has been homogenized and the stresses computed using a single strategy and redefining the boxes of inversion, which still remain arbitrarily chosen by the authors on the basis of the type of focal mechanism. The main features observed in map (A) are confirmed, although this approach is much more precise, at the scale of the belt. (C) New seismotectonic stress map by Mathey et al. (2021). This map is based on the inversion of about 2200 newly computed focal mechanisms (corresponding Frolich's diagram in the inset). This new approach uses an automatic strategy of inversion, with a sliding window scanning the overall study area, in which stress inversion is performed only if the number of focal mechanisms in the considered window reach a given threshold (fixed at 10 for this map). Indeed, the obtained stress field is displayed on a regular grid, only if the number of data is sufficient for a given inversion, and does not depend on author's choice of inversion's boxes as in (A) and (B). This map is the most accurate view of the actual state of stress in the Western and Central Alps. It appears that the transcurrent stress field is predominant in the northern and western foreland of the belt, whereas the extensional stress field is prevalent in the high chain. Strictly compressive areas almost disappear, except South of Torino in the Pô plain. Another important observation concerns the direction of extensional axes in the core of the arc, which appears to be slightly deflected from strictly orogen-perpendicular extension (see Mathey et al. (2021) for details and exhaustive discussion).*

4.3.1. *Seismotectonic synthesis*

The first observable classically used to infer active tectonics in a given region is the overall seismicity and mechanical information brought by the focal mechanisms that can be computed (e.g. Scholz 1990). In the Western and Central Alps, several teams work on such analysis at the local/regional scale (e.g. Eva et al. 1998; Sue et al. 1999; Kastrup et al. 2004; Reiter et al. 2018). Nevertheless, the last large-scale seismotectonic synthesis has been published by Delacou et al. (2004), who compiled 389 focal mechanisms of earthquakes recorded between 1969 and 2000 with local magnitudes (Ml) spanning from 0.7 to 6.0 to characterize the alpine ongoing deformation. Recent works have also been achieved on the overall alpine seismicity (e.g. Potin 2016; Baucé et al. 2019) in terms of earthquake detection and re-localization, or on specific crises, such as the Ubaye swarm (e.g. De Barros et al. (2019) and references herein), allowing us to take a step forward in the alpine seismicity analysis. Furthermore, an ongoing study aims to provide a high-resolution overview of the current alpine deformation, with a new database of more than 2200 focal mechanisms (Mathey et al. 2021). The first level of analysis is the mode of deformation that arises from the type of mechanisms (regarding the three extreme poles: extension, shortening, strike-slip, using the Frohlich's diagram), and the distribution and regionalization of the corresponding P/B/T axes; the second one is the stress field that comes from focal mechanisms inversion; the third one is the associated strain rate, computed using the Kostrov's methodology (Kostrov 1974; modified by Molnar (1983)). On this basis, comparisons and correlations between deformation-related features and the alpine crustal and lithospheric structures can be investigated.

4.3.1.1. *Regionalization of the deformation*

In terms of deformation mode, several difficulties arise to characterize the alpine tectonism. Indeed, these methods are based on focal mechanism compilations (Delacou et al. (2004), see background in Figure 4.6) and related statistic methodologies, and thus rely on the global number, the quality and the repartition of the focal mechanisms used to characterize a given region. In the alpine context, the seismicity's distribution is highly inhomogeneous (Figure 4.6) and remains moderate in terms of energy released. Notwithstanding the inhomogeneity of both the seismic distribution and the focal mechanisms' repartition, statistical characterization of the mode of deformation throughout the whole Western and Central Alps has been proposed by Delacou et al. (2004), on the basis of a 389 focal mechanism compilation, using a regionalization and interpolation of the P- and T- axes' dips. A major improvement of this kind of analysis has recently been developed by Mathey et al. (2021), who developed a more robust and thorough statistical regionalization based on more than 2,200 newly computed focal mechanisms. At the first order, both studies allow us to recognize a continuous zone of extension in the higher zones of the chain, all along the belt from the Southern Valais to the north of the Argentera massif, also developed eastward in the Piemont zone (Figure 4.6). The second main characteristic is the existence of small discontinuous areas with a compressive/transpressive regime localized along the outer periphery of the Alpine chain. Actually, these compressional zones appear to be much more limited in the more robust analysis of Mathey et al. (2021), which provides a precise view of the overall tectonics scheme. A third important feature is the prevalence of transcurrent tectonics all around the alpine arc. The P- and T-axes trajectory fields show radial fan-shaped distributions of T-axes in the core of the belt, and radial fan-shaped distributions of P-axes in its periphery and foreland, as already proposed by Fréchet (1978), Pavoni (1980) and Ménard (1988). Indeed, this kind of regionalization provides a clear view of the current tectonics around the belt of the Alpine arc, with an extensional zone in the core of the belt, correlated at the first order with the topography (and thus the orogenic root), and surrounded by areas characterized by transcurrent tectonics, and some scarce shortening areas, pointing out a so-called "tectonic contrast" across the belt (Sue et al. 2007).

4.3.1.2. *Current stress field inversion*

The current state of stress is computed using inversion methods of focal mechanisms with a similar theoretical background as the one used for paleostress inversion from faults/strias databases (see, for instance, Gephart and Forsyth (1984); Delacou et al. (2004) and references herein for details on the methodology). Such stress maps have been displayed in the Alpine realm since the end of the 1990s (Maurer et al. 1997; Eva et al. 1998; Sue et al. 1999; Baroux et al. 2001; Kastrup et al. 2004; Delacou et al. 2004; Bauve et al. 2014; Mathey et al. 2021), with an increasing accuracy. They generally rely on focal mechanism inversion selected on a basis of so-called "homogeneous zones", that is, presenting more or less compatible mechanisms (uniform deformation modes). However, such an approach still presents a bias on the choice and the geographic limits of the given inversion zones. Nonetheless, first regional stress inversion (Figure 4.6(a)) then large-scale synthesis such as the one published by Delacou et al. (2004), exhibit at the first order a clearly contrasted state of stress around the Alpine arc, with an orogen-perpendicular extensional stress field in the core of the arc, surrounded by transcurrent state of stress at the periphery of the orogen (Figure 4.6(b)). The transcurrent stress in the outer zone of the Western Alps is locally modulated by a minor reverse component (transpression), such as in front of the Belledonne massif or in the Pô plain. Principal stress directions are coherent all along the backbone of the arc, with an orogen-perpendicular pattern of $\sigma3$ in the high chain rotating progressively from the N-S direction in the Valais, to the E-W direction in the Briançon region, and the WSW-ENE direction North of the Argentera massif. Similarly, the $\sigma1$ axes also draw a fan pattern more or less perpendicular to the belt in the outer zones evolving from the NNW–SSE direction in Northern Switzerland to the NW–SE in front of Belledonne and the SW–NE in front of the Digne nappe, and everywhere compatible with dextral movement in the longitudinal alpine strike.

This overall scheme has been reassessed recently by Mathey et al. (2021; Figure 4.6(c)) using a new set of more than 2,200 focal mechanisms computed in the alpine realm. The huge quantity of data analyzed in this work allows us to provide a much more accurate view of the current stress field. However, they also allow us to provide a stress map based on a regular gridding of the overall area, without human-selected zones of inversion, made possible thanks to the large number of focal mechanisms. Indeed, the corresponding stress field confirms the first order extension in the core of the Alpine arc, but it provides extensional axes systematically deflected with respect to the orogen-perpendicular extension found by previous authors, of about 20°–30° clockwise (Figure 4.6(c)), which may lead to alternative and/or modulated geodynamic interpretations. At the periphery of the orogen, this new stress map largely confirms the predominant transcurrent stress field, with globally radial $\sigma 1$ scheme, and minimizes the impact of shortening/transpressive zones with respect to previous publications.

4.3.1.3. *Seismic strain quantification*

The third level of analysis that may arise from focal mechanisms datasets is the quantification of the seismic strain, obtained via the Kostrov's method (Kostrov 1974; Molnar 1983). This approach is based on the summation of the moment tensor of each focal mechanism. The moment tensor combines the mechanical information of a given mechanism (formalized as a nine-component tensor), with the energy information composed of the seismic moment (scalar value) or estimated via the magnitude and conversion laws between magnitude and moment. This summation brings a strain rate tensor for a given region where the moment tenors are summed. Actually, this method theoretically asks for a completed seismic catalog, as the main part of the resulting strains tensor arises from the larger/largest quake(s). In the alpine context with potentially very long return times for the characteristic earthquake of each subarea, and relatively ill-defined magnitudes and thus related seismic moments, the quantitative evaluation of the seismic strain rate remains poorly constrained and must be used with care. Nevertheless, such attempts have been led at the regional scale (Alps-Rhone valley transitional zone: Hannouz et al. submitted) and Alpine scale (Sue et al. 2007b). Notwithstanding the uncertainties of such an approach in a region of moderate seismicity such as the Alps, these works provided seismic strain values in the order of 0.1–0.4 mm/yr over areas of typical size of 30–50 km, in particular in the Southern Western Alps, where the extensional strain axes are confirmed to trend radially to the backbone of the orogen. Indeed,

Sue et al. (2007) pointed out a discrepancy between the seismic strain rates obtained (some 10^{ths} of mm/yr) and the 1–2 mm/yr of extension across the Alps seen by early GPS studies (Calais et al. 2002). Since these publications, new geodetic studies haves been published in the alpine realm (see section 4.3.2) showing geodetic strain rates of c.a. 0.5 mm/yr of geodetic strain across a baseline of 50 km (Walpersdorf et al. 2015, 2018). Indeed, taking into account uncertainties on the geodetic and seismic strains' quantifications, it seems that both kinds of observations are now converging toward comparable strain rates, at least in the SW Alps.

4.3.2. *Geodetic strain field*

Complementarily to the seismotectonic approaches synthetized in the previous section, increasing efforts on GPS (then GNSS) surveying of the Alps sensus lato have been developed since the end of the 1990s, regarding both permanent network and campaign studies. Although the measured deformations remain very small and barely larger than the associated uncertainties, they constrain the present-day deformation schemes at several scales: overall plate tectonic context, alpine realm (i.e. the orogen's deformation itself), and local/regional specific deformations. GNSS results can be analyzed in terms of horizontal deformations versus vertical deformations.

4.3.2.1. *Horizontal geodetic strain field*

The first level of analysis concerns the boundary conditions at the limits of the Alps, that is to say the relative motion of the Apulo-Adriatic plate with respect to the European plate. According to various datasets and strategies of analysis, all the publications dealing with GPS/GNSS deformation at the plates scale displayed the counterclockwise rotational component of Adria with respect to stable Europe (e.g. Calais et al. 2002; Serpelloni et al. 2005, 2007; D'Agostino et al. 2008; Nocquet 2011; Nguyen et al. 2016; Masson et al. 2019). However, depending on the various positions of the pole of rotation calculated, the plate models provide different kinds of boundary conditions across the Alps, most of them resulting in a small extensional component, together with a right-lateral strike slip component, in the 0.1–1 mm/yr range. Looking closer at the alpine arc itself, the first results by Vigny et al. (2002) and Calais et al. (2002) show a major orogen-perpendicular extension in the core of the belt, but their data relied only on some years of measurements, and indeed stay close to still large uncertainties. More locally, such extensional deformation has been measured in the Briançon area across a repeatedly measured survey network, showing a convergence of the strain rates and yielding a deformation rate of about 16 nanostrain/yr, corresponding to c.a. 0.5 mm/yr across this region about 50 km wide (Sue et al. 2000; Walpersdorf et al. 2015). It is important to note that most of the overall extension measured across the whole Western Alps could be concentrated in the internal area (overall Briançonnais and Piemontais), which is also the most active area in terms of seismicity (see Figure 4.6). Recently, an updated synthesis combining all the available GPS/GNSS data available in the Alpine realm and surroundings in a new solution has been published by Walpersdorf et al. (2018). This study integrates permanent GPS sites and survey GPS networks, and compares several computing strategies. Major results rose up from this work, establishing a reference geodetic analysis in the Alps so far. First, regarding individual stations, the signal recorded on the horizontal component of the deformation would be of the order of 0.2 mm/yr, which means that 25–35 years of measurements are necessary to get significant local deformation. Consequently, a pretty good correlation begins to arise between permanent and survey sites. The resulting overall strain tensor obtained across the belt indicates a roughly E-W extension of about 3 nanostrain/yr and N-S shortening of about 1 nanostrain/yr (corresponding to some tenths of mm/yr). Geodetic cross-sections across the arc show that the extensional deformation is concentrated over the orogenic relief, whereas a small shortening can be seen in the foreland. These cross-sections also indicate a right-lateral strike-slip component lateral to the arc (Walpersdorf et al. 2018). To go further in the geodetic analysis of the horizontal strain, a strategy of inversion of the velocity's differences between all pairs of stations in the network has been applied (see the methodology in Masson et al. (2014)), so that the signal to noise ratio could be improved. The main result is a strain map covering the bend of the Western and Central Alps as a whole (Figure 4.7), with a regular 0.5° × 0.5° grid. This strategy provides an accurate strain field, which is dominated by extensional geodetic strain in the core of the Alpine arc, rotating from the N-S direction to the North to E-W direction to the South. Smaller shortening strain is observed at the periphery of the orogen. Note that the maximum of deformation is obtained in the E-W extension in the Briançon area (see Walpersdorf et al. (2015) for a specific regional analysis). It is very important to underline the remarkable coherency between the seismotectonic strain/stress fields (see above) and the geodetic horizontal strain fields, both in terms of regionalization of the deformation and directions of the strain axes.

4.3.2.2. *Vertical geodetic strain field*

Given the observations developed above, in particular concerning the prevalence of extensional tectonics in the core of the Alps, it is of the utmost importance to constrain the vertical motions that can be active within the belt, to allow proper geodynamic interpretations. Historically, the first study on vertical deformation in the Alpine realm was published by Schaer and Jeanrichard (1974), restricted within the

Swiss borders, as it relies on national leveling data. This work, re-used by many subsequent studies (e.g. Jouanne et al. 1995), established an overall uplift of the Swiss Alps reaching 2 mm/yr, with positive gradient from the molasse basin where the zero has been fixed toward the internal zones and the highest part of the belt (South Valais, Grisons). This scheme has been largely confirmed later on, using GNSS observations (Serpelloni et al. 2013; Nocquet et al. 2016; Nguyen et al. 2016; Walpersdorf et al. 2018; Masson et al. 2019; Sternai et al. 2019). Thus, the first integrated alpine-scale GNSS vertical field was published by Serpelloni et al. (2013), showing uplift rates in the 2–3 mm/yr concentrated in the North-Western Alps, and establishing a strong relationship between the alpine topography and the vertical uplift rates. Nocquet et al. (2016) provided a more accurate vertical velocity field for the Western Alpine arc, with a maximum uplift rate of about 2 mm/yr in-between the Mont-Blanc and Vanoise areas, and established a good coherency between GNSS and leveling data. This work also showed that the uplift is concentrated over the orogen, especially in the North of the Western Alps, and significantly decreases toward the forelands and toward the south of the belt, with values lower than 0.5 mm/yr south of a line running from the Viso massif to the Valensole basin. Moreover, it confirms and enlarges the results of Walpersdorf et al. (2015) on the Brinçonnais area. Recently, Sternai et al. (2019) re-assessed and updated the GNSS vertical motions' field by optimizing the number of available stations and the duration of observation, applying standard quality estimators (see, for example, Sternai et al. (2019); Serpelloni et al. (2018) for details). They obtained a similar pattern of deformation as the previous studies for the Western Alps, with comparable vertical velocities of about 2 mm/yr in the Northwestern part of the Alpine arc, and established a second maxima of uplift in the South Valais area (Figure 4.7). Notwithstanding the computational strategies and accuracy of the results of each given publication, it is worth noting that all of them provide very similar results both in terms of overall deformation scheme and vertical uplift values (up to 2–2.5 mm/yr), which underlines the reproducibility and the robustness of these observations. Though they show a positive correlation between uplift rates and topography, and provide a vertical strain field reaching 2 mm/yr at several places of the belt, the horizontal strain field only reaches about 0.2 mm/yr across the belt (see above). Indeed, the vertical deformation of the Alps seems to be one order of magnitude higher than the horizontal one. This observation strongly constrains the geodynamic implications (see below). Actually, a strong qualitative and quantitative coherency is found between seismological activity and horizontal geodetic strain field, whereas a clear decorrelation is observed between the actual maximum of seismicity and maximum of uplift (Figure 4.7).

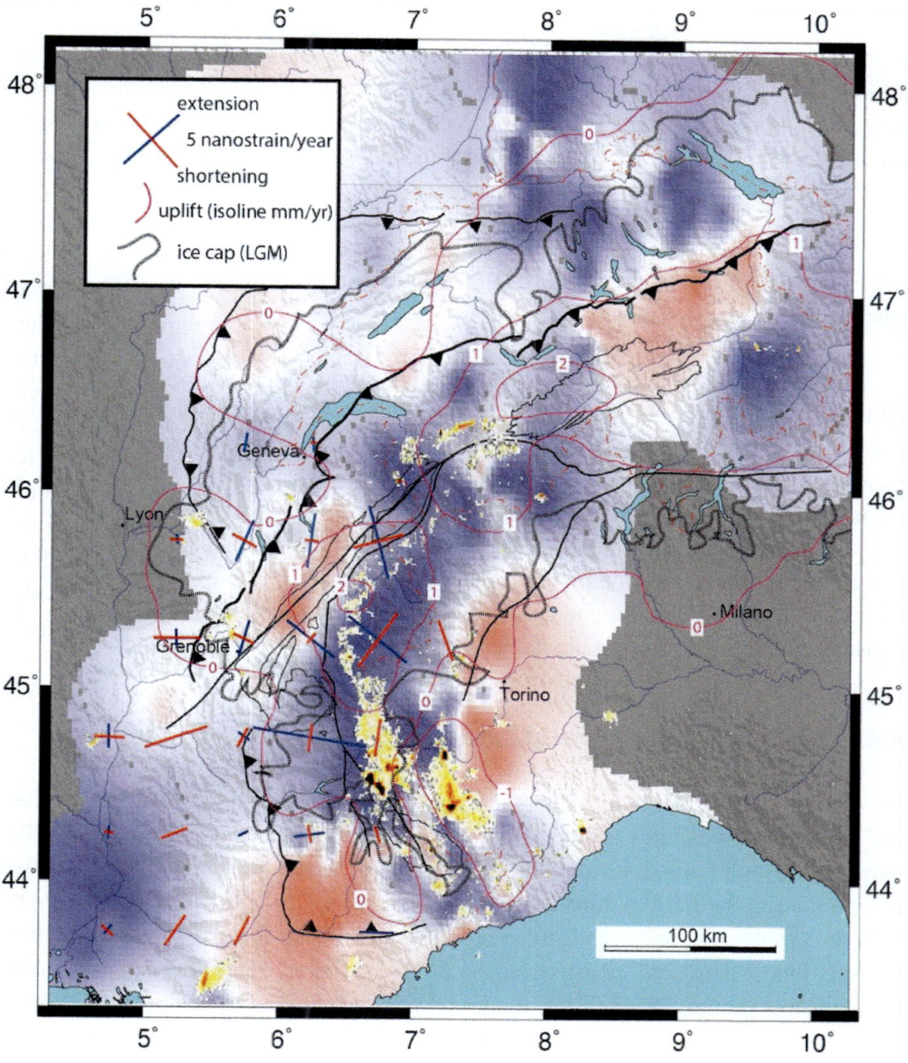

Figure 4.7. *Synthetic map of the geodetic horizontal strain together with the vertical motions (modified from Walpersdorf et al. (2018); Sternai et al. (2019). Same color code as Figure 4.6)*

COMMENT ON FIGURE 4.7.– *The extension of the ice cap for the LGM is plotted with a gray dotted line (from Ehlers and Gibbard 2004). The background is the deformation mode from seismotectonics and the density of earthquakes (see Figure 4.6). Uplift rates are nicely correlated with the LGM's ice cap, which support the interpretation that part of the current uplift arises from GIA processes. Geodetic horizontal extension is pretty well correlated with the seismotectonics extension (see Figure 4.6) at the alpine scale. However, extension is also correlated with uplift rates in the northern part of the alpine arc, whereas there is a decorrelation between extension and uplift to the south, which indicates that extensional tectonics in the Alpine arc may be only partly ruled by processes inducing coeval uplift (see text for discussion).*

4.4. Interpretations and discussion: some insights on the alpine geodynamic processes

Notwithstanding the importance of collisional and indentation processes in the Alps from the Oligocene, which led to important shortening and crustal thickening (e.g. Schmid and Kissling 2000), the late Alpine tectonics is imprinted by two phases of late extension at the scale of the belt observed on the field, as well as through the seismicity and the geodetic deformation for the current tectonics. Up until now, the relationships between the orogen-parallel extension (globally Miocene) and the orogen-perpendicular extension (mainly Plio-quaternary), and the evolution from one to the other are not clearly established. However, we can consider them as two distinct phenomena at the first order.

4.4.1. *Miocene orogen-parallel extension*

The Neogene brittle deformation analyzed along the internal zones of the Western and Central Alps exhibits a preeminent and homogeneous pattern of orogen-parallel extension, which turns direction with the bending of the alpine arc, and which can be followed up to the Eastern Alps (see Bertrand and Sue (2017)). It developed coevally with the propagation of the collisional processes toward the external zones of the belt. Indeed, this extensional phase developed in the inner zones during an overall syn-collision context at the global alpine scale. Such an extension would have participated to the accommodation of the shortening across the belt and would imply lateral tectonic flow, favored by boundary conditions at the tips of the chain, such as the Pannonian basin opening to the east and related lateral extrusion (Royden 1993; Ratschbacher et al. 1991a, 1991b; Decker et al. 1993; Peresson and Decker 1997; Sperner et al. 2002; Bertrand et al. 2015). A similar model could barely be invoked to the Southwestern tip of the Alps with the opening of the Ligurian Sea, as the arc-shape geometry of the Western Alps largely interferes with a typical extrusion model. A consequence of this coeval compression in the outer zones, plus back-thrusting in the easternmost ones (Tricart and Sue 2006) and parallel extension in the inner zones, is the enhancement of the exhumation processes, seen in particular with low temperature thermochronology (review in Fox et al. (2015)). In this frame, orogen-parallel extension may have accommodated the thickening of the crustal prism by re-equilibration of the overall architecture of the orogen.

However, orogen-parallel extension has been observed in other orogenic contexts, such as the Himalaya (Larson et al. (2020) and references herein). The interpretation proposed by Larson et al. (2018) emphasizes the role of alternative phases of coupling/uncoupling of major crustal discontinuities (main thrusts), which may activate lateral lower crustal flow, inducing lateral extension in the upper crust. Such conceptual models could be invoked in the alpine context, although the size of the structures and the importance of crustal overthickening are not directly comparable. Nonetheless, the idea of lower crustal or sub-crustal and lithospheric lateral fluxes as driving engines for the observed upper-crustal parallel extension in an active orogens raises the matter of coupled/uncoupled structural levels. Indeed, lower crustal and lithospheric flow, which still remains a matter of debate, largely depends on the large-scale plate tectonic configuration (review in Kästle et al. (2020)). In that framework, Bertrand and Sue (2017) underlined the correlation observed between the SKS seismic waves anisotropy's trajectories (Qorbani et al. 2015) and the directions of brittle extension all along the backbone of the Alps. The SKS splitting comes from olivine's lattice preferred orientation and reflects the (upper) mantle flow. Assuming that the observed correlation between the two directional fields reflects actual geodynamic relations, the Neogene orogen parallel extension in the entire Alps could be at least partly controlled by the mantle dynamics under the orogen, and by the Panonian and Apennine slabs retreat processes (e.g. Doglioni 1991; Faccenna et al. 2004; Jolivet et al. 2008; Molli et al. 2010). Beyond these deep-seated processes, the Neogene orogen parallel extension results from mechanical interactions between active plate collision, taking into account the counterclockwise rotation component of Adria with respect to Europe, and the overall buoyancy within the 3D orogenic prism. This complex system asks for further modeling to better understand the acting and interacting geodynamic processes.

4.4.2. *Plio-quaternary and current orogen-perpendicular extension*

The Western and Central Alps display a weird pattern of active deformation characterized by high uplift rates and upper crustal extension. Active orogen-perpendicular extension deduced from seismotectonics is similar to the last brittle deformation phase observed in the field using structural geology, with Pliocene age clues. Indeed, this tectonic context may have developed, at least during the whole Plio-Quaternary period, as the current seismotectonic scheme works in continuity with the latest faulting event, especially in the Western Alps (e.g. Sue et al. 2007).

In detail, the prevalence of uplift in terms of strain, together with the tectonic contrast characterized by an overall orogen-perpendicular extension in the core of the Alps along the high chain, and local compression and/or transpression at the periphery of the orogen, lead to the proposal of geodynamic models ruled by buoyancy forces impacting the global isostasy of the belt (e.g. Molnar and Lyon-Caen 1988; England and Molnar 1990; Turcotte and Schubert 2002; Sue et al. 2007; Watts 2009; Gvirtzman et al. 2016; Sternai et al. 2019). This class of model based on intrinsic forces would interact with extrinsic forces ruled by plate conditions at the limit of the system, mainly the counterclockwise rotation of Apulia with respect to Europe (Anderson and Jackson 1987; Collombet et al. 2002; Battaglia et al. 2004; Nocquet and Calais 2004; D'Agostino et al. 2008; Rolland et al. 2012), as no shortening is observed by GPS across the arc (Calais et al. 2002; Walpersdorf et al. 2018). However, in the classical model, surface uplift is due to collision and associated thickening of the crust is discarded by the absence of shortening (at least in the upper brittle crust). Simple so-called post-orogenic collapse model implying subsidence (e.g. Selverstone 2004) is also discarded as the belt is uplifting with uplift rates reaching up to 2.5 mm/yr (Sternai et al. 2019). Actually, concomitant extension and uplift ask for a proper mechanical model. Numerical modeling has been developed, for instance, by Delacou et al. (2005), Champagnac et al. (2007), Sue et al. (2007), Gardi et al. (2010), Vernant et al. (2013) and Sternai et al. (2019), in which it has been shown that isostasy-related uplift mechanically drives upper crustal extension over the uplifting area, with shortening located at its borders, which reflects the tectonic contrast observed in the Western/Central Alps well.

At this stage, it is worth noting that the correlation between uplift and extension in the map is only partial. Indeed, extension and uplift are pretty well correlated in the Central and Northwestern Alps (Figure 4.7), whereas the extension found in the Southwestern Alps takes place in a zone of much slower uplift or even subsidence for the Southeastern seismic arc (Piemont) (Figure 4.7). Another interesting point to keep in mind is the decorrelation between the most uplifting zone (Northwestern Alps) and the most active area in terms of seismicity (Southwestern branch of the Alpine arc), where geodesy also establishes the higher horizontal strain rates (Walpersdorf et al. 2018; Figure 4.7). The overall neotectonic scheme of the Alps probably arises from complex interactions between the plate motions, mainly ruled by the counterclockwise rotation of Apulia, and intrinsic buoyancy forces in the orogen's inner structure, which are discussed hereafter. Two classes of geodynamic processes may impact the overall isostatic equilibrium and associated gravity potential adjustment (GPA) of an orogen: (i) external processes linked to glacier melting (GIA) and erosional unloading, in the framework of tectonic-climate interactions (e.g. Molnar and England 1990; Champagnac et al. 2012), and (ii) deep-seated processes linked to the overall architecture of the crustal and lithospheric roots and to mantle dynamics.

i) External processes

The impact of the GIA in the Alps has been discussed since the 1990s (e.g. Gudmundsson 1994; Stocchi et al. 2005; Barletta et al. 2006) and modelized using visco-elastic rheology (Turcotte and Schubert 2002). Notwithstanding the precision on the rheological parameters, these models predicted uplift rates in the order of the actually measured geodetic ones, that is to say in the 0.5–2 mm/yr range. Chery et al. (2016) proposed that the entire Western Alps' uplift can be explained by GIA, using ad hoc viscosity lateral contrasts. Mey et al. (2016) also argued that the GIA over the entire Alps account for 90% of the uplift field constrained mainly using old leveling measurements. However, this model comprises several important discrepancies with the actual updated GNSS-related vertical field (Serpelloni et al. 2013; Nocquet et al. 2016; Walpersdorf et al. 2018; Sternai et al. 2019). Recently, Sternai et al. (2019) published an updated quantitative review of the different geodynamic processes responsible for the vertical deformation in the Alps. They re-assessed the available modeling and obtained an evaluation of about 10–30% of the overall uplift due to quaternary GIA in the Western and Central Alps, plus eventually 10–20% due to the present-day glacier shrinkage (Barletta et al. 2006). Erosional unloading, eventually enhanced by glacier melting (e.g. Koppes and Montgomery 2009), and associated re-distribution of sediment masses in intra-mountain or peripheral basins, belongs to the same family of external processes as GIA, taking into account a 1-to-3 ratio in terms of relative density between ice and rocks. Indeed, millimeter erosion rate over interglacial period would produce unloading similar to the ice melting itself, and would induce comparable GPA. Sediment budgets in peripheral orogens' basins, hence erosion in the mountains belts, document a global increase of erosion from the Pliocene onward (e.g. Herman et al. 2013). In the Alpine realm, this erosion increases and then enhanced deposition (Kuhlemann et al. 2002; Mancin et al. 2009; Fox et al. 2015) leads to an estimated erosion rate of about 0.6–0.9 mm/yr (Hinderer 2001; Wittmann et al. 2007; Sternai et al. 2013). Notwithstanding the spatial repartition of sediments within the valleys or intra-mountain basins or outside of the belt, which is still a matter of debate (e.g. Brocard et al. 2003; Korup and Schlunegger 2007; Valla et al. 2010; van den Berg et al. 2012; Rolland et al. 2017), we can note a pretty good correlation between uplift rates and erosion rates at the first order (Schlunegger and Hinderer 2001; Wittmann et al. 2007). Indeed, Champagnac et al. (2007) provided an elastic model of the erosion unloading using the so-called "Geophysical Relief" (Small and Anderson 1998) as a proxy for erosion. They obtained a maximum uplift of about 500 m in the Northwestern Alps, for a million-year time scale, well correlated with the leveling-measured uplift in Switzerland (originally Schaer and Jeanrichard (1974); see also Serpelloni et al. (2013); Sternai et al. (2019)). The improved model of Sternai et al. (2012, 2019) implements a reconstructed pre-glacial topography to precisely evaluate the erosion distribution. They obtain an erosion-related uplift at about 250 m maximum for the same time scale. Sternai et al.'s (2019) review provides a complete updated discussion on erosion rate in the Alps and establishes that the proportion of erosion-related uplift at the scale of the Western and Central Alps reaches 10%–20% of the total GNSS-observed uplift.

ii) Deep-seated processes

The second class of geodynamic processes that can impact the overall isostasy of the orogen (static and dynamic) refers to the deep structure of its root and to the underlying mantle's dynamics. The Moho isobaths under the Western/Central Alps present a complex structure with three imbricated specific Mohos (European, Ligurian, and Apulian Mohos: Waldhauser et al. 2002; Di Stefano et al. 2009; Schreiber et al. 2010; Wagner et al. 2012; Spada et al. 2013). However, at a first level of analysis, the residual topography arising from the Moho depth in the Western/Central Alps remains moderate (some hundreds of meters), that is to say that the belt is somehow compensated (Sternai et al. 2019), which is not the case for the Eastern Alps. Yet, these observations would only impact the static equilibrium and related GPA of the lithospheric structure. From a dynamic point of view, a Pliocene slab break-off under the Western Alps has been proposed from tomographic studies (e.g. Lippitsch et al. 2003; Kissling et al. 2006; Kästle et al. 2018). The removal of such dense material, eventually associated with crustal delamination, would induce large-scale uplift of the belt (e.g. Molnar and Lyon Caen 1988; Figure 4.8). It also invoked the interpretation of a recent thermochronology-related dataset in the Western Alps (Baran et al. 2014; Fox et al. 2015). Such a slab tear-off would induce a rapid uplift, probably affecting large areas (Duretz et al. 2014). It could explain up to 50% of the current uplift and account for the actual lateral variations in the uplift field observed from the Western to the Eastern Alps (e.g. Serpelloni et al. 2013; Nocquet et al. 2016; Sternai et al. 2019). Nevertheless, the slab break-off under the Western Alps is still a matter of debate, and there is no consensus on that topic in the community. Indeed, Zhao et al. (2016) interpreted a continuous steep slab under the Western Alps from a new seismic tomography and showed a region of low viscosity in the mantle 200 km under the Western Alps. This zone of low viscosity, therefore more buoyant material, could account for the residual uplift modelized by Nocquet et al. (2016) in the NW Alps, in addition to the GIA and erosion-related uplift. The quantification of the mantle flow and related "dynamic topography" (e.g. Flament et al. 2013; Faccenna et al. 2014) process is still quite difficult in the Alps (see the attempt of Sternai et al. (2019)). However, it could account for up to 20% of the actual uplift. The complexity of the mantle and lithospheric structure under the Alps prevents us from providing a more accurate modeling of the different processes that may account for the observed deformation (extension, uplift). Nevertheless, it is worth noting that several correlations rise up between the deformation fields and the deep structures, which indeed rule part of the current upper crustal deformation. In that frame, the ongoing Alparray experiment (Hetényi et al. 2018) should provide first-order information, allowing us to better constrain the deep orogen's structure and dynamics. However, the decorrelation observed between the uplifting zone and extensional tectonics in the southern part of the arc, together with the extensional stress axes computed by Mathey et al. (2021), which are deflected to the South from strictly orogen-perpendicular extension, would result from interactions between buoyancy forces in the Alpine's root and rotational plate tectonic at the limits of the system.

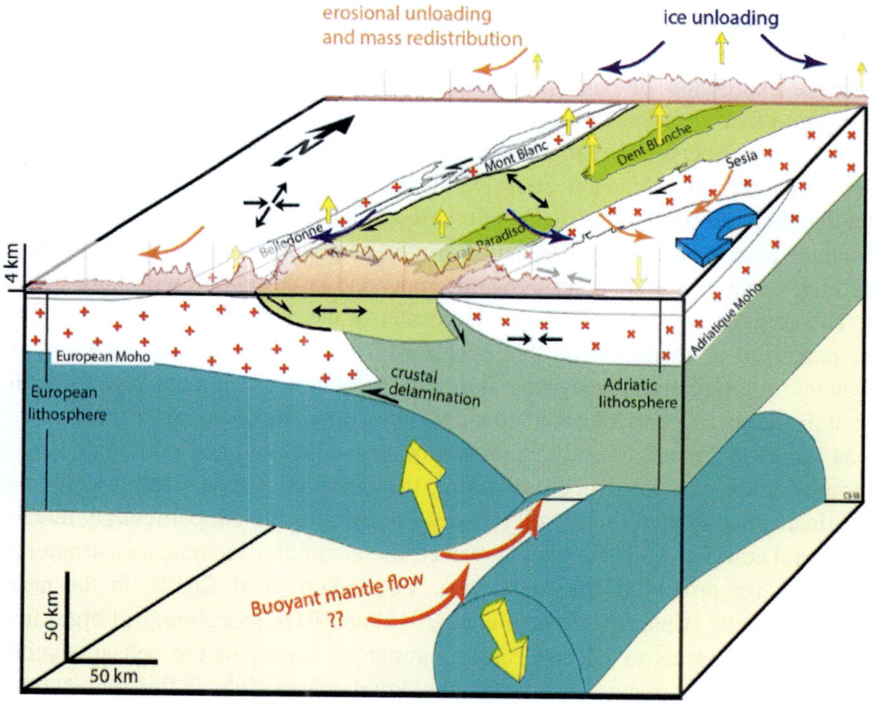

Figure 4.8. *Synoptic 3D block diagram of the Western Alps at the lithospheric scale (modified from Sue et al. 1999)*

COMMENT ON FIGURE 4.8.– *External processes such as GIA and erosional unloading appear as first-order driving mechanisms to explain the actual tectonic scheme combining uplift and extension in the high chain, but concomitant deep-seated processes such as slab break-off, mantle flow, or crustal delamination are necessary to explain the entire observed uplift in the overall frame of the orogen's isostatic equilibrium (see text for discussion).*

4.5. Concluding remarks

This chapter presents an up-to-date overview of the extensional tectonics in the Western and Central Alps from the Neogene onward. The late alpine faulting analyzed in the field displays two distinct phases: a Miocene orogen-parallel extension, then an orogen-perpendicular extension, which developed during the Pliocene and is still seismically active. The Alpine foreland is characterized by transcurrent and/or transpressive state of stress. The Miocene parallel extension in the internal zones is coeval with the propagation of the shortening toward the outer zones of the orogen, and to the opening of the Ligurian Sea to the south. It could be the surface expression of deep processes at the lithosphere scale. Beyond fault analysis from field data, seismotectonics strain/stress fields allow us to constrain the still active orogen-perpendicular extension and can be compared with respect to the geodetically inferred horizontal strain field and vertical uplift rates. It is worth noting that the vertical motions in the belt represent the most important deformations in absolute value. At the first order, active orogen-perpendicular extension is located in the high chain (therefore thick orogenic root) and is correlated with high uplift rates. This scheme is valid in the north and center of the Arc, but fades to the south, where extension is characterized in a sector with no vertical motion. In terms of geodynamic interpretation, buoyancy is forced to play a preeminent role through external processes, such as GIA and erosional unloading, and deep-seated processes, such as slab detachment, crustal delamination, or mantle flow and related dynamic topography. Indeed, the balance controlling the interplay between intrinsic forces and extrinsic plate-related forces still remains poorly constrained, particularly due to the structural complexity of the belt. However, these complex tectonic and structural schemes ask for proper 3D modeling (e.g. Vouillamoz et al. 2012). In the near future, the ongoing Alparray (Hetényi et al. 2018) and RGF (see: http://rgf.brgm.fr/) programs will allow us to improve the 3D structural model of the belt at several imbricated scales, and to develop proper 3D/4D-dynamic models. In this framework, some of the remaining challenges to address will concern (i) the problem of orogen-parallel extension, still poorly interpreted and/or modelized in the Western Alps, and the timing of the shift toward the orogen-perpendicular extension; (ii) the quantification of the mechanical processes explaining the complex 3D relationship between the observed extension and uplift; (iii) the modeling and quantification of the balance between intrinsic buoyancy engine and extrinsic plate motions.

4.6. Acknowledgments

I would like to thank the editors Rosenberg and Bellahsen for their patience and fruitful remarks. I owe much to the careful and constructive reviews by Ortner and Bistacchi. Long-lasting scientific exchanges in the frame of many projects and discussions with colleagues within the "alpine community", especially in the frame of the Argand Alpine Workshop series, were very important throughout the evolution of the works synthetized in this paper. Special thanks to Pierre, Jean-Daniel, Bastien, Naomi, Peter, Andrea, Audrey, Jean-Mathieu, Pietro, Margot, ... and Ombeline. I would like to dedicate this work to the memory of Martin.

4.7. References

Allanic, C. (2012). Kinematics, age and dynamic of the brittle deformation within the Lepontine dome (Central Alps). PhD Thesis, University of Orléans.

Amstutz, A. (1954). Pennides dans l'Ossola et problème des racines. *Arch. Sci. Genève*, 7, 411.

Anderson, H. and Jackson, J. (1987). Active tectonics of the Adriatic region. *Geophys. J. Int.*, 91(3), 937–983.

Angelier, J. (1989). From orientation to magnitudes in paleostress determinations using fault slip data. *Journal of Structural Geology*, 11, 37–50.

Angelier, J. (1990). Inversion of field data in fault tectonics to obtain the regional stress – A new rapid direct inversion method by analytical means. *Geophys. J. Int.*, 103, 363–376.

Angelier, J. and Goguel, J. (1979). Sur une méthode simple de détermination des axes principaux des contraintes pour une population de failles. *C. R. Acad. Sci.*, 282, 307–310.

Angelier, J. and Mechler, P. (1977). Sur une méthode graphique de recherche des contraintes principales également utilisable en tectonique et sismologie : la méthode des dièdres droits. *Bull. Soc. Géol. Fr.*, 7(19), 1309–1318.

Argand, E. (1916). Sur l'arc des Alpes occidentales. *Eclogae Geologicae Helveticae*, 14, 145–191.

Argand, E. (1924). La tectonique de l'Asie. *Compte-rendu du XIIIème Congrès Géologique International*, 1922(1), 171–372.

Ault, A.K., Frenzel, M., Reiners, P.W., Woodcock, N.H., Thomson, S.N. (2016). Record of paleofluid circulation in faults revealed by hematite (U-Th)/He and apatite fission-track dating: An example from Gower Peninsula fault fissures, Wales. *Lithosphere*, 8, 379–385.

Ballèvre, M. and Merle, O. (1993). The Combin fault: Compressional reactivation of a late Cretaceous–early Tertiary detachment fault in the Western Alps. *Schweizrische Mineralogische Petrografische Mitteilung*, 73, 205–227.

Ballèvre M., Lagabrielle Y., Merle O. (1990). Tertiary ductile normal faulting as a consequence of lithospheric stacking in the western Alps. In *Deep Structure of the Alps*, Roure, F., Heitzmann, P., Polino, R. (eds). Société Géologique de France, Paris.

Baran, R., Friedrich, A.M., Schlunegger, F. (2014). The late Miocene to Holocene erosion pattern of the Alpine foreland basin reflects Eurasian slab unloading beneath the western Alps rather than global climate change. *Lithosphere*, 6(2), 124–131.

Barletta, V.R., Ferrari, C., Diolaiuti, G., Carnielli, T., Sabadini, R., Smiraglia, C. (2006). Glacier shrinkage and modeled uplift of the Alps. *Geophys. Res. Lett.*, 33(14).

Baroux, E., Béthoux, N., Bellier, O. (2001). Analyses of the stress field in southeastern France from earthquake focal mechanisms. *Geophys. J. Int.*, 145, 336–348.

Battaglia, M., Murray, M.H., Serpelloni, E., Burgman, R. (2004). The Adriatic region: An independent microplate within the Africa-Eurasia collision zone. *Geophys. Res. Lett.*, 31(9).

Bauve, V., Plateaux, R., Rolland, Y., Sanchez, G., Bethoux, N., Delouis, B., Darnault, R. (2014). Long-lasting transcurrent tectonics in SW Alps evidenced by Neogene to present-day stress fields. *Tectonophysics*, 621, 85–100.

Beaucé, E., Frank, W.B., Paul, A., Campillo, M., van der Hilst, R.D. (2019). Systematic detection of clustered seismicity beneath the Southwestern Alps. *Journal of Geophysical Research: Solid Earth*, 124, 11,531–11,548.

Behrmann, J.H. (1988). Crustal-scale extension in a convergent orogen: The Sterzing-Steinach mylomte zone in the Eastern Alps. *Geodinamica Acta*, 2(2), 63–73.

van den Berg, F., Schlunegger, F., Akçar, N., Kubik, P. (2012). 10Be-derived assessment of accelerated erosion in a glacially conditioned inner gorge, Entlebuch, Central Alps of Switzerland. *Earth Surf. Process. Landforms*, 37(11), 1176–1188.

Bertrand, A. and Sue, C. (2017). Reconciling late faulting over the whole Alpine belt: From structural analysis to geochronological constrains. *Swiss J. Geosci.*, 110, 565–580.

Bertrand, A., Rosenberg, C.L., Garcia, S. (2015). Fault-slip analysis and late exhumation of the Penninic units of the Tauern Window, Eastern Alps. *Tectonophysics*, 649, 1–17.

Beucher, R., Sue, C., Tricart, P. (2017). Orogen-parallel brittle extension as a major tectonic imprint in the Neogene evolution of the south-western Alpine arc. *Int. J. Earth Sci.*, 106, 2973–2990.

Bilau, A. (2019). Signature géochimique des fluides associés au rejeu extensif du front pennique (Alpes Occidentales) : implications pour la datation des failles. Masters Thesis, University of Grenoble Alpes, Grenoble.

Bilau, A., Rolland, Y., Schwartz, S., Godeau, N., Guihou, A., Deschamps, P., Brigaud, B., Noret, A., Dumont, T., Gautheron, C. (2021). Extensional reactivation of the Penninic frontal thrust 3 Myr ago as evidenced by U–Pb dating on calcite in fault zone cataclasite. *Solid Earth*, 12, 237–251.

Bistacchi, A. and Massironi, M. (2000). Post-nappe brittle tectonics and kinematic evolution of the north-western Alps: An integrated approach. *Tectonophysics*, 327, 267–292.

von Blanckenburg, F. (1992). Combined high-precision chronometry and geochemical tracing using accessory minerals: Applied to the Central-Alpine Bergell intrusion (central Europe). *Chemical Geology*, 100(1–2), 19–40.

von Blanckenburg, F. and Davies, J.H. (1995). Slab breakoff: A model for syncollisional magmatism and tectonics in the Alps. *Tectonics*, 14, 120–131.

Bolognesi, F. and Bistacchi, A. (2016). Weakness and mechanicalanisotropy of phyllosilicate-rich cataclasites developed after mylonites of a low-angle normal fault (Simplon Line, Western Alps). *J. Struct. Geol.*, 83, 1–12.

Bott, M.H. (1959). The mechanism of oblique slip faulting. *Geol. Mag.*, 96(2), 109–117.

Brandon, M.T., Roden-Tice, M.K., Graver, J.I. (1998). Late Cenozoic exhumation of the Cascadia accretionary wedge in the Olympic Mountains, northwest Washington State. *Geological Society of America Bulletin*, 110, 985–1009.

Brocard, G.Y., Van Der Beek, P.A., Bourlès, D.L., Siame, L.L., Mugnier, J.L. (2003). Long-term fluvial incision rates and postglacial river relaxation time in the French Western Alps from 10Be dating of alluvial terraces with assessment of inheritance, soil development and wind ablation effects. *Earth Planet. Sci. Lett.*, 209(1–2), 197–214.

Calais, E., Nocquet, J.M., Jouanne, F., Tardy, M. (2002). Current strain regime in the Western Alps from continuous Global Positioning System measurements. 1996–2001. *Geology*, 30, 651–654.

Campani, M., Mancktelow, N., Seward, D., Rolland, Y., Müller, W., Guerra, I. (2010). Geochronological evidence for continuous exhumation through the ductile-brittle transition along a crustal-scale low-angle normal fault: Simplon Fault Zone, central Alps. *Tectonics*, 29(3).

Carey, E. and Brunier, B. (1974). Analyse théorique et numérique d'un modèle mécanique élémentaire appliqué à l'étude d'une population de failles. *Comptes Rendus de l'Académie des Sciences*, 279, 891–894.

Champagnac, J.D., Sue, C., Delacou, B., Burkhard, M. (2003). Brittle orogen-parallel extension in the internal zones of the Swiss Alps (south Valais). *Eclogae Geologicae Helveticae*, 96, 325–338.

Champagnac, J.D., Sue, C., Delacou, B., Burkhard, M. (2004). Brittle deformation in the inner northwestern Alps: From early orogen-parallel extrusion to late orogen-perpendicular collapse. *Terra Nova*, 16, 232–242.

Champagnac, J.D., Sue, C., Delacou, B., Tricart, P., Allanic, C., Burkhard, M. (2006). Miocene lateral extrusion in the inner western Alps revealed by dynamic fault analysis. *Tectonics*, 25, TC3014. doi: 10.1029/2004TC001779.

Champagnac, J.D., Molnar, P., Anderson, R.S., Sue, C., Delacou, B. (2007). Quaternary erosion-induced isostatic rebound in the Western Alps. *Geology*, 35, 195–198.

Champagnac, J.D., Molnar, P., Sue, C., Herman, F. (2012). Tectonics, climate, and mountain topography. *J. Geophys. Res.*, 117, B02403.

Chery, J., Genti, M., Vernant, P. (2016). Ice cap melting and low-viscosity crustal root explain the narrow geodetic uplift of the Western Alps. *Geophys. Res. Lett.*, 43(7), 3193–3200.

Ciancaleoni, L. and Marquer, D. (2008). Late Oligocene to early Miocene lateral extrusion at the eastern border of the Lepontine dome of the Central Alps (Bergell and Insubric areas, eastern central Alps). *Tectonics*, 27, TC4008.

Collombet, M., Thomas, J.C., Chauvin, A., Tricart, P., Bouillin, J.P., Gratier, J.P. (2002). Counterclockwise rotation of the Western Alps since the Oligocene: New insights from paleomagnetic data. *Tectonics*, 21(4).

Coward, M. and Dietrich, D. (1989). Alpine tectonics: An overview. In *Alpine Tectonics*, Coward, M., Dietrich, D., Park, R. (eds). Geological Society London Special Publications, Oxford.

D'Agostino, N., Avallone, A., Cheloni, D., D'anastasio, E., Mantenuto, S., Selvaggi, G. (2008). Active tectonics of the Adriatic region from GPS and earthquake slip vectors. *J. Geophys. Res.*, 113(B12).

De Barros, L., Baques, M., Godano, M., Helmstetter, A., Deschamps, A., Larroque, C., Courboulex, F. (2019). Fluid-induced swarms and coseismic stress transfer: A dual process highlighted in the aftershock sequence of the 7 April 2014 earthquake (ml 4.8, Ubaye, France). *Journal of Geophysical Research: Solid Earth*, 124, 3918–3932.

De Graciansky, P.C., Robert, R.D., Tricart, P. (2011). *The Western Alps, from Rift to Passive Margin to Orogenic Belt: An Integrated Geoscience Overview*. Elsevier, Amsterdam.

Decker, K., Meschede, M., Ring, U. (1993). Fault slip analysis along the northern margin of the Eastern Alps (Molasse, Helvetic nappes, North and South Penninic flysch, and the Northern Calcareous Alps). *Tectonophysics*, 223, 291–312.

Delacou, B., Sue, C., Champagnac, J.D., Burkhard, M. (2004). Present-day geodynamics in the bend of the western and central Alps as constrained by earthquake analysis. *Geophys. J. Int.*, 158(2), 753–774.

Delacou, B., Sue, C., Champagnac, J.D., Burkhard, M. (2005). Origin of the current stress field in the Western/Central Alps: Role of gravitational reequilibration constrained by numerical modelling. *Geol. Soc. London Spec. Publ.*, 243, 295–310.

Dewey, J.F., Pitman, W.C.I., Ryan, W.B.F., Bonnin, J. (1973). Plate tectonics and the evolution of the Alpine system. *Bulletin of the Geological Society of America*, 84, 137–180.

Di Stefano, R., Kissling, E., Chiarabba, C., Amato, A., Giardini, D. (2009). Shallow subduction beneath Italy: Three-dimensional images of the Adriatic-European-Tyrrhenian lithosphere system based on high-quality P wave arrival times. *J. Geophys. Res.*, 114(B5).

Doglioni, C. (1991). A proposal for the kinematic modelling of W-dipping subductions – Possible applications to the Tyrrhenian-Apennines system. *Terra Nova*, 3, 423–434.

Duretz, T., Gerya, T., Spakman, W. (2014). Slab detachment in laterally varying subduction zones: 3D numerical modeling. *Geophys. Res. Lett.*, 41. doi: 10.1002/2014GL059472.

England, P. and Molnar, P. (1990). Surface uplift, uplift of rocks, and exhumation of rocks. *Geology*, 18, 1173–1177.

Eva, E. and Solarino, S. (1998). Variations of stress directions in the Western Alpine arc. *Geophys. J. Int.*, 135, 438–448.

Faccenna, C., Piromallo, C., Crespo-Blanc, A., Jolivet, L., Rossetti, F. (2004). Lateral slab deformation and the origin of the western Mediterranean arcs. *Tectonics*, 23, TC1012.

Faccenna, C., Becker, T., Miller, M.S., Serpelloni, E., Willett, S.D. (2014). Isostasy, dynamic topography, and the elevation of the Apennines of Italy. *Earth and Planetary Science Letters*, 407, 163–174.

Flament, N., Gurnis, M., Müller, R.D. (2013). A review of observations and models of dynamic topography. *Lithosphere*, 5(2), 189–210.

Fleitout, L. and Froidevaux, C. (1982). Tectonics and topography for lithosphere containing density heterogeneities. *Tectonics*, 1, 21–56.

Fox, M., Herman, F., Kissling, E., Willett, S.D. (2015). Rapid exhumation in the Western Alps driven by slab detachment and glacial erosion. *Geology*, 43(5), 379–382.

Fréchet, J. (1978). Sismicité du sud-est de la France et une nouvelle méthode de zonage sismique. Postgraduate Thesis, Université de Grenoble.

Fréchet, J. and Pavoni, N. (1979). Etude de la sismicité de la zone briançonnaise entre Pelvoux et Argentera (Alpes occidentales) à l'aide d'un réseau de stations portables. *Eclog. Geol. Helv.*, 72(3), 763–779.

Frohlich, C. (1992). Triangle diagrams: Ternary graphs to display similarity and diversity of earthquake focal mechanisms. *Physics of the Earth and Planetary Interiors*, 75.1–3, 193–198.

Gardi, A., Baize, S., Scotti, O. (2010). Present-day vertical isostatic readjustment of the Western Alps revealed by numerical modelling and geodetic and seismotectonic data. *Geological Society, London, Special Publications*, 332(1), 115–128.

Gephart, J.W. (1990). Stress and the direction of slip on fault planes. *Tectonics*, 9, 845–858.

Gephart, J.W. and Forsyth, W.D. (1984) An improved method for determining the regional stress tensor using earthquake focal mechanism data: Application to the San-Fernando earthquake sequence. *J. Geophys. Res.*, 89, 9305–9320.

Grosjean, G., Sue, C., Burkhard, M. (2004). Late neogene extension in the vicinity of the Simplon fault zone (central Alps, Switzerland). *Eclogae Geologicae Helveticae*, 97, 33–46.

Gudmundsson, G.H. (1994). An order-of-magnitude estimate of the current uplift-rates in Switzerland caused by the Würm Alpine deglaciation. *Eclogae Geologicae Helvetiae*, 87(2), 545–557.

Gvirtzman, Z., Faccenna, C., Becker, T.W. (2016). Isostasy, flexure, and dynamic topography. *Tectonophys*, 683, 255–271.

Handy, M.R., Schmid, S.M., Bousquet, R., Kissling, E., Bernoulli, D. (2010). Reconciling plate-tectonic reconstructions of Alpine Tethys with the geological–geophysical record of spreading and subduction in the Alps. *Earth Sci. Rev.*, 102(3), 121–158.

Hannouz, E. (2019). Quantification des taux de déformation sismique dans la zone de transfert tectonique des Alpes à la vallée du Rhône. Masters Thesis, University of Grenoble Alpes, Grenoble.

Hannouz, E., Sue, C., Walpersdorf, A., Mazzotti, S., Baize, S., Lemoine, A., Mathey, M. (submitted). Seismic strain quantification in the foreland of the Western Alps and implications on seismic hazard. *Natural Hazard*.

Herman, F., Seward, D., Valla, P.G., Carter, A., Kohn, B., Willett, S.D., Ehlers, T.A. (2013). Worldwide acceleration of mountain erosion under a cooling climate. *Nature*, 504(7480), 423–426. doi: 10.1038/nature12877.

Herwegh, M., Berger, A., Baumberger, R., Wehrens, P., Kissling, E., (2017). Large-scale crustal-block-extrusion during late alpine collision. *Scientific Reports*, 7(1), 413.

Hetényi, G., Molinari, I., Clinton, J., Bokelmann, G., Bondár, I., Crawford, W.C., Dessa, J.-X., Doubre, C., Friederich, W., Fuchs, F. et al. (2018). The AlpArray seismic network: A large-scale European experiment to image the Alpine orogeny. *Surveys in Geophysics*, 39, 1009–1033.

Hinderer, M. (2001). Late quaternary denudation of the Alps, valley and lake fillings and modern river loads. *Geodinamica Acta*, 14(4), 231–263.

Jolivet, L., Augier, R., Faccenna, C., Negro, F., Rimmele, G., Agard, P., Robin, C., Rossetti, F., Crespo-Blanc, A. (2008). Subduction, convergence and the mode of backarc extension in the Mediterranean region. *BSGF*, 179, 525–550.

Jomard, H., Cushing, E.M., Palumbo, L., Baize, S., David, C., Chartier, T. (2017). Transposing an active fault database into a seismic hazard fault model for nuclear facilities – Part 1: Building a database of potentially active faults (BDFA) for metropolitan France. *Nat. Hazards Earth Syst. Sci.*, 17, 1573–1584.

Jouanne, F., Ménard, G., Darmendrail, X. (1995). Present-day vertical displacements in the north-western Alps and southern Jura Mountains: Data from leveling comparisons. *Tectonics*, 14(3), 606–616.

Kästle, E.D., El-Sharkawy, A., Boschi, L., Meier, T., Rosenberg, C.L., Bellahsen, N., Cristiano, L., Weidle, C. (2018). Surface wave tomography of the Alps using ambient-noise and earthquake phase velocity measurements. *J. Geophys. Res.*, 123, 1770–1792.

Kästle, E.D., Rosenberg, C.L., Boschi, L., Bellahsen, N., Meier, T., El-Sharkawy, A. (2020). Slab break-offs in the Alpine subduction zone. *Int. J. Earth Sci.*, 109, 587–603.

Kastrup, U., Zoback, M.L., Deichmann, N., Evans, K., Giardini, D. (2004). Stress field variations in the Swiss Alps and the Northern Alpine foreland derived from inversion of fault plane solutions. *J. Geophys. Res.*, 109.

Kissling, E., Schmid, S.M., Lippitsch, R., Ansorge, J., Fügenschuh, B. (2006). Lithosphere structure and tectonic evolution of the Alpine arc: New evidence from high-resolution teleseismic tomography. *Geol. Soc., London, Memoirs*, 32(1), 129–145.

Koppes, M.N. and Montgomery, D.R. (2009). The relative efficacy of fluvial and glacial erosion over modern to orogenic timescales. *Nat. Geosci.*, 2, 644–647.

Korup, O. and Schlunegger, F. (2007). Bedrock landsliding, river incision, and transience of geomorphic hillslope-channel coupling: Evidence from inner gorges in the Swiss Alps. *J. Geophys. Res.*, 112(F3).

Kostrov, B. (1974). Seismic moment and energy of earthquakes and seismic flow of rock. *Izv. Acad. Sci. USSR Phys. Solid Earth*, 1, 23–40.

Kralik, M., Klima, K., Riedmuller, G. (1987). Dating fault gouges. *Nature*, 327, 315–317.

Kralik, M., Clauer, N., Holnsteiner, R., Huemer, H., Kappel, F. (1992). Recurrent fault activity in the Grimsel test site (GTS, Switzerland), revealed by Rb–Sr, K-Ar and tritium isotope techniques. *Journal of the Geological Society of London*, 149, 293–301.

Kuhlemann, J., Frisch, W., Székely, B., Dunkl, I., Kazamér, M. (2002). Post-collisional sediment budget history of the Alps: Tectonic versus climatic control. *Int. J. Earth Sci.*, 91, 818–837.

Lacombe, O. (2012). Do fault slip data inversions actually yield "paleostresses" that can be compared with contemporary stresses? A critical discussion. *Comptes Rendus Géoscience*, 344, 159–173.

Larson, K.P., Kellett, D.A., Cottle, J.M., Camacho, A., Brubacher, A.D. (2020). Mid-Miocene initiation of E-W extension and recoupling of the Himalaya. *Terra Nova*, 32, 151–158.

Lippitsch, R., Kissling, E., Ansorge, J. (2003). Upper mantle structure beneath the Alpine orogen from high-resolution teleseismic tomography. *J. Geophys. Res.*, 108.

Lyons, J.B. and Snellenberg, J. (1971). Dating faults. *Geological Society of America Bulletin*, 82, 1749–1752.

Malusá, M.G., Polino, R., Zattin, M., Bigazzi, G., Martin, S., Piana, F. (2005). Miocene to present differential exhumation in the Western Alps: Insights from fission track thermochronology. *Tectonics*, 24, TC3004.

Mancin, N., Di Giulio, A., Cobianchi, M. (2009). Tectonic vs. climate forcing in the Cenozoic sedimentary evolution of a foreland basin (Eastern Southalpine system, Italy). *Basin Res.*, 21(6), 799–823.

Mancktelow, N.S. (1985). The Simplon line: A major displacement zone in the western Lepontine Alps. *Eclogae Geologicae Helvetiae*, 78/1, 73–96.

Mancktelow, N.S. (1987). Quartz textures from the Simplon fault zone, Southwest Switzerland and North Italy. *Tectonophysics*, 135/1–3, 133–153.

Mancktelow, N.S. (1990). The Simplon fault zone. *Beitr. Geol. Karte CH*, 163, 74.

Mancktelow, N.S. and Pavlis, T.L. (1994). Fold-fault relationships in low-angle detachment systems. *Tectonics*, 13/3, 668–685.

Marrett, R. and Allmendinger, R.W. (1990). Kinematic analysis of fault-slip data. *Journal of Structural Geology*, 12, 973–986.

Masson, F., Lehujeur, M., Ziegler, Y., Doubre, C. (2014). Strain rate tensor in Iran from a new GPS velocity field. *Geophysical Journal International*, 197(1), 10–21.

Masson, C., Mazzotti, S., Vernant, P., Doerflinger, E. (2019). Extracting small deformation beyond individual station precision from dense Global Navigation Satellite System (GNSS) networks in France and western Europe. *Solid Earth*, 10(6), 1905–1920.

Mathey M., Sue, C., Potin, B., Pagani, C., Bodin, T., Husson, L., Hannouz, E., Baise, S., Walpersdorf, A. (2021). Present-day geodynamics of the Western Alps: New insights from earthquake mechanisms. *Solid Earth*, 12(7), 1661–1681.

Maurer, H., Burkhard, M., Diechmann, N., Green, A.G. (1997). Active tectonism in the Central Alps: Contrasting stress regimes north and south of the Rhone Valley. *Terra Nova*, 9, 91–94.

McDougall, I. and Harrison, M.T. (1999). *Geochronology and Thermochronology by the 40Ar/39Ar Method*. Oxford University Press, Oxford.

Ménard, G. (1988). Structure et cinématique d'une chaîne de collision : les Alpes occidentales et centrales. Doctoral Thesis, Université de Grenoble.

Mercier, D. (1977). Les modalités de la transgression jurassique dans la zone briançonnaise (région de Briançon, Haute Alpes) : étude stratigraphique et sédimentologique. Postgraduate Thesis, Université Paris VI.

Mey, J., Scherler, D., Wickert, A.D., Egholm, D.L., Tesauro, M., Schildgen, T., Strecker, M.R. (2016). Glacial isostatic uplift of the European Alps. *Nature Commun* 7, 13382. doi: 10.1038/ncomms13382.

Molli, G., Crispini, L., Malusa, M., Mosca, P., Piana, F., Federico, L. (2010). Geology of the Western Alps-Northern Apennine junction area: A regional review. In *The Geology of Italy: Tectonics and Life Along Plate Margins*, Beltrando, M., Peccerillo, A., Mattei, M., Conticelli, S., Doglioni, C. (eds). *Journal of the Virtual Explorer*, 36, 1441–8142.

Molnar, P. (1983). Average regional strain due to slip on numerous faults of different orientations. *J. Geophys. Res.*, 88, 6430–6432.

Molnar, P. and England, P. (1990). Late Cenozoic uplift of mountain ranges and global climate change: Chicken or egg? *Nature*, 346(6279), 29–34.

Molnar, P. and Lyon-Caen, H. (1988). Some simple physical aspects of the support, structure, and evolution of mountain belts. *Geol. Soc. Am. Bull.*, 218, 179–207.

Nguyen, H.N., Vernant, P., Mazzotti, S., Khazaradze, G., Asensio, E. (2016). 3D GPS velocity field and its implications on the present-day postorogenic deformation of the western Alps and Pyrenees. *Solid Earth Discussions*, 7(5), 1349.

Nicolas, A., Hirn, A., Nicolich, R., Polino, R., Group ECW (1990). Lithospheric wedging in the Western Alps inferred from the ECORS-CROP traverse. *Geology*, 18, 587–590.

Nocquet, J.M. (2011). Rapport de la géodésie spatiale à la connaissance de la géodynamique en Europe et en Méditerranée. Habilitation à diriger des recherches, Université Nice Sophia Antipolis.

Nocquet, J.M. and Calais, E. (2004). Geodetic measurements of crustal deformation in the Western Mediterranean and Europe. *Pure Appl. Geophys.*, 161(3), 661–681.

Nocquet, J.M., Sue, C., Walsperdorf, A., Tran, T., Lenotre, N., Vernant, P., Cushing, M., Jouanne, F., Masson, F., Baize, S. et al. (2016). Present-day uplift of the western Alps. *Sci. Rep.*, 6, 28404.

Ortner, H. (2003). Local and far field stress – Analysis of brittle deformation in the western part of the Northern Calcareous Alps, Austria. *Geologisch-Paläontologische Mitteilungen Innsbruck*, 26, 109–131.

Ortner, H., Reiter, F., Acs, P. (2002). Easy handling of tectonic data: The programs TectonicVB for Mac and TectonicsFP for Windows (TM). *Computers & Geosciences*, 28, 1193–1200.

Pavoni, N. (1961). Faltung durch horizontal verschiebung. *Ecl. Geol. Helv.*, 54, 515–534.

Pavoni, N. (1980). Crustal stresses inferred from fault-plane solutions of earthquakes and neotectonic deformation in Switzerland. *Rock Mech. Suppl.*, 9, 63–68.

Peresson, H. and Decker, K. (1997). Far-field effects of late Miocene subduction in the Eastern Carpathians: e-W compression and inversion of structures in the Alpine-Carpathian-Pannonian region. *Tectonics*, 16, 38–56.

Perrone, G., Cadoppi, P., Balestro, G., Tallone, S. (2011). Post-collisional tectonics in the Northern Cottian Alps (Italian Western Alps). *International Journal of Earth Sciences*, 100, 1349–1373.

Petit, J.P. (1987). Criteria for the sense of movement on fault surfaces in brittle rocks. *Journal of Structural Geology*, 9, 597–608.

Pfiffner, O.A. and Burkhard, M. (1987). Determination of paleostress axis orientations from fault, twin and earthquake data. *Annales Tectonicae*, 1(1), 48–57.

Platt, J.P. (1986). Dynamic of orogenic wedges and the uplift of high pressure metamorphic rocks. *Geological Society of America Bulletin*, 97, 1037–1053.

Potin, B. (2016). Les Alpes occidentales : tomographie, localisation de séismes et topographie du Moho. PhD Thesis, University Grenoble Alpes.

Qorbani, E., Kurz, W., Bianchi, I., Bokelmann, G. (2015). Correlated crustal and mantle deformation in the Tauern Window, Eastern Alps. *Austrian Journal of Earth Sciences*, 108, 161–173.

Ramsay, J.G. and Huber, M.I. (1983). *The Techniques of Modern Structural Geology*, Volume 2. Academic Press, London.

Ratschbacher, L., Frisch, W., Linzer, H.G., Merle, O. (1991a). Lateral extrusion in the Eastern Alps, Part 1: Boundary conditions and experiments scaled for gravity. *Tectonics*, 10(2), 257–271.

Ratschbacher, L., Frisch, W., Linzer, H.G., Merle, O. (1991b). Lateral extrusion in the Eastern Alps, Part 2: Structural analysis. *Tectonics*, 10(2), 257–271.

Reiter, F., Freudenthaler, C., Hausmann, H., Ortner, H., Lenhardt, W., Brandner, R. (2018). Active seismotectonic deformation in front of the Dolomites indenter, Eastern Alps. *Tectonics*, 37/12, 4625–4654.

Roberts, N.M.W. and Walker, R.J. (2016). U-Pb geochronology of calcite-mineralized faults: Absolute timing of rift-related fault events on the northeast Atlantic margin. *Geology*, 44, 531–534.

Rolland, Y., Cox, C.F., Corsini, M. (2009). Constraining deformation stages in brittle–ductile shear zones from combined field mapping and 40Ar/39Ar dating: The structural evolution of the Grimsel Pass area (Aar Massif, Swiss Alps). *Journal of Structural Geology*, 31, 1377–1394.

Rolland, Y., Lardeaux, J.M., Jolivet, L. (2012). Deciphering orogenic evolution. *J. Geodyn.*, 56, 1–6.

Rolland, Y., Petit, C., Saillard, M., Braucher, R., Bourlès, D., Darnault, R., Cassol, D., ASTER Team (2017). Inner gorges incision history: A proxy for deglaciation? Insights from Cosmic Ray Exposure dating (10Be and 36Cl) of river-polished surfaces (Tinée River, SW Alps, France). *Earth Planet. Sci. Lett.*, 457, 271–281.

Rosenberg, C.L., Berger, A., Bellahsen, N., Bousquet, R. (2015). Relating orogen width to shortening, erosion, and exhumation during Alpine collision. *Tectonics*, 34, 1306–1328.

Rothé, J.P. (1941) Les séismes des Alpes françaises en 1938 et la sismicité des Alpes occidentales. *Ann. Inst. Phys. Globe Strasbourg*, 3, 1–105.

Royden, L.H. (1993). The tectonic expression slab pull at continental convergent boundaries. *Tectonics*, 12, 303–325.

Schaer, J.P. and Jeanrichard, F. (1974). Mouvements verticaux anciens et actuels dans les Alpes suisses. *Eclogae Geologicae Helvetiae*, 67(1), 101–119.

Schlunegger, F. and Hinderer, M. (2001). Crustal uplift in the Alps: Why the drainage pattern matters. *Terra Nova*, 13(6), 425–432.

Schmid, S. and Froitzheim, N. (1993). Oblique slip and blockrotation along the Engadine line. *Eclogae Geologicae Helvetiae*, 86, 569–593.

Schmid, S.M. and Kissling, E. (2000). The arc of the western Alps in the light of geophysical data on deep crustal structure. *Tectonics*, 19, 62–85.

Schmid, S.M., Fugenshuh, B., Kissling, E., Schuster, R. (2004). Tectonic map and overall architecture of the Alpine orogen. *Eclogae Geologicae Helvetiae*, 97, 93–117.

Schmid, S.M., Kissling, E., Diehl, T., van Hinsber, D., Molli, G. (2017). Ivrea mantle wedge, arc of the Western Alps, and kinematic evolution of the Alps–Apennines orogenic system. *Swiss J. Geosci.*, 110(2), 581–612.

Scholtz C.H. (1990). *The Mechanics of Earthquakes and Faulting*. Cambridge University Press, Cambridge.

Schreiber, D., Lardeaux, J.M., Martelet, G., Courrioux, G., Guillen, A. (2010). 3-D modelling of Alpine mohos in southwestern Alps. *Geophys. J. Int.*, 180(3), 961–975.

Schwartz, S. (2002). La zone Piémontaise des Alpes Occidentales : un paléo-complexe de subduction. BRGM Thesis, Claude Bernard Lyon 1.

Selverstone, J. (1988). Evidence for east-west crustal extension in the Eastern Alps: Implications for the unroofing history of the Tauern Window. *Tectonics*, 7(1), 87–105. doi: 10.1029/TC007i001p00087.

Selverstone, J. (2004). Are the Alps collapsing? *Annu. Rev. Earth Planet. Sci.*, 33, 113–132

Serpelloni, E., Anzidei, M., Baldi, P., Casula, G., Galvani, A. (2005). Crustal velocity and strain-rate fields in Italy and surrounding regions: New results from the analysis of permanent and non-permanent GPS networks. *Geophysical Journal International*, 161(3), 861–880.

Serpelloni, E., Vannucci, G., Pondrelli, S., Argnani, A., Casula, G., Anzidei, M., Baldi, P., Gasperini, P. (2007). Kinematics of the western Africa-Eurasia plate boundary from focal mechanisms and GPS data. *Geophysical Journal International*, 169, 1180–1200.

Serpelloni, E., Faccenna, C., Spada, G., Dong, D., Williams, S.D. (2013). Vertical GPS ground motion rates in the Euro-Mediterranean region: New evidence of velocity gradients at different spatial scales along the Nubia-Eurasia plate boundary. *J. Geophys. Res.*, 118, 6003–6024.

Serpelloni, E., Pintori, F., Gualandi, A., Scoccimarro, E., Cavaliere, A., Anderlini, L., Belardinelli, M.E., Todesco, M. (2018). Hydrologically induced karst deformation: Insights from GPS measurements in the Adria-Eurasia plate boundary zone. *J. Geophys. Res.*, 123, 4413–4430.

Small, E.E. and Anderson, R.S. (1998). Pleistocene relief production in Laramide mountain ranges, western United States. *Geology*, 26(2), 123–126.

Spada, M., Bianchi, I., Kissling, E., Agostinetti, N.P., Wiemer, S. (2013). Combining controlled-source seismology and receiver function information to derive 3-D Moho topography for Italy. *Geophys. J. Int.*, 194(2), 1050–1068.

Sperner, B., Ratschbacher, L., Ott, R. (1993). Fault-striae analysis: A Turbo pascal program for graphical presentation and reduced stress tensor calculation. *Computers & Geosciences*, 19, 1361–1388.

Stampfli, G.M., Mosar, J., Marquer, D., Marchant, R., Baudin, T., Borel, G. (1998). Subduction and obduction processes in the Swiss Alps. *Tectonophysics*, 296, 159–204.

Steck, A. and Hunziker, J. (1994). The tertiary structure and thermal evolution of the Central Alps – Compressional and extensional structures in an orogenic belt. *Tectonophysics*, 238, 229–254 .

Sternai, P., Herman, F., Champagnac, J.-D., Fox, M., Salcher, B., Willett, S.D. (2012). Preglacial topography of the European Alps. *Geology*, 40(12), 1067–1070.

Sternai, P., Herman, F., Valla, P.G., Champagnac, J.-D. (2013). Spatial and temporal variations of glacial erosion in the Rhône valley (Swiss Alps): Insights from numerical modeling. *Earth Planet. Sci. Lett.*, 368, 119–131.

Sternai, P., Sue, C., Husson, L., Serpelloni, E., Becker, T.W., Willett, S.D., Faccenna, C., Di Giulio, A., Spada, G., Jolivet, L. et al. (2019). Present-day uplift of the European Alps: Evaluating mechanisms and models of their relative contributions. *Earth-Sci. Rev.*, 190, 589–604.

Stocchi, P., Spada, G., Cianetti, S. (2005). Isostatic rebound following the Alpine deglaciation: Impact on the sea level variations and vertical movements in the Mediterranean region. *Geophys. J. Int.*, 162(1), 137–147.

Sue, C. (1998). Dynamique actuelle et récente des Alpes occidentales internes – Approche structurale et sismologique. PhD Dissertation, Université Joseph Fourier, Grenoble.

Sue, C. and Tricart, P. (1999). Late-Alpine brittle extension above the frontal pennine thrust near Brianc ̧on western Alps. *Eclogae Geologicae Helveticae*, 92, 171–181.

Sue, C. and Tricart, P. (2002). Widespread post-nappe normal faulting in the Internal western Alps: A new constraint on arc dynamic. *Journal of the Geological Society of London*, 159, 61–70.

Sue, C. and Tricart, P. (2003). Neogene to ongoing normal faulting in the inner western Alps: A major evolution of the alpine tectonics. *Tectonics*, 22, 1–25.

Sue, C. and Schmid, S.M. (2017). From ocean formation to mountain evolution in Alpine-type orogens. *Swiss Journal of Geosciences*, 110(2), 417–418.

Sue, C., Thouvenot, F., Fréchet, J., Tricart, P. (1999). Widespread extension in the core of the western Alps revealed by earthquake analysis. *J. Geophys. Res.*, 104(B11), 25611–25622.

Sue, C., Martinod, J., Tricart, P., Thouvenot, F., Gamond, J.F., Frechet, J., Marinier, D., Glot, J.P., Grasso, J.R. (2000). Active deformation in the inner Western Alps inferred from comparison between 1972-classical and 1996-GPS geodetic surveys. *Tectonophysics*, 320, 17–29.

Sue, C., Grasso, J.R., Lahaie, F., Amitrano, D. (2002). Mechanical behavior of western alpine structures inferred from statistical analysis of seismicity. *Geophys. Res. Lett.*, 29, 65–69.

Sue, C., Delacou, B., Champagnac, J.D., Allanic, C., Tricart, P., Burkhard, M. (2007a). Extensional neotectonics around the bend of the Western/Central Alps: An overview. *Int. J. Earth Sci.*, 96, 1101–1129.

Sue, C., Delacou, B., Champagnac, J.D., Allanic, C., Burkhard, M. (2007b). Aseismic deformation in the Alps: GPS vs. seismic strain quantification. *Terra Nova*, 19(3), 182–188.

Tapponnier, P. (1977). Evolution tectonique du système alpin en Méditerranée : poinçonnement et écrasement rigide plastique. *Bulletin de la Société Géologique de France*, 7, 437–460.

Termier, P. (1903). Les nappes des Alpes orientales et la synthèse des Alpes. *Bulletin de la Société Géologique de France*, 3, 711–766.

Thouvenot, F., Fréchet, J., Jenatton, L., Gamond, J.F. (2003). The Belledonne Border Fault: Identification of active seismic strike-slip fault in the Western Alps. *Geophys. J. Int.*, 155, 174–192.

Tricart, P. (1984). From passive margin to continental collision: A tectonic scenario for the Western Alps. *American Journal of Science*, 284, 97–120.

Tricart, P. and Sue, C. (2006). Faulted backfold versus reactivated backthrust: The role of inherited structures during late extension in the frontal Piémont nappes east of Pelvoux (Western Alps). *Int. J. Earth Sci. (Geol. Rundsch.)*, 95, 827–840.

Tricart, P., Lardeaux, J.M., Schwartz, S., Sue, C. (2006). The late extension in the inner western Alps: A synthesis along the south-Pelvoux transect. *Bull. Soc. Geol. Fr.*, 177, 299–310.

Turcotte, D.L. and Schubert, G. (2002). *Geodynamics*. Cambridge University Press, Cambridge.

Turner, F.J. (1953). Nature and dynamic interpretation of deformation lamellae in calcite of three marbles. *American Journal of Science*, 251, 276–298.

Twiss, R.J. and Unruh, J.R. (1998). Analysis of fault slip inversions: Do they constrain stress or strain rate? *Journal of Geophysical Research Solid Earth*, 103, 12205–12222.

Valla, P., van der Beek, P.A., Lague, D. (2010). Fluvial incision into bedrock: Insights from morphometric analysis and numerical modeling of gorges incising glacial hanging valleys (Western Alps, France). *J. Geophys. Res.*, 115.

Van der Pluijm, B.A., Hall, C.M., Vrolijk, P.J., Pevear, D.R., Covey, M.C. (2001). The dating of shallow faults in the Earth's crust. *Nature*, 412, 172–175.

Vernant, P., Hivert, F., Chery, J., Steer, P., Cattin, R., Rigo, A. (2013). Erosion-induced isostatic rebound triggers extension in low convergent mountain ranges. *Geology*, 41, 467–470.

Vigny, C., Chery, J., Duquesnoy, T., Jouanne, F., Ammann, J., Anzidei, M., Avouac, J.P., Barlier, F., Bayer, R., Briole, P. et al. (2002). GPS network monitors the Western Alps' deformation over a five-year period: 1993-1998. *J. Geodesy.*, 76, 63–76.

Virlouvet, B., Tricart, P., Villemin, T. (1996). Blocs basculés tardi-alpins dans les nappes briançonnaises de Haute-Durance (Alpes occidentales, France) et évolution néotectonique des zones alpines internes. *C. R. Acad. Sci. Paris*, 322, 475–481.

Vouillamoz, N., Sue, C., Champagnac, J., Calcagno, P. (2012). 3D cartography modeling of the Alpine Arc. *Tectonophysics*, 579, 131–143.

Wagner, M., Kissling, E., Husen, S. (2012). Combining controlled-source seismology and local earthquake tomography to derive a 3-D crustal model of the western Alpine region. *Geophys. J. Int.*, 191(2), 789–802.

Waldhauser, F., Lippitsch, R., Kissling, E., Ansorge, J. (2002). High-resolution teleseismic tomography of upper-mantle structure using an a priori three-dimensional crustal model. *Geophys. J. Int.*, 150(2), 403–414.

Wallace, R.E. (1951). Geometry of shearing stress and relation to faulting. *The Journal of Geology*, 59, 118–130.

Walpersdorf, A., Sue, C., Baize, S., Cotte, N., Bascou, P., Beauval, C., Collard, P., Daniel, G., Dyer, H., Grasso, J-R. et al. (2015). Coherence between geodetic and seismic deformation in a context of slow tectonic activity (SW Alps, France). *J. Geodyn.*, 85, 58–65.

Watts, A.B. (2009). *Isostasy and Flexure of the Lithosphere*. Cambridge University Press, Cambridge.

Wittmann, H., von Blanckenburg, F., Kruesmann, T., Norton, K., Kubik, P. (2007). Relation between rock uplift and denudation from cosmogenic nuclides in river sediment in the Central Alps of Switzerland. *J. Geophys. Res.*, 112(F4).

Yamada, R., Tagami, T., Nishimura, S., Ito, H. (1995). Annealing kinematics of fission tracks in zircon: An experimental study. *Chemical Geology (Isotope Geoscience Section)*, 122, 249–258.

Yamaji, A. (2000). The multiple inverse method: A new technique to separate stresses from heterogeneous fault-slip data. *J. Structural Geology*, 22, 441–452.

Zhao, L., Paul, A., Malusà, M.G., Xu, X., Zheng, T., Solarino, S., Guillot, S., Schwartz, S., Dumont, T., Salimbeni, S., Aubert, C. (2016). Continuity of the Alpine slab unraveled by high-resolution P wave tomography. *J. Geophys. Res.*, 121(12), 8720–8737.

Zwingmann, H. and Mancktelow, N. (2004). Timing of Alpine fault gouge. *Earth and Planetary Science Letters*, 223, 415–425.

PART 3

Regional Alpine Studies from Extension to Collision

5

Tectonic Evolution of the Eastern Alps

Walter KURZ[1], **Harald FRITZ**[1], **Andreas WÖLFLER**[2]

[1] *Institute of Earth Sciences, University of Graz, NAWI Graz Geocenter, Austria*

[2] *Institute for Geology, Leibniz University Hannover, Germany*

5.1. Introduction

The Alps, and the Eastern Alps in particular, expose nappe systems resulting from two orogenies. Formation of Austroalpine and Southalpine Nappe Systems is related to the closure of an embayment of the Meliata ocean, as part of the Neotethys, into the Adriatic plate from Late Jurassic times onwards, followed by Cretaceous continental collision. The formation of Penninic and Sub-Penninic Nappe Systems is related to the closure of the Penninic (Piemont-Ligurian and Valais) ocean(s) between the Adriatic plate and Europe. The latter also affected the north(western) Adriatic continental margin (Lower Austroalpine Nappe System).

Large parts of the Eastern Alps comprise nappe systems that originate from the (south) eastern continental margin of the former Adriatic plate. The Paleo- and Eo-Alpine plate tectonic cycles, which formed the present Austroalpine nappe complex, may be subdivided into two distinct phases: (1) the (ES)E-ward subduction and closure of the Meliata oceanic basin, resulting in the assembly of nappe systems in the Northern Calcareous Alps (NCA), including its basement, that is, the Greywacke Zone. After closing of the Meliata oceanic basin during the Late Jurassic, the Cretaceous orogeny in the Eastern Alps encompasses the collision between (south) easternmost parts of the Adriatic plate, including the Austroalpine continental crust,

Geodynamics of the Alps 1,
coordinated by Claudio L. ROSENBERG and Nicolas BELLAHSEN. © ISTE Ltd 2023.

and a continental Tisza fragment to the east. (2) Southward intracontinental subduction within the Adriatic plate resulted in the Eo-Alpine imbrication of the Central Austroalpine nappe complexes. These units were affected by pronounced Eo-Alpine metamorphim, increasing from greenschist facies conditions in the northern parts to amphibolite and eclogite facies conditions in the southernmost parts. Continuous underplating was accompanied by extension in the internal parts of the orogen, and exhumation of high-pressure metamorphic rocks during the Late Cretaceous. Coeval extension in the upper plate resulted in the formation of the Gosau sedimentary basins.

The southward subduction of the Penninic ocean, and the subsequent collision of the European margin with the Adriatic Plate (Neo-Alpine cycle) resulted in nappe assembly and high-pressure metamorphism of parts of Penninic and Sub-Penninic nappes during Late Eocene to Early Oligocene times. Although accompanied by massive crustal thickening, this evolution within the Lower European Plate is hardly reflected in the upper plate.

The present structure of the Eastern Alps is characterized by a system of fault zones that developed during Late Oligocene to Miocene times, related to orogen-parallel escape of Austroalpine units towards east, a process also termed lateral extrusion. Intramontane basin formation and distinct signatures of faulting activity indicate orogen-parallel faulting either prior or subsequent to Sub-Penninic nappe assembly. During Oligocene to Middle Miocene times extrusion-related faulting continuously propagated from the western toward the central eastern part of the Eastern Alps. Sub-Penninic high-pressure assemblages of the Lower European Plate, exposed within the Tauern Window, were exhumed contemporaneously during oblique convergence between the Adriatic and European plates.

5.2. The Eastern Alps: Overview

The Alpine orogen was formed during the convergence of African and European plates, which more or less continuously converged since Late Jurassic times (e.g. Schmid et al. 2008). The geology of the Alpine–Mediterranean area is complex, however, because of the existence of more than one oceanic realm and several microplates between Africa and Europe, as well as the interplay between shortening processes and lateral movements (Schuster et al. 2013). This makes it challenging to determine the plate tectonic arrangement through time. Models of Alpine tectonics have developed rapidly during the last few decades, mainly as a result of modern structural, stratigraphic, petrological and geochronological investigations which, together with deep reflection seismic profiling and tomographic studies, have provided new insights into the present-day structures. However, contrasting interpretations on the evolution of the Alpine orogen still remain, further complicated by the use of different nomenclatures and tectonic subdivisions of nappe systems.

The Alpine orogen can be subdivided into tectonic units reflecting the Mesozoic to Paleogene paleogeography. From north to south, respectively, from bottom to the top, the Eastern Alps are formed by the following plate tectonic elements (Figures 5.1 and 5.2).

The European plate is represented by the Helvetic and Ultrahelvetic nappes and the Sub-Penninic unit. Relics of the former Penninic Oceans (Piemont-Ligurian and Valais Ocean) are represented by or included into the overlaying Lower and Upper Penninic nappes which form the Rhenodanubian Flysch belt and the Sub-Penninic nappes of the Engadine, Tauern and Rechnitz Windows. The Middle Penninic nappes are the remnant of the Iberian-Briançonnais microcontinent particularly exposed in the Engadine Window (Figure 5.1). In the western part of the Alps, this microcontinent separated the Penninic Ocean into the Piemont-Ligurian and Valais oceanic basins, both being part of the so-called Alpine Tethys (e.g. Stampfli 2000; Schmid et al. 2004, 2008). The term "Alpine Tethys", however, seems to be misleading in terms of Alpine tectonics and paleogeography, as the evolution of the related oceans (Penninic/Piemont/Ligurian/Valais) was never related to the Tethys, but to the Atlantic. Adria-derived fragments tectonically above the Penninic units are separated by the Periadriatic Lineament (PAL) into a northern Austroalpine and a southern Southalpine domain (Figures 5.1 and 5.3). Remnants of the Neotethys oceanic realm (Meliata, Vardar Ocean) occur in a few outcrops in the easternmost part of the Eastern Alps (Mandl and Ondrejickova 1993).

Opening of the Neotethyan Meliata-Maliac Ocean started in the Triassic at ~220 Ma. The Piemont-Ligurian and Valais Oceans are kinematically linked to the opening of the Atlantic Ocean. Opening of the Piemont-Ligurian Ocean resulted in the separation of the Adriatic Plate (including the future Southern Alps) from the European continent in Middle to early Late Jurrasic time at ~180 to 156 Ma. Simultaneously, the Triassic Meliata Ocean is being subducted south-eastward, in present-day orientation, below the upper plate Jurassic parts of the Neotethys. Later, the Briançonnais microcontinent, that is, the eastern tip of the Iberia block, was separated from Europe in conjunction with the opening of the Valais Ocean in Early Cretaceous times. In Santonian times (84 Ma), the two branches of the Alpine Tethys (Valais and Piemont-Ligurian Oceans) connected with younger, that is, Cretaceous-age branches of the Neotethys (see Figure 5.2 and Schmid et al. 2008).

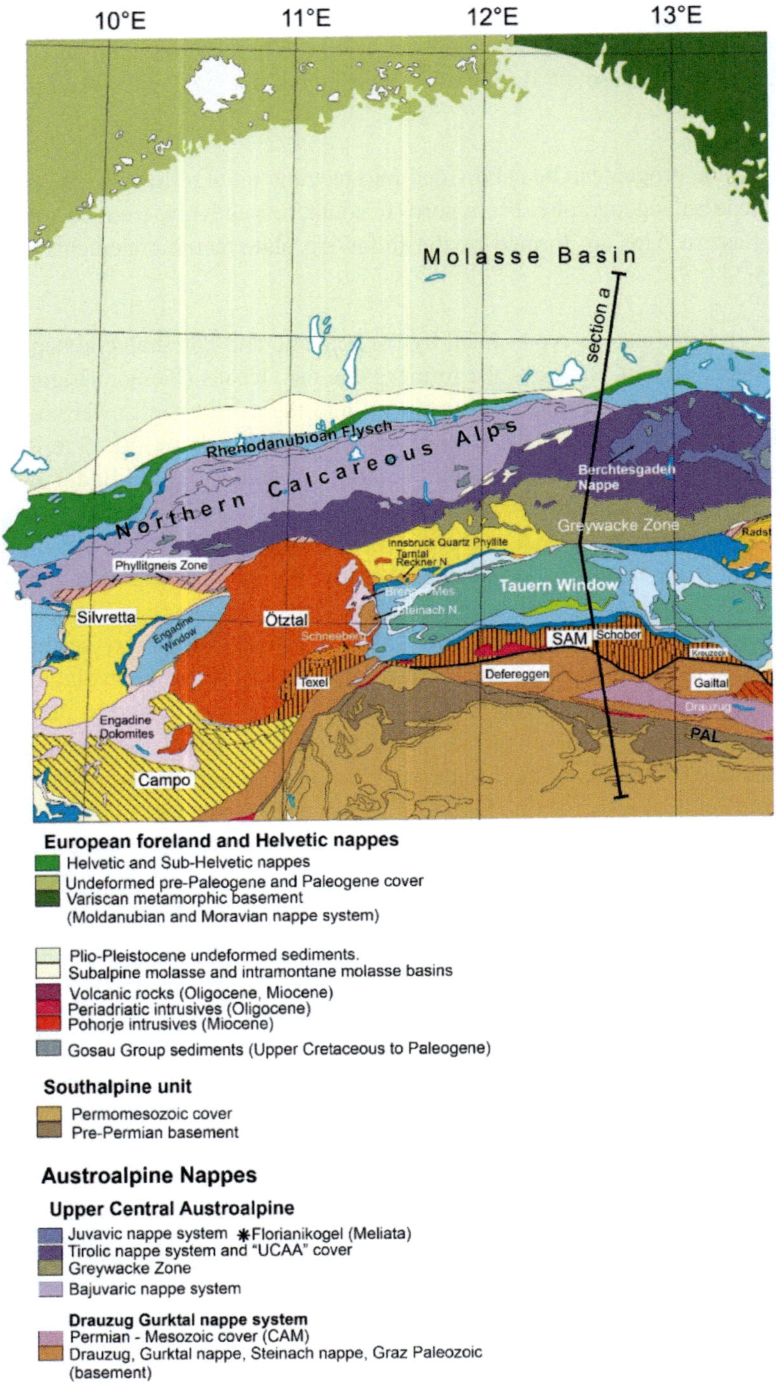
10°E
11°E
12°E
13°E
Molasse Basin
section a
Rhenodanubioan Flysch
Northern Calcareous Alps
Berchtesgaden Nappe
Greywacke Zone
Innsbruck Quartz Phyllite
Tarntal
Reckner N
Phyllitgneis Zone
Tauern Window
Silvretta
Engadine Window
Ötztal
Schneeberg
Steinach N.
SAM
Schober
Kreuzeck
Defereggen
Texel
Gailtal
Drauzug
PAL
Engadine Dolomites
Campo
European foreland and Helvetic nappes
Helvetic and Sub-Helvetic nappes
Undeformed pre-Paleogene and Paleogene cover
Variscan metamorphic basement (Moldanubian and Moravian nappe system)
Plio-Pleistocene undeformed sediments.
Subalpine molasse and intramontane molasse basins
Volcanic rocks (Oligocene, Miocene)
Periadriatic intrusives (Oligocene)
Pohorje intrusives (Miocene)
Gosau Group sediments (Upper Cretaceous to Paleogene)
Southalpine unit
Permomesozoic cover
Pre-Permian basement
Austroalpine Nappes
Upper Central Austroalpine
Juvavic nappe system ✱Florianikogel (Meliata)
Tirolic nappe system and "UCAA" cover
Greywacke Zone
Bajuvaric nappe system
Drauzug Gurktal nappe system
Permian - Mesozoic cover (CAM)
Drauzug, Gurktal nappe, Steinach nappe, Graz Paleozoic (basement)

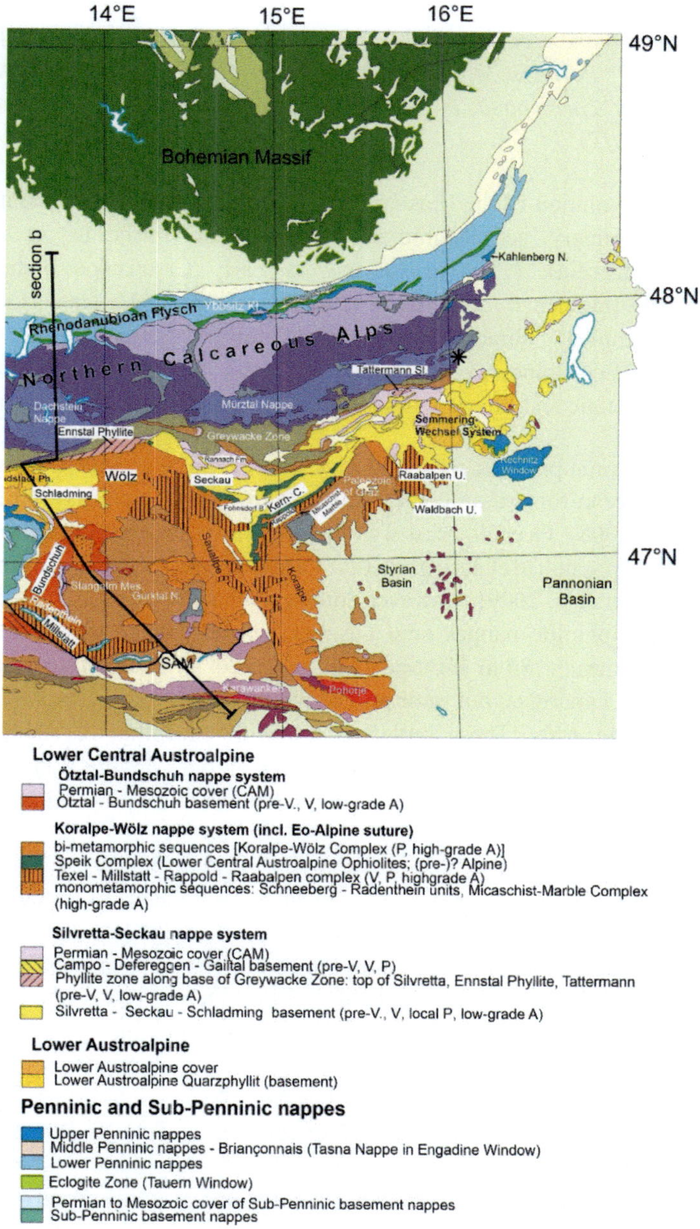

Figure 5.1. *Tectonic map of the Eastern Alps including the tectonic structure with nappe system subdivision and the metamorphic evolution of several units*

COMMENT ON FIGURE 5.1.– *SAM: Southern border of Alpine Metamorphism (according to Hoinkes et al. 1999); PAL: Periadriatic fault; V: Variscan Metamorphism; P: Permian Metamorphism; A: Eo-Alpine Metamorphism. Colors (legend) basically correspond to the map "Tectonic framework of the Alps" (Bousquet et al. 2012).*

The tectonic evolution of the (Eastern) Alps can be generally grouped into three phases (for a summary, see Schmid et al. 2008; van Hinsbergen et al. 2020; and references therein): (1) Late Jurassic to Early Cretaceous obduction of Meliata-derived fragments onto the Adriatic margin; (2) Late Cretaceous (Cenomanian/Turonian) intracontinental subduction of parts of the Adriatic microcontinent (Eo-Alpine event); and (3) latest Cretaceous to Paleogene subduction of Penninic and Sub-Penninic (European) crust (Neo-Alpine event).

The oldest tectonic phase is related to inversion of the Neotethys Ocean margin. The Neotethys Ocean formed an embayment into the Adriatic plate (see Figure 5.2 in Schmid et al. 2008). Oceanic crust was formed during the westward propagation of the ocean in Triassic (Meliata), Jurassic (Vardar Ocean) and also in Cretaceous times (Ustaszewski et al. 2009). However, intraoceanic subduction processes and the emplacement of ophiolite nappes onto the Adriatic margin are documented since Middle Jurassic times (~180 to 160 Ma) (e.g. Schmid et al. 2008). These nappes are widespread in the Dinarides, but outlayers of these nappes were also present in the eastern part of the Alps. These outlayers are with exception of small Meliata fragments at the Eastern margin of the Alps (Florianikogel, Figure 5.1) completely eroded, but redeposited material is present in the Upper Cretaceous sediments of the Gosau Group (Schuster et al. 2007).

In the Eastern Alps, the Eo-Alpine event was related to intracontinental subduction and collision within the Austroalpine unit (e.g. Kurz and Fritz 2003). Maximum burial depth and peak metamorphic conditions were reached in the early Late Cretaceous (Turonian; ca. 92 Ma) (Thöni 2006). During this event, parts of the Austroalpine unit were in a lower plate position, whereas other parts were in an upper plate position.

The Penninic Oceans, recently referred to as "Alpine Tethys" (Stampfli 2000; Schmid et al. 2004), were closed during the (Neo-)Alpine phase – from the Late Cretaceous onwards, followed by a Paleogene collisional event. Europe acted as the lower, and the Adriatic plate (Austroalpine and Southalpine unit) as the tectonic upper plate.

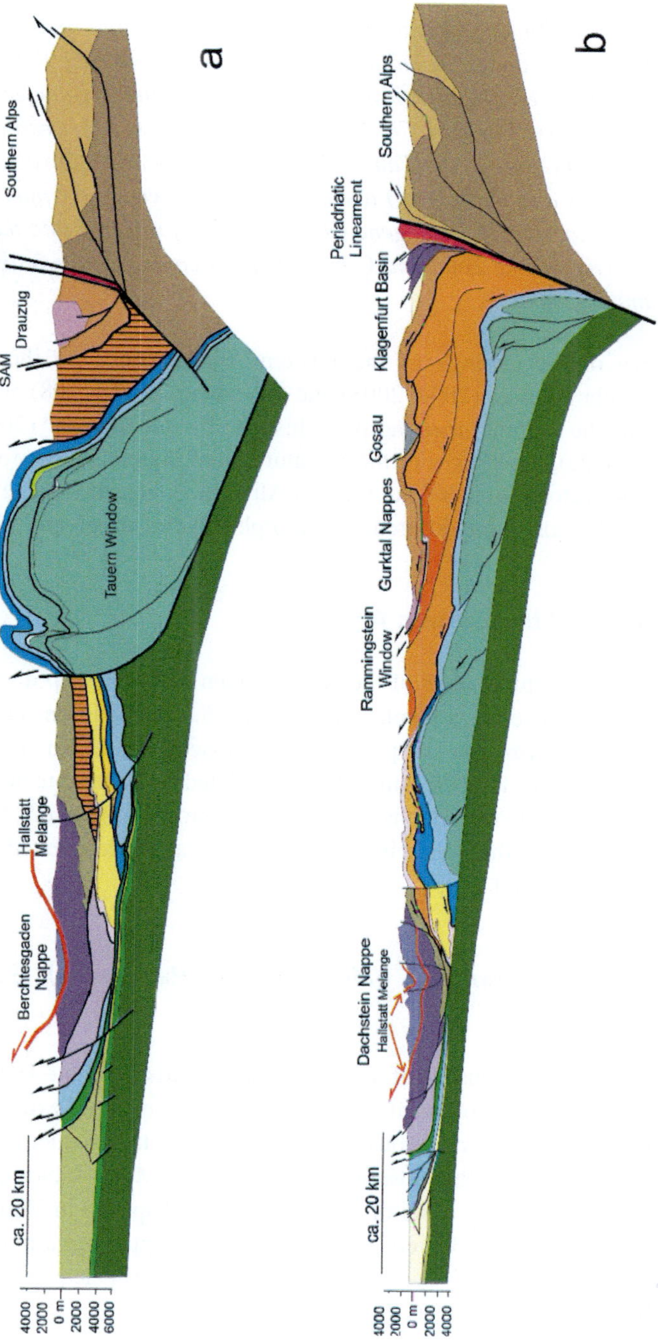

Figure 5.2. *Schematic geological cross-sections expanded from maps and profiles provided by the Austrian Geological Survey (https://www.geologie.ac.at/)*

COMMENT ON FIGURE 5.2.– *For profile location and legend, see Figure 5.1(a): Profile across central Tauern Window. The red line and half-arrow indicate emplacement of Juvavic nappes including Hallstatt Mélange during Paleo-Alpine (Late Jurassic) tectonics. Black half-arrows display motion during younger tectonic activities. SAM: Southern Limit of Eo-Alpine Metamorphism. (b) Profile across Austroalpine units to the east of the Tauern Window. Paleo-Alpine Juvavic nappes at Dachstein (red lines and half-arrow) are divided into Lower and Upper Juvavic nappes. The Late Cretaceous to Paleogene Gosau Basin is projected into the profile line. Mention that Greywacke Zone and Gurktal nappes contain individual tectonic units that are summarized here.*

This summary of the geology of the Eastern Alps expands on the tectonic interpretation by Schmid et al. (2004, 2008) and Froitzheim et al. (2008) and on the recent review of the Alpine metamorphic history by Oberhänsli (2004) and Bousquet et al. (2012a), together with the literature cited therein. In the following, an overview of the tectonic units of the Eastern Alps, including their plate tectonic origin, and an introduction to the different metamorphic cycles are given.

5.3. Tectonic units of the Eastern Alps

The subdivision of tectonic units of the Alps is taken from the map displayed in Figure 5.1 and the cross-sections displayed in Figure 5.2. In addition, the map in Figure 5.1 includes information on pre-Alpine (Variscan), Permian and Alpine metamorphism for distinct tectonic units. Details related to several tectonic units will be discussed in separate paragraphs below. From bottom to top (from N to S, or NW to SE, respectively), the following tectonic units are incorporated into the Eastern Alps (Figures 5.1 and 5.2).

5.3.1. *Units derived from the (Mesozoic to Paleogene) European continent*

The European continent consists of a deeply eroded Variscan (Late Devonian to Carboniferous) metamorphic continental crust, rich in plutonic rocks (north of the Alpine front), covered by Carboniferous to Eocene sedimentary sequences. It dips beneath the Alps and contains the Late Eocene to Neogene Molasse basin, forming the northern peripheral foreland basin of the orogen (see Chapter 5 in Volume 3 by Ortner et al.) (Figure 5.1). The Helvetic and Ultrahelvetic nappes are a thin-skinned fold and thrust belt that formed exclusively of detached cover sequences. In the Eastern Alps, they are present only as thin slices.

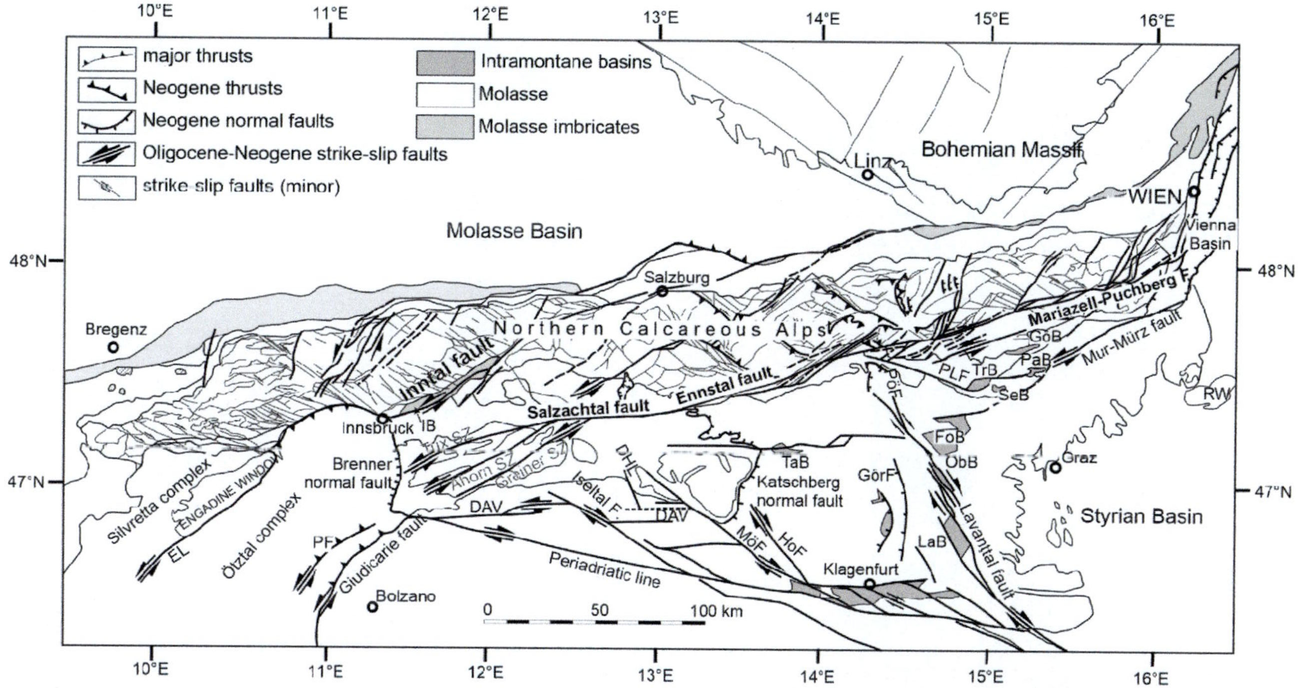

Figure 5.3. *Tectonic map of the Eastern Alps displaying the Paleogene to Neogene fault systems (after Linzer et al. 2002)*

COMMENT ON FIGURE 5.3.– *DAV: Defereggen-Anthols-Vals fault; DHL: Döllach-Heiligenblut line; EL: Engadine line; GörF: Görtschitztal Fault; HoF: Hochstuhl fault; MöF: Mölltal fault; PF: Peijo fault; PLF: Palten–Liesing fault; PöF: Pöls fault; RW: Rechnitz Window; IB: Inntal Basin; GöB: Göriach basin; PaB: Parschlug basin; SeB: Seegraben basin; FoB: Fohnsdorf basin; LaB: Lavanttal Basin; ObB: Obdach basin; TrB: Trofaiach Basin; TaB: Tamsweg Basin.*

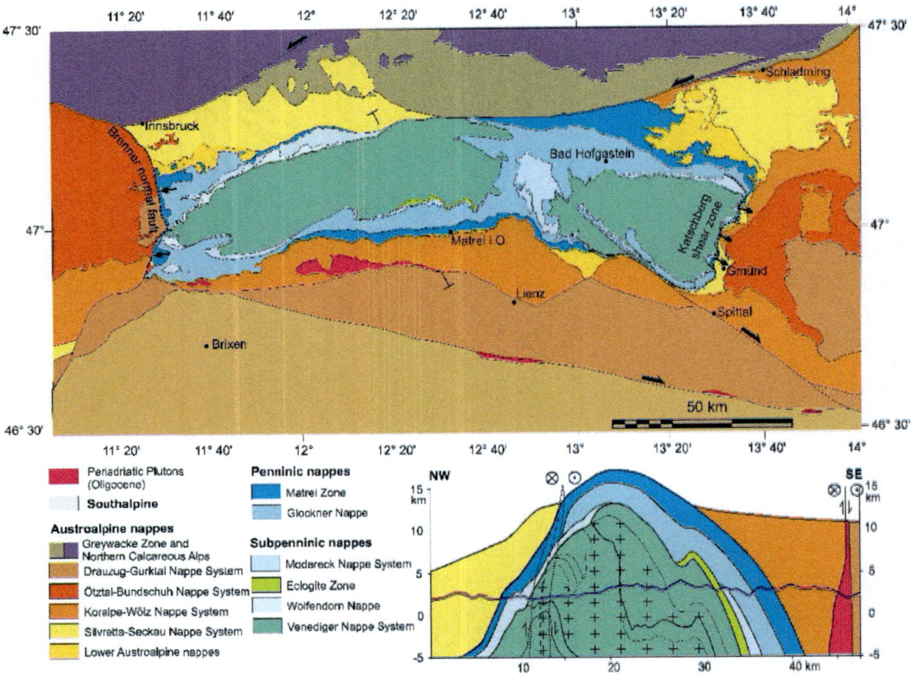

Figure 5.4. *Tectonic map and cross section of the Tauern Window and adjacent Austroalpine units (after Kurz et al. 1998; Schmid et al. 2013)*

The Sub-Penninic nappes (Figures 5.1 and 5.4) represent the distal European margin, forming ductilely deformed basement and cover nappes, which lost contact with their lithospheric mantle and served as the basement for the Helvetic nappes (Lammerer 1986; Schmid et al. 2013). They form, contrary to many earlier publications who consider the Sub-Penninic nappes as part of the Iberian-Briançonnais microcontinent, the Venediger Nappe System and the Modereck Nappe System (Figure 5.4). This interpretation is based on the conclusion that the crustal material of the Venediger Nappe System was not separated from the European margin by an oceanic basin (e.g. Froitzheim et al. 1996; Kurz et al. 2001), and on the nature of the cover sequence of the Venediger Nappe, that is inferred to be of Helvetic origin (Frisch 1975; Vesela et al. 2008, 2011).

The eclogitic Sub-Penninic units (Modereck Nappe System and Eclogite Zone of the Tauern Window) contain material derived from the southern European continental margin and transitional crust toward the Penninic Ocean and developed in a Paleogene subduction and accretion channel (Engi et al. 2001; Kurz and Froitzheim 2002; Kurz 2005, 2006; Schmid et al. 2013; Groß et al. 2020).

5.3.2. *Penninic nappes*

The Penninic nappes (Figures 5.1 and 5.4) comprise three paleogeographic elements: the Piemont-Ligurian Ocean, the Briançonnais microcontinent and the Valais Ocean (Schmid et al. 2004). The Penninic nappes can therefore be subdivided into the Upper, Middle and Lower Penninic nappes, whereby each unit consists mainly of one of the paleogeographic elements mentioned above. Here, we comment that this subdivision holds for the Engadine Window (Figure 5.1), where all three nappes are exposed. Further east, within the Tauern and Rechnitz Windows, this subdivision is questionable.

The Piemont-Ligurian Ocean opened in Late Jurassic times (Ratschbacher et al. 2004; Schmid et al. 2004). Rifting triggered exhumation of the subcontinental mantle of the Adriatic microplate (Froitzheim and Manatschal 1996). These mantle rocks are overlain by Jurassic radiolarites, aptychus limestones and Cretaceous calcareous turbiditic metasediments. The Briançonnais microcontinent was a part of the European distal margin until it was cut off by the opening of the Valais Ocean. The Valais oceanic crust comprises Cretaceous ophiolites overlain by Cretaceous to Eocene calcareous turbiditic metasediments: toward the east the Valais Ocean merged into the Piemont-Ligurian Ocean, thus forming a single oceanic basin east of the present-day Engadine Window (e.g. Stampfli 1994; Froitzheim et al. 1996; Kurz 2006; Schmid et al. 2013). Although this situation in the eastern part of the Eastern Alps makes any subdivision of the Piemont-Ligurian from the Valais basin somewhat artificial (Kurz et al. 2001), characteristic successions, analyses of the source areas of the clastic sedimentary successions and the age and chemical characteristics of the ophiolitic rocks allow the differentiation of elements from the northern or southern part of this joint oceanic basin.

The Lower Penninic nappes consist predominantly of material from the Valais oceanic province and from the northern parts of the joint oceanic basin. The Lower Penninic nappes make up large parts of the Central Alps and the central part of the Engadine Window. The lower nappes of the Rhenodanubian Flysch Zone, which are present along the northern margin of the Eastern Alps (Figure 5.1), represent a continuation of the Central Alpine Valais basin sediments into the Eastern Alps (Kurz et al. 1998); they comprise Cretaceous flyschoid sediments deposited in a basin along the European margin. The Glockner Nappe System of the Tauern Window, as well as the nappes of the Rechnitz Window Group, consisting of calcareous flyschoid metasediments and metaophiolites, is thought to be a southern continuation of the lower nappes of the Rhenodanubian Flysch Zone. Parts of the Glockner Nappe System contain high-pressure/low-temperature (HP/LT) mineral assemblages (e.g. Cornelius and Clar 1934; Dachs and Proyer 2001; Frank et al. 1987a; Groß et al. 2020, 2021), indicating that at least some of the oceanic units were subducted to great depth and exhumed during the Neo-Alpine orogeny. The term Neo-Alpine refers to the evolution related to subduction of the Penninic Ocean and collision of the southern European margin and Adria.

All units and nappe contacts within the Tauern Window, including the aforementioned HP assemblages in the Sub-Penninic nappes, were overprinted by a Barrow-type, high-temperature–medium pressure metamorphic event, the so-called “Tauernkristallisation” (Sander 1914; see also Droop 1985; Hoinkes et al. 1999; Dachs et al. 2005) in Late Oligocene time (Favaro et al. 2015; Cliff et al. 2015; Rosenberg et al. 2018; and references therein). This event varies in metamorphic grade from peak amphibolite facies conditions in the two basement subdomes at either end of the Tauern Window (Figure 5.4) to greenschist facies in the central Tauern Window (Oberhänsli et al. 2004; Bousquet et al. 2012a).

The Middle Penninic nappes are mainly derived from the Briançonnais microcontinent and are common in the Western and Central Alps. The easternmost nappes that include material from the Briançonnais microcontinent are represented by the Tasna nappe of the Engadine Window (Figure 5.1).

Rocks derived from the Piemont-Ligurian Ocean and the accretionary wedge along the southern margin of the oceanic basin toward the Adriatic microcontinent make up the Upper Penninic nappes. They are widespread in the Western and Central Alps and continue into the Apennines. In the Eastern Alps, the Upper Penninic nappes form the uppermost tectonic elements in the Engadine and Tauern Windows (e.g. the Arosa Zone within the Engadine Window and along the western margin of the Eastern Alps, the Matrei Zone and Reckner Complex within the Tauern Window). In the northern part of the Eastern Alps, the Ybbsitz Klippen Belt (Decker 1990) (Figure 5.1) is a remnant of the Piemont-Ligurian Ocean, containing the typical sequence of serpentinites, Jurassic radiolarites and aptychus limestones. It is in contact with the Kahlenberg nappe of the Rhenodanubian Flysch Zone, which is also interpreted to be an Upper Penninic nappe (Faupl and Wagreich 1992).

5.3.3. *Adriatic margin*

The Adriatic microcontinent consists of Cadomian continental crust (Neubauer 2002) with Paleozoic metasedimentary sequences and with magmatic rocks related to rifting and subduction processes lasting until the Carboniferous (for summary, see von Raumer and Neubauer 1993; von Raumer et al. 2002, 2013). During the Variscan orogeny, large parts of this crust were affected by metamorphic overprint and synorogenic magmatism. Post-orogenic Permian to Carboniferous sediments were deposited locally, and were incorporated into several Alpine nappe systems. Parts of the post-Variscan Molasse are preserved, for example, within the Drauzug-Gurktal Nappe System (Figures 5.1 and 5.2). In Permian times, these units were affected by lithospheric extension expressed by basaltic magmatic underplating, intense acidic magmatism and related high-temperature/low-pressure (HP/LT) metamorphism (Schuster and Stüwe 2008). More than 3 km of Permo-Mesozoic sediments were subsequently deposited on top of the thermally subsiding microcontinent, which formed a broad carbonate shelf toward the Meliata Ocean in the southeast. Parts of these sequences are preserved within the NCA and the Southern Alps. From the Jurassic onwards, Adria was bordered to the north, in present-day orientation, by a passive continental margin facing toward the Piemont-Ligurian Ocean. In the Eastern Alps, the Adriatic microcontinent is represented by the Austroalpine and Southalpine units, being separated by the Periadriatic lineament.

The Austroalpine units form a complex nappe stack of crustal material, which can be subdivided into several nappe systems. The subdivision and nappe system terms described in this paragraph basically follow Schmid et al. (2004). The subdivision and tectonic implications, however, are discussed below (see section 5.5).

The Lower Austroalpine unit derived from the Adriatic continental margin toward the Piemont-Ligurian Ocean and was affected by extension and nappe stacking during the opening and closing of this oceanic realm, respectively (Schmid et al. 2004). It overlies the Penninic nappes of the Eastern Alps. The Pre-Alpine basement of the Lower Austroalpine is characterized by greenschist-facies quartz phyllites from Early Paleozoic protoliths (e.g. the Radstadt Quartz Phyllite), and by Late Variscan Granitoids (e.g. "Grobgneis" of the Semmering Nappe System) (Figure 5.1), as well as higher grade metamorphic rocks intruded by these granitoids (Schuster et al. 2004). Permian cover sequences are associated in some areas with thick bimodal volcanic flow rocks. The Lower and Middle Triassic sequences are lithologically similar to the basal sequences in the NCA, but much thinner. The Upper Triassic sequences show a transitional Alpine (Hauptdolomit) to Germanic

(Keuper) facies. Testimonies of Jurassic extension tectonics are Jurassic sedimentary breccias, shed from submarine fault scarps. Sedimentary breccias are widespread within Lower and Middle Jurassic sequences, also containing basement clasts. The sediment succession reaches up into the early Late Cretaceous (flysch) (e.g. Froitzheim and Manatschal 1996). During closure of the Penninic Ocean, slivers of Lower Austroalpine rocks were mixed with oceanic sediments and fragments of oceanic crust within an accretionary wedge between the Penninic and the Austroalpine, for example, in the Matrei Zone of the Tauern Window and the Arosa Zone (Frisch et al. 1987; Ring 1992) (Figure 5.1). These sedimentary-tectonic mélanges are generally considered as Penninic. Austroalpine slivers in these zones may represent olistoliths that were gliding from the continental slope into the trench, or emplaced as extensional allochthons that formed during the breakup of the Penninic ocean (e.g. Froitzheim and Manatschal 1996).

In the Eastern Alps, the Lower Austroalpine mainly occurs in four complexes (Figure 5.1): (1) the Semmering and Wechsel nappes; (2) the Radstadt nappes along the northeastern margin of the Tauern Window; (3) the Tarntal Mesozoic along the northwestern margin of the Tauern Window, and (4) the western margin of the Eastern Alps with the Err and Bernina nappes.

The Austroalpine nappes situated above the Lower Austroalpine represent an Eo-Alpine nappe pile. Its lowermost unit, the Silvretta-Seckau Nappe System (Figures 5.1 and 5.2), consists of a Pre-Alpine metamorphic and magmatic basement and remnants of Carboniferous to Triassic cover sequences. During the Eo-Alpine event, it was overprinted by sub-greenschist to greenschist facies conditions (Schuster et al. 2013; Pfingstl et al. 2015).

To the north, the Silvretta-Seckau Nappe System is overlain by the nappes of the Greywacke Zone (Figures 5.1 and 5.2), which consists of Paleozoic sedimentary sequences and basement units metamorphosed during Late Cretaceous, Eo-Alpine greenschist facies metamorphism, and by the Juvavic, Tirolic and Bajuvaric Nappe Systems (Figures 5.1 and 5.2). The latter form the NCA, comprising unmetamorphosed to lowermost greenschist facies metamorphic Permian to Mesozoic, partly Lower Cretaceous, sediments deposited on the shelf facing originally toward the Meliata Ocean, with the sequences of the Juvavic Nappe System representing the most distal shelf.

To the south, the Silvretta-Seckau Nappe System is overlain by the Koralpe-Wölz Nappe System (Figures 5.1 and 5.2). It consists of polymetamorphic nappes with a Permian to Triassic HT/LP and an Eo-Alpine LT/HP metamorphic overprint (Schuster et al. 2004). The Koralpe and Wölz Nappe Systems show internally entirely different lithologies. The Wölz Nappe System mainly comprises Eo-Alpine amphibolite facies metapelites and carbonates, whereas the Koralpe Nappe System includes Eo-Alpine eclogite facies metapelites, eclogites, and meta-granitoides.

The Ötztal-Bundschuh Nappe System (Figures 5.1 and 5.2) shows a similar lithological composition as the Silvretta-Seckau Nappe System, but is positioned on top of the Koralpe-Wölz Nappe System. The overlying Drauzug-Gurktal Nappe System is made up of a Variscan metamorphic basement, anchizonal to greenschist facies Paleozoic metasedimentary sequences and by unmetamorphosed Permian to Triassic sediments (Rantitsch and Russegger 2000). The base of the Drauzug nappe is formed by a normal fault (Hoke 1990), being part of a fault system known as "Southern Limit of Alpine Cretaceous Metamorphism" (SAM) (Hoinkes 1990) (Figures 5.1 and 5.2). Within the Ötztal-Bundschuh and Drauzug-Gurktal Nappe Systems the Eo-Alpine metamorphic grade decreases upwards from amphibolite facies at the base to diagenetic conditions at the top of the nappe pile (Schuster et al. 2013).

The Upper Cretaceous to Paleogene sediments of the Gosau Group (Figure 5.1) were deposited in syn- to postorogenic basins with respect to the Eo-Alpine orogenic event (Faupl and Wagreich 2000).

The Southalpine unit shows minor Alpine thrusting-related deformation. Deformation is mainly localized along its margins, with southward backthrusting in its northern part. Its major part is in contact with a subcontinental Adriatic lithosphere; this contact is visible at the surface in the Ivrea Zone in the Western Alps (e.g. Schmid et al. 1996). In the southeast, the Southalpine unit continues into the External Dinarides, being affected by Eocene (Dinaric) and Middle Miocene to recent backthrusting.

5.3.4. *Meliata unit*

Rare remnants of the Meliata oceanic basin, as part of the Neotethys, are preserved in the Eastern Alps (Figure 5.1) (Kozur and Mostler 1992). These include serpentinites, basic volcanic rocks and Triassic deep-water sediments, partly redeposited in Jurassic metasediments. These rocks can be correlated with those from the Meliata Zone in the Western Carpathians (Mandl 2000). The Meliata Zone occurs as tiny klippen within the eastern part of the Eastern Alps between the Tirolic and Juvavic Nappe Systems (Figure 5.1). They show a sub-greenschist facies metamorphic imprint but no indications of subduction-related HP/LT metamorphism. Material from the Meliata oceanic basin is also present as detritus in

Cretaceous sediments of Austroalpine units (Faupl and Wagreich 2000) and in the Haselgebirge, an evaporite tectonite at the base of the Juvavic Nappe System (Schorn et al. 2013).

5.3.5. *Eocene to Miocene magmatism*

Intrusions along the Periadriatic fault (Figures 5.1 and 5.2) comprise calkalkaline tonalities, granodiorites and granites, and minor alkaline basaltic dikes (see also Chapter 1 in Volume 3 by Berger et al.). Intrusions are closely associated with contemporaneous strike-slip movements along the Periadriatic fault. They are (Eocene to) Oligocene in age and were for a long time considered to be related to the breakoff of the Penninic oceanic lithosphere from the distal European margin (Davis and von Blanckenburg 1995). Recent studies, however, suggest the existence of a continuous slab, at least in the (eastern) Central Alps (Ji et al. 2019). Magmatism is therefore assumed to be related to slab steepening when subduction was still active. The Pohorje pluton (Figure 5.1) west of Maribor is not belonging to the Periadriatic intrusives sensu strictu. It is Miocene in age and related to the Pannonian magmatism in the course of the extensional tectonics, which leads to the development of the Pannonian Basin (Fodor et al. 2008). Locally upwelling asthenosphere in the junction of the Periadriatic and Pöls-Lavanttal fault systems (Figure 5.3), due to a slab window within the eastward subducting Adriatic Plate, could be the trigger for this intrusion; additionally, pluton emplacement may have been facilitated by the thinned lithosphere beneath the adjacent Styrian Basin (Wölfler et al. 2011).

Eocene to Miocene magmatism within the Alps is described in detail in another chapter of this book (see Chapter 1 in Volume 3 by Berger et al.).

5.4. Distribution and timing of metamorphism within the Eastern Alps

In the following paragraphs, a brief summary of the metamorphic events within the Eastern Alps, following Schuster et al. (2013), is provided (see also Figure 5.1). In general, four major metamorphic cycles can be recognized in the Eastern Alps since Paleozoic times.

5.4.1. *Variscan collisional event (Late Devonian to Carboniferous)*

The Variscan event is induced by the Late Devonian to Carboniferous collision of Gondwana, Laurussia and the intervening microplates (e.g. Avalonia, Galatia), during the formation of the Pangea supercontinent (e.g. von Raumer et al. 2002; Kroner et al. 2008; Mandl et al. 2018).

$^{40}Ar/^{39}Ar$ ages of detrital white mica of about 380 Ma are the oldest remnants of the Variscan event within the Alpine realm (Handler et al. 1997). A collision-related LT/HP imprint with formation of eclogites occurred prior to 350 Ma (Thöni 2006) and the thermal metamorphic peak was reached at about 340 Ma at medium pressure conditions (ca. 25°C/km). During exhumation the sillimanite stability field at a mean geothermal gradient of 35°C/km was crossed in the medium to high-grade metamorphic units (Tropper and Hoinkes 1996). Typical Variscan cooling ages are in the range of 310–290 Ma (Thöni 1999). Due to the reworking during the Alpine tectonometamorphic cycle not much is known about the tectonic style of the Variscan orogen within the present-day Alps. Top to the south-directed ductile shearing, in present-day orientation, was recognized in some of the pre-Alpine basement units and Variscan thrust tectonics within Palaeozoic sequences is proved within parts of the low-grade metamorphic, pre-Alpine units, for example, within the Gurktal Nappe System (Flügel and Neubauer 1984).

5.4.2. *Permian extensional event*

The Permian event in the Alps followed in the wake of the Variscan tectonic evolution. Lithospheric extension affected the northern part of the Adriatic microplate and caused magmatism and metamorphism in the Austroalpine, Southalpine and also in the Western Carpathian realm. The onset of the Permian HT/LP event may have initiated when crustal thickness was thinned below normal, and thus cannot be related to gravitational collapse of the Variscan orogen.

Evidence for active thinning in Permian time is the formation of graben systems filled by clastic sediments (e.g. the Collio Graben in the Southern Alps), intense magmatic activity, as the widespread Permian intrusions preserved in Austroalpine basement units and the Permian acidic volcanic rocks in the Southern Alps (e.g. the Bozen Quarzporphyr) and high temperature metamorphism (Schuster and Stüwe 2008). High-grade (sillimanite and andalusite grade) conditions are known particularly well from the Silvretta and Koralpe regions (Figure 5.1), and the Ivrea Zone (the only region in the Alps with Permian kinzingites), but exemplary evidence is given south of the Tauern Window (Figure 5.1). The metamorphic imprint reached a geothermal gradient of up to 45°C/km. Peak metamorphic conditions were reached at about 280–260 Ma and were accompanied by the formation of pegmatitic dikes in upper amphibolites facies metamorphic rocks. After that the continental crust was continuously subsiding and the lithosphere cooled down to the steady-state geotherm of ca. 25°C/km at about 200 Ma (Schuster and Stüwe 2008).

5.4.3. *Alpine subduction and collision (Late Jurassic to Neogene) (Eo-Alpine and Neo-Alpine event)*

Convergence through closure of the Meliata Ocean, resulting in first, mainly west-directed thrusts within marine carbonate sequences and synorogenic basin formation, initiated already in the Late Jurassic (e.g. Gawlick and Missoni 2019). However, early nappe stacking was of thin-skinned tectonic style and associated with minor metamorphic imprint. The convergence between the African, Adriatic and Eurasian Plate since Early Cretaceous time caused shortening in the Alpine realm. This led to the formation of the high-grade metamorphosed Alpine orogenic belt by "tectonic progradation" (Frisch 1979) from south (-east) to north (-west). This process started with Eo-Alpine intracontinental subduction within the Adriatic plate during Early Cretaceous time (Kurz and Fritz 2003; Janak et al. 2004), resulting in the formation of the Eo-Alpine eclogites, which are preserved within the Texel, Schober, Kreuzeck mountains, and within the Koralpe, Saualpe and the Pohorje (Figure 5.1). This was followed by the Late Cretaceous-Eocene subduction of the Piemont-Ligurian and Valais Ocean and prograding continental collisional events that continued until recent times (Neo-Alpine). High-pressure metamorphism, that affected the related continental margins, is associated with both collisional events. Detailed reviews of the metamorphic conditions and the timing of metamorphism in the individual units, together with maps showing the distribution of the metamorphic grade, are given by Frey et al. (1999), Oberhänsli et al. (2004) and Bousquet et al. (2012a) (Figure 5.1). Recent plate tectonic reconstructions and interpretations are provided by Handy et al. (2010), van Hinsbergen et al. (2020) and Schmid et al. (2020).

In Early Cretaceous time, a southeast-directed subduction zone developed which was at least in part situated within the continental lithosphere of the Adriatic plate (e.g. Janak et al. 2006). Along this subduction zone, the Austroalpine nappes formed an orogenic wedge, which was continuously growing during the Cretaceous. During this tectonic event, which is referred to as the Eo-Alpine tectonometamorphic event, the northwesternmost part of the Adriatic plate, represented by the main part of the Austroalpine nappes (Lower Central Austroalpine *sensu* Janak et al. 2004), acted as the tectonic lower plate, whereas the main part of the Adriatic plate including some Austroalpine nappes (Upper Central Austroalpine *sensu*; Janak et al. 2004) and the Southalpine unit were part of the upper plate. Today, the structure of the Eo-Alpine suture is obscured by Cenozoic tectonics. Nappe stacking related to the initial Eo-Alpine subduction was N- to NW-directed (Kurz and Fritz 2003). Large parts of the sedimentary sequences of the Austroalpine unit were stripped off from their basement, which was buried and metamorphosed up to eclogite facies conditions (Koralpe-Wölz Nappe System). Geochronological data suggest peak metamorphic conditions of the HP metamorphic event at about 92 Ma (Thöni 2006) and a subsequent medium pressure overprint during exhumation of the Koralpe-Wölz high pressure rocks. This Eo-Alpine high-pressure belt mainly consists of pre-Mesozoic basement rocks, as well as meta-pelites and meta-carbonates of widely unknown precursor sedimentary deposition age. Recent U-Pb studies on detrital zircons from these meta-pelites have shown that, at least in some parts of the Koralpe-Wölz Nappe System, the youngest zircons show maximum ages in the range of 308–312 Ma (Frank et al. 2018, 2019), indicating a post-Variscan sedimentation age. Parts of the basement were already eclogitized during the Variscan orogeny (Hauke et al. 2019). In the eastern part of the Eastern Alps, exhumation of the deeply buried units occurred by N- or NW-directed thrusting and S- to SE-directed extensional tectonics, resulting in the formation of a metamorphic extrusion wedge. The latter shows a lower part with an inverted metamorphic field gradient and an upper part with an upright metamorphic field gradient. Cooling ages, mainly $^{40}Ar/^{39}Ar$ mineral ages, are in the range of 90 to 70 Ma (Dallmeyer et al. 1996; Hoinkes et al. 1999; Thöni 1999).

Subduction of the Penninic ocean initiated at late Early Cretaceous to Late Cretaceous times (e.g. Ratschbacher et al. 2004; Schmid et al. 2004, 2008). After its closure, the European continent entered the subduction zone in the Paleogene at about 55 Ma (e.g. Kurz et al. 2008; Schmid et al. 2013). These processes are here referred to as the Neo-Alpine event, which is the main event in the Western Alps. It resulted in the formation of the Penninic and Sub-Penninic nappes. In the Eastern Alps, the Neo-Alpine event caused a HP/LT metamorphism in the Penninic and Sub-Penninic nappes. The ages proposed for HP metamorphism within the Sub-Penninic nappes of the Tauern Window are between 45 and 33 to 32 Ma (Ratschbacher et al. 2004; Glodny et al. 2005, 2008; Kurz et al. 2008; Nagel et al. 2013) which were followed within a very short time interval by medium pressure thermal overprint at about 30 Ma ("Tauernkristallisation") (Sander 1914; see also Höck and Miller 1980; Cliff et al. 1985; Droop 1985; Dachs et al. 2005; Glodny et al. 2008; Groß et al. 2020, 2021). Typical cooling ages ($^{40}Ar/^{39}Ar$ and Rb/Sr mineral ages, fission track ages) post-dating the "Tauernkristallisation" are in the range of 25–10 Ma (Luth and Willingshofer 2008; Bertrand et al. 2017). In the Tauern Window, exhumation of these units is related to north-south shortening, crustal thickening, resulting erosion as well as E–W extension since Miocene time (see discussion and references in Rosenberg et al. 2018). Adjacent to the Tauern Window, the Austroalpine unit shows a tectonic and very low to low-grade Neo-Alpine metamorphic overprint (Hoinkes et al. 1999).

Subsequent exhumation and cooling of the underlying Penninic and Sub-Penninic nappes within windows (Selverstone 1988; Behrmann 1988; Genser and Neubauer 1989; Fügenschuh et al. 1997) and lateral extrusion of the orogen in the Miocene (Ratschbacher et al. 1989) was accompanied by a system of normal and strike-slip faults and related intra-orogenic Molasse basins, as, for example, the Styrian, Fohnsdorf and Klagenfurt Basins (Figure 5.3). These Miocene structures and related tectonics are described and discussed in a separate paragraph in the following.

5.5. Austroalpine paleogeography and tectonics

Arising from studies in the 1960s, the classical subdivision into Lower Austroalpine (LAA), Middle Austroalpine (MAA) and Upper Austroalpine (UAA) nappes was established by Tollmann (1959, 1961, 1963, 1973, 1977). The subdivision is based on the distribution of Permian to Triassic sedimentary facies elements now allocated in different positions within the Austroalpine nappe pile. Triassic facies distribution with the variation from open marine pelagic sediments (Hallstatt Facies), over shallow marine to lagoonal facies seamed by reefs (Dachstein and Hauptdolomit Facies), to siliciclastic/carbonatic shelf sediments (Central Austroalpine Facies) has been interpreted to reflect deposition on the south-eastward sloping margin of the Adriatic plate toward the Meliata Ocean (Mandl 2000). During Alpine thrusting, the Permian and Mesozoic sediments were incorporated into the Alpine nappe pile with the Hallstatt Facies occupying the uppermost tectonic position (Juvavic nappes), the Dachstein and Hauptdolomit Facies the intermediate tectonic positions (Tirolic and Bajuvaric nappes) and near continental facies (Hauptdolomit and Keuper Facies), the so-called Central Austroalpine Mesozoics (CAM), in lowermost position (Figures 5.1, 5.2 and 5.5). Juvavic nappes, characterized by reddisch pelagic limestones and salt deposits (Schorn and Neubauer 2014; Leitner and Spötl 2017; Granado et al. 2018), are rootless thin-skinned nappes sheared off from the distal Adriatic margin. Tirolic nappes of the NCA, or Noric Tirolic nappe if basement/cover contacts are preserved, contain Permian clastics and Upper Triassic formations of the Dachsteinkalk facies resting on Paleozoic basement (Noric nappe as part of the Greywacke Zone) (Neubauer et al. 1994). The Bajuvaric Nappe System (Figures 5.1 and 5.5) comprises Permian to Mesozoic sediment sequences including Upper Triassic Hauptdolomite facies platform carbonates (Tollmann 1976a, 1976b) that rest, in Tollmann's model, on Paleozoic phyllites (Landeck Quartzphyllite). All these nappe systems were summarized as Upper Austroalpine nappes.

The Middle Austroalpine cover sediments (Central Alpine Mesozoic cover: CAM) (Figures 5.1 and 5.5) comprise Upper Permian to Triassic sedimentary sequences with basal clastics (often referred to as Alpine Verrucano) and

Hauptdolomite Facies carbonates being typical for the Upper Triassic evolution. They were deposited onto a highly diverse crytalline basement. The Lower Austroalpine occupies the deepest structural position and was, in Triassic paleogeographic reconstructions, located farthest from the Maliata margin (Mandl 2000), close to the future Penninic ocean. Permian/Triassic cover sediments comprise CAM facies (Alpine Verrucano and Keuper facies) when resting on crystalline basement of the Semmering-Wechsel nappes system (Figure 5.1). Occurrences structurally above the Penninic units of the Tauern Window, that is, Radstadt Nappe System and slices to the south and north of the Tauern Window, contain additionally thick Jurassic carbonate sequences deposited on the subsiding Adriatic continental margin adjacent to the Penninic ocean.

Although the structural and paleogeographic position of the Lower Austroalpine is not debated, the subdivision of Austroalpine nappes into Upper and Middle Austroalpine was never entirely accepted (e.g. Frank 1987), especially by geologists working in the western part of the Eastern Alps and the Central Alps. In particular, recognition of widespread Alpine high-grade metamorphism (e.g. Frank 1987; Hoinkes et al. 1999) torpedoed the simple picture of a Variscan (or older) crystalline basement ("Altkristallin") and a Permian to Mesozoic cover found in distinct nappes. Findings of Cretaceous high-temperature/high-pressure rocks, in particular ca. 90 Ma old eclogites (Thöni 2006), require at least the implementation of major Alpine tectonic elements within the "Altkristallin". In addition to this, the position of Bajuvaric units was challenged. Frank (1987) proposed that parts of the NCA (Bajuvaric Lechtal nappe) are in primary contact with the underlying Silvretta basement (Figures 5.1 and 5.2). According to this presumption, the Silvretta nappe and its Bajuvaric cover in the NCA would thus be part of the MAA in terms of the subdivision defined by Tollmann (1959, 1961). Here, we comment that the interpretation by Frank (1987) is seriously challenged and needs to be confirmed. Presumption of a primary sedimentary contact between Silvretta basement and its Bajuvaric cover would have dramatic implications for the Austroalpine tectonostratigraphy. These new findings and interpretations were integrated into the "Tectonic map and overall architecture of the Alpine orogen" by Schmid et al. (2004) and "Tectonic framework of the Alps" (Bousquet et al. 2012b). Both groups of authors present a subdivision of the Austroalpine nappes based on the structural position of tectonic units metamorphosed during the Eo-Alpine event at ca. 90 Ma. Lower Austroalpine nappes remain untouched occupying the deepest structural position. The former Middle Austroalpine is divided into nappe systems below (Silvretta-Seckau) and above (Öztal-Bundschuh) the high-pressure Koralpe-Wölz Nappe System. The Drauzug-Gurktal Nappe System occupies the highest tectonic position and the nappes of the NCA including the Greywacke Zone (Figures 5.1 and 5.5) are considered as "far travelled Austroalpine Nappes". Here, we comment that this subdivision is largely based on the distribution of units affected by Eo-Alpine high-grade metamorphism. By contrast, Tollmann's model is based on the distribution of Triassic facies elements, being emplaced and distributed within the Austroalpine nappe edifice from the Late Jurassic onwards (ca. 150 Ma: Gawlick and Missoni 2019). Both views cannot be compared and have their justification when specific items are addressed.

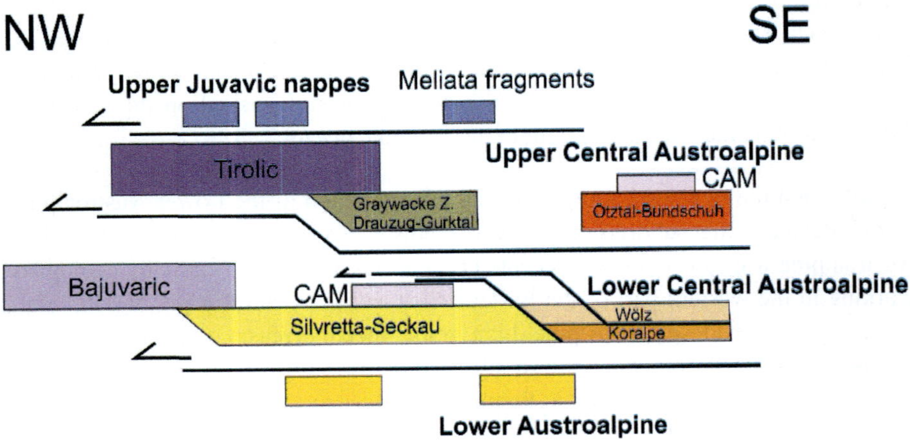

Figure 5.5. *Subdivision and correlation of Austroalpine nappes after Janák et al. (2004, 2006)*

Consequent expansion of the model of Schmid et al. (2004) by implementing an Eo-Alpine high-pressure wedge and presuming a sedimentary contact between the Silvretta basement and a Bajuvaric cover results in a tectonostratigraphic succession that is non-intuitive and gives rise to new questions. Sediments of the Bajuvaric Nappe System, and consequently also those exposed in the NCA, would occupy a relatively low tectonostratigraphic position within the Austroalpine nappe pile. The Ötztal Bundschuh Nappe System, including its Permian to Mesozoic cover, would occupy a tectonic position above the Tirolic and Juvavic nappes of the NCA. Since this situation is hard to retrace, we favor the model proposed by Janak et al. (2004, 2006) and Nagel (2006) as it is more descriptive and less interpretative (Figure 5.5). Accordingly, the Austroalpine is subdivided into a Lower Austroalpine, a Lower Central Austroalpine (LCAA) and an Upper Central Austroalpine (UCAA) unit (Figures 5.1, 5.2, and 5.5). The UCAA comprises Juvavic, Tirolic and Bajuvaric nappes of the NCA as well as the low-grade metamorphosed Paleozoic units (i.e. Drauzug-Gurktal Nappe System; Graz Paleozoic, Greywacke Zone). The LCAA comprises poly-metamorphic pre-Alpine basement units including their Permian to

Mesozoic cover that may be further subdivided into units described by Schmid et al. (2004), that is, Silvretta-Seckau, Koralpe-Wölz and Ötztal-Bundschuh systems.

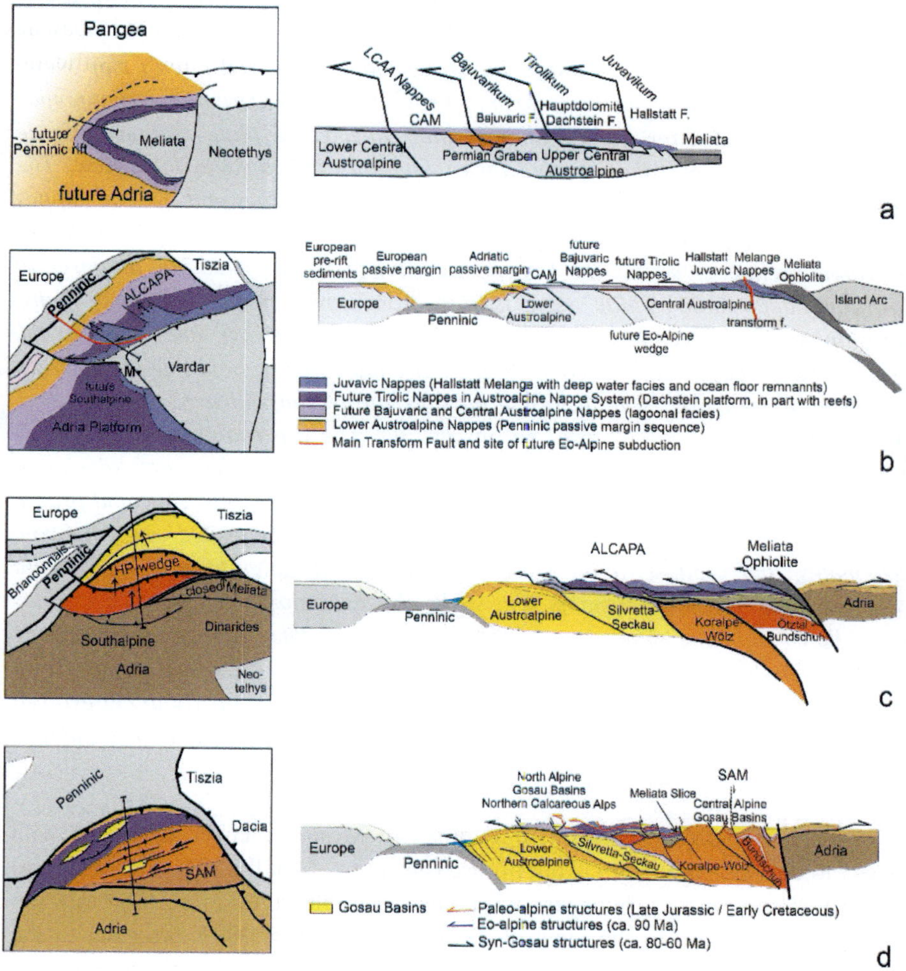

Figure 5.6. *Evolutionary steps of Austroalpine nappe tectonics*

COMMENT ON FIGURE 5.6.– *For colors, see, unless indicated, the legend in Figure 5.1. Paleogeographic reconstructions (left row) modified after Schmid et al. (2004, 2008) and Handy et al. (2010). Profile lines of sections (right row) are indicated. (a) Late Triassic (Norian, ca. 215 Ma) reconstruction showing distribution of sedimentary facies realms on the Meliata margin. CAM: Central Austroalpine Mesozoics. Expiring orange color in map view denotes terrigenous Keuper Facies.*

The thinned crust below the inferred Permian Graben is the site of the future Eo-Alpine subduction zone. Future nappes are given in italics. (b) Late Jurassic (ca. 150 Ma), Paleo-Alpine reconstruction. Adria is the site of opposing plate motions, the opening Penninic and the closing Meliata oceans. Large transfer systems crossing Adria are the site of future Eo-Alpine subduction (red line in map view) and divide Adria into a northern domain (ALCAPA: Alps-Carpathians-Pannonia) and a southern domain (future Southalpine). Interfering plate motions release a system of Jurassic strike-slip and thrust faults as indicated in map view. The cross-section shows Jurassic emplacement of Meliata ophiolites and Juvavic nappes including the Hallstatt Mélange. The position of future nappes and the future Eo-Alpine wedge is given. (c) Cenomanian (ca. 95 Ma) situation at onset of Eo-Alpine deformation. Map view is adit from cover nappes showing major Autroalpine basement domains. Top-north motion (arrows) and site of the high-pressure wedge (Koralpe–Wölz System) are indicated by bold line. Also shown is the site of the closed Meliata ocean passing into a trans-Adriatic transfer zone. (d) Late Cretaceous (ca. 75 Ma) reconstruction. The Austroalpine nappe pile is completed and extension related Gosau-type sediments are deposited (deviating from Figure 5.1, Gosau sediments are shown in yellow). The map view and profile shows schematically North Alpine Gosau sediments, deposited on Mesozoic carbonates, and Central Alpine Gosau sediments. The latter formed coeval with exhumation of basement domains is a strike-slip–extensional setting. The profiles show Paleo-Alpine (Jurassic to Early Cretaceous) thrusts in red, Eo-Alpine structures in blue and syn-Gosauic structures in black. Note that compressive structures at the base of the Austroalpine nappe pile (Lower Austroalpine) evolved coeval with extensional structures in upper tectonic levels. SAM: Southern Limit of Alpine Cretaceous Metamorphism.

The paleogeographic position of Permian to Mesozoic cover sediments with respect to the Meliata margin is out of discussion (e.g. Mandl 2000) (Figures 5.6(a) and (b)). Disputes arise, however, when their possible substratum, the basement units, are considered. In the view of Tollmann, the entire NCA originate to the south of the MAA/LCAA basement nappes (e.g. Silvretta-Seckau Nappe System), a hypothesis first proposed by Tollmann (1959) and also known as "Fernschubhypothese" – "far traveled nappe hypothesis" (e.g. Tollmann 1959, 1976b; Auer and Eisbacher 2003). If the Bajuvaric nappes, however, are part of the Silvretta-Seckau Nappe System (terminology of Schmid et al. 2004) or LCAA nappe complex (terminology of Janak et al. 2004), the sediment sequences of Bajuvaric Facies can be regarded as para-autochtohonous, whereas only Tirolic and Juvavic nappes are far-traveled. Furthermore, the Eo-Alpine eclogites would occupy an intra-continental suture located north (west) of the Meliata suture, that is, between the Silvretta-Seckau and Ötztal-Bundschuh nappes (Kurz and Fritz 2003; Janak et al.

2004). In this view, a two-phase evolution of the Austroalpine nappe complex is required, starting with the consumption of the Meliata ocean during Late Jurassic times (Dal Piaz et al. 1995; Stampfli and Borel 2002; Kurz and Fritz 2003; Frisch and Gawlick 2003; Cavazza et al. 2004) and subsequent continental collision during Early Cretaceous times (Kurz and Fritz 2003; Frisch and Gawlick 2003; Janak et al. 2004) (Figures 5.6(b) and (c)). The location of the intracontinental suture could be situated along a former Permian rift or transform system that evolved at high angle to the Meliata embayment. Remnants of this domain of elevated thermal gradient are preserved, for example, in Permian gabbros and peridodites, extending from the Koralpe in the east to the Ivrea Zone in the west, transformed to eclogites during Eo-Alpine high-pressure metamorphism (Thöni and Jagoutz 1993). This belt coincides roughly with the Permian high-temperature metamorphic belt including Permian granulites (Schuster and Stüwe 2008) (Figures 5.6(a) and (c)).

5.5.1. *Kinematics and timing of Austroalpine nappe stacking and extension*

The time span considered here expands from the onset of Austroalpine thrusting through to the consumption of the Meliata Ocean in the Middle/Late Jurassic (ca. 160 Ma, e.g. Gawlick and Missoni 2019; Putiš et al. 2019), over Turonian (ca. 90 Ma) subduction of continental fragments as exemplified by formation of eclogite embedded in units derived from continental crust (e.g. Thöni 1999) to Late Cretaceous/Paleocene exhumation of previously buried units, and extension as evident from the onset of Central Alpine Gosau Basin (Figures 5.1 and 5.6(d)) formation starting in the Santonian (ca. 80 Ma: e.g. Neubauer et al. 1995). Metamorphic imprint and structures evolving over this period of at least 80 Myr can be seen to result from a continuous process with temporally and spatially different evolution (Dallmeyer et al. 1998). Here, we subdivide this time span into distinct periods, described as Paleo-Alpine, Eo-Alpine and Post Eo-Alpine evolution steps, and justify this with different prevailing kinematics, tectonic style and metamorphic conditions. In short, the Paleo-Alpine evolution is characterized by thin-skinned tectonics (for details, see below). The Eo-Alpine evolution is of thick-skinned tectonic style associated with high-grade metamorphism. The Post Eo-Alpine period is characterized by extension and deposition of clastic deep-water sediments (Gosau) in upper crustal segments of the central Eastern Alps.

Basically three attempts exist to explain Eo-Alpine thrust geometries within the Eastern Alps. A relatively long-standing model of continuous nappe stacking interprets the Austroalpine nappe stack as a forward propagating thrust system active throughout the Cretaceous (approximately between 140 and 70 Ma). It is largely based on cooling age distributions, formation ages of syntectonically formed mica (Dallmeyer et al. 1998) and few datings which allow for better assumptions on thrust tectonics (Ratschbacher 1986; Fritz 1988). Geochronological data indicate that ages decrease from the uppermost tectonic units (formerly known as Upper Austroalpine sensu Tollmann: ca. 130–95 Ma) over middle tectonic units (former Middle Austroalpine: 90–80 Ma) to lower tectonic units (Lower Austroalpine: ca. 80–70 Ma). This has been interpreted as foreward and downward propagation of thrusting with thin-skinned nappe stacking in upper tectonic domains and progressive downward involvement of thick-skinned nappes. It has also been observed that Cretaceous nappe stacking previous to ca. 90 Ma took place at low-grade metamorphic conditions with top-to-the W (NW to SW) kinematics, whereas younger thrusts (<90 Ma) show high-grade metamorphisms and top-to-the N kinematics (e.g. Krohe 1987; Kurz and Fritz 2003) (Figure 5.6). This continuous nappe stacking model, however, explains the distribution of metamorphic assemblages only in parts. Existence of an Eo-Alpine high-pressure wedge bordered by lower grade Austroalpine units below and above cannot be explained by this attempt. Furthermore, no statements are made upon the site of Alpine suture zones, that is, the Meliata Ocean margin and/or intra-Adriatic subduction.

A second approach came up with the recognition of an Eo-Alpine high-pressure wedge (also described as Koralpe-Wölz System) (Schmid et al. 2004), best documented by ca. 90 Ma old eclogites within the Koralpe and Saualpe and the Schobergruppe (Thöni and Jagoutz 1992, 1993; Thöni and Miller 1996; Sölva et al. 2001; Thöni 2002; Janak et al. 2004, 2006; Hauke et al. 2019) (Figure 5.1). This model is based on metamorphic gradients with respect to the high-pressure wedge. An inverted field gradient occurs toward Austroalpine low-grade metamorphic units in the footwall (Silvretta-Seckau Nappe System) and a normal field gradient toward low-grade metamorphic units in the hanging-wall (Ötztal-Bundschuh Nappe System) of this high-pressure wedge. This distribution of Eo-Alpine metamorphism (Figure 5.1) is the key argument for the tectonic subdivision provided by Schmid et al. (2004). However, the pre-Eo-Alpine metamorphic and tectonic evolution is not well covered by that model.

A third model was shortly outlined by Kurz and Fritz (2003) and explains the formation of the Austroalpine nappe stack by a two-phase discontinuous evolution. A first phase being related to the closure of the Meliata Ocean (Paleo-Alpine event), resulted in Jurassic to Early Cretaceous thin-skinned tectonics as outlined above. Gabbro-bearing slices at the base of the Juvavic nappes and the Hallstatt salt mélange are remnants of the Meliata suture. Associated nappe kinematics were top-to-the W (e.g. Ratschbacher 1986; Fritz 1988; Ratschbacher and Neubauer 1989; Fritz et al. 1991; Linzer et al. 1995; Neubauer et al. 2000; Frisch and Gawlick 2003)

(Figure 5.6(a)). For the accreted Hallstatt mélange, the salt-bearing assemblages having plastic-viscous rheology at low temperatures play a major role. Evaporites acted as detachment horizons for W-directed thrusting before their propagation into strike-slip faults (Habermüller 2005; Schorn and Neubauer 2014).

5.5.1.1. *Paleo-Alpine evolution*

During the Permian, the lithosphere along the (south) eastern Adriatic margin was thinned along an embayment of the Neotethys ocean (usually described as Meliata), resulting in formation of a passive margin. From ca. 240 Ma onwards, the sag stage (Lein 1987; Mandl 2000; Stüwe and Schuster 2010) is well documented by the Triassic platform carbonates within the present-day NCA. Early Austroalpine nappe stacking (Paleo-Alpine) was triggered by the closing Meliata-Hallstatt ocean. In the course of thrusting, the Permian and Mesozoic sediments were incorporated into the Alpine nappe pile with distal, oceanward portions occupying the uppermost tectonic position and proximal, near continental portions in lower tectonic position. Thrusting evolved through foreward propagation of thin-skinned nappes when cover sediments (including occasionally remnants of ophiolites and pre-Alpine basement) where scrapped off from the formerly extended margin. This style is most evident in the nappe pile exposed in the NCA, where a low-grade metamorphic fold-and-thrust belt of cover nappes was assembled without significant involvement of basement units (Figure 5.6(a)). A complication arises from the fact that the uppermost tectonic units, the Juvavic nappes, show duplication with Upper and Lower Juvavic nappes separated by Permian evaporites (the so-called Haselgebirge). This can be explained by out-of-sequence thrusting or, that Upper Juvavic nappes derived from the opposite Meliata shelf (Southalpine/Dinaride shelf), as proposed by Schweigl and Neubauer (1997). These ductile thrusts represent the Early Cretaceous accretion stage post-dating the subduction and extrusion of Meliata rocks around 160 Ma (Spötl et al. 1996, 1998; Frisch and Gawlick 2003; Gawlick and Schlagintweit 2006; Schorn et al. 2013).

Simultaneously, the Penninic Ocean opened leaving the Adriatic plate in a domain of opposing plate motions (divergent motion in the Northwest and convergent motion in the Southeast) and transected by transform fault systems (Figures 5.6(a–c)). A complex structural interference pattern evolved through simultaneous activity of Jurassic–Early Cretaceous transform faults transecting entire Adria and thrust faulting related to convergence and closure of the Meliata Ocean. Here, we admit that Jurassic transform faults were never unambiguously identified, but manifold independent arguments suggest existence of such fault systems. Most arguments are built up upon correlation of Mesozoic Facies distribution and correlation of dispersed major tectonic units that suggest significant lateral movement (see below for details).

The best arguments for Jurassic to Early Cretaceous thrusting and emplacement of ophiolites arise from the former northern Neotethys branch (the Meliata margin exposed in the Western Carpathians of Slovakia) and the southern Neotethys or West-Vardar margin in the Dinarides. The closure of the Meliata Ocean evolved through intra-oceanic subduction between ca. 180 and 150 Ma (Schmid et al. 2008). Fragments from the continental margin and the oceanic margin were subducted to high-pressure conditions, incorporated into Jurassic flysch sediments and accreted onto the Adriatic margin (Gemeric Superunit in the Western Carpathians) at ca. 150–130 Ma (Putis et al. 2019). Blueschist formation is dated at ca. 160–140 Ma ($^{40}Ar/^{39}Ar$ plateau ages from phengite in blueschist assemblages: Faryad and Henjes-Kunst 1997; Dallmeyer et al. 2008). Ophiolite obduction onto the wider Adriatic shelf in the Dinaride branch of the Neotethys is well defined from stratigraphic data. The ocean started to close around the Early/Mid Jurassic times with the onset of an east-dipping intraoceanic subduction (Schmid et al. 2008; Gawlick and Missoni 2019). An intra-oceanic subduction is supported by the existence of a Middle Jurassic metamorphic sole in the Vardar Zone of the Dinarides (Dimo-Lahitte et al. 2001). Onset of ophiolite obduction occurred in the Middle Jurassic and mélange formation proceeded until the early Late Jurassic (Pamič et al. 1998; Tari and Pamič 1998; Gawlick and Missoni 2019).

Closure of the Meliata-Hallstatt ocean within Austroalpine nappes is likewise recorded from Jurassic facies distribution. Subsidence of the drowning carbonate platform culminated in deposition of Callovian to Oxfordian radiolarites within Tirolic units (Ruhpolding Group; ca. 160 Ma) that is succeeded by shallow marine facies (Plassen Formation). This scenario is interpreted to reflect uplift due to basin inversion (Gawlick et al. 1999; Frisch and Gawlick 2003; Gawlick and Missoni 2019). By the Late Bathonian to Oxfordian (165–155 Ma) propagating ophiolite obduction and imbrication of the former Neotethys passive margin resulted in formation of the Hallstatt Mélange (Frisch and Gawlick 2003; Schorn et al. 2013), now exposed at the base of the Juvavic nappes. Trench-like basins, collisional breccias and sedimentary mélanges formed in front of the propagating nappe stack (Gawlick et al. 1999; Frisch and Gawlick 2003). Within the Austroalpine only indirect evidence exists for the presence of a Middle Jurassic intraoceanic subduction as reported from the Vardar Zone of northern Greece (Michail et al. 2016). The siliciclastic Rossfeld Formation, an Early Cretaceous basin in front of the approaching Hallstatt imbricate zone with flyschoid facies, contains detrital material from an eroded island arc (Gawlick and Schlagintweit 2006; Krische and Gawlick 2015). Indirect evidence for presence of a metamorphic sole arises from eroded material deposited in post-collisional Gosau basins that contain pebbles of amphibolite that gave Sm/Nd garnet crystallization ages of ca. 163±11 Ma and Ar/Ar hornblende plateau ages of 189 ± 6 Ma and 174 ± 5 Ma (Gruber et al. 1992;

Schuster et al. 2007). Meliata equivalents are rare in the Eastern Alps but exposed in small outcrops of Middle Triassic (Anisian) deep-water sediments, mafic magmatic rocks and serpentinites at the eastern end of the Alps (Florianikogel: Kozur and Mostler 1992) (Figure 5.1). Meliata remnants also scatter along the base of the Juvavic nappes, in particular as gabbros, meta-dolerites and blueschist fragments within the Haselgebirge (Hallstatt Facies including salt deposits) (Schorn et al. 2013; Leitner and Finger 2019). These tectonic units were defined as Hallstatt mélange or Hallstatt nappes (Frisch and Gawlick 2003) and thus contain the Meliata suture (Figures 5.1 and 5.6(a) and (b)). The Haselgebirge itself was not subducted but contains slices of subducted material. The Paleo-Alpine (Jurassic to Early Cretaceous) thermal effect during the accretion phase was mild. Within the Hallstatt mélange, temperatures hardly exceeded Argon retention temperatures of biotite in the range of 300 ± 25°C (for a summary, see Schorn et al. 2013).

Unlike to the northern and southern margins of the Neotethys embayment where Jurassic convergence was largely orthogonal toward foreland units, the western, Austroalpine sector displays a combination of transform and thrust tectonics. Middle Jurassic/Early Cretaceous large-scale transform tectonics is argued from paleogeographic facies fitting and correlation of basement units (Schmidt et al. 1991; Haas et al. 1995; Frank and Schlager 2006). Large-scale sinistral Jurassic transform faulting, later overprinted by younger movement, was first proposed by Schmidt et al. (1991) and Haas et al. (1995) who recognized that the Drauzug and the western Transdanubian Mountains of Hungary appear as exotic blocks when Triassic Facies are compared with their surroundings. Paleogeographic facies fitting suggests that the Drauzug (Figure 5.1) and the Transdanubian Mountains, together with the South Alpine realm, were displaced by 300–400 km to the east along transform faults during Middle Jurassic to Early Cretaceous times. Another argument arises from appearance of exotic units below the Noric-Tirolic nappe in the eastern Greywacke Zone (Figure 5.1). The Paleozoic evolution of the Veitsch nappe differs significantly from all other Paleozoic occurrences in the Eastern Alps but shows similarities with successions in the Slovakian Western Carpathians. This exotic domain has been defined as laterally distributed tectonic sliver, the Veitsch-Nötsch-Ochtina terrane (Neubauer and Vozarova 1990; Ebner et al. 2008), probably juxtaposed during the Paleo-Alpine event against other Austroalpine units.

Due to the low-grade metamorphic overprint and the absence of suitable geochronometers available in the past, the Paleo-Alpine event is not well dated. Hints for a metamorphic event older than the well-known Eo-Alpine metamorphic phase around 90 Ma arise from cooling age data. Juvavic and Tirolic nappes and the Greywacke Zone together with the Drauzug-Gurktal Nappe systems experienced cooling below Rb/Sr biotite (ca. 320 ± 40°C) and Ar/Ar muscovite retention temperatures (ca. 425 ± 30°C) between 145 and 95 Ma (Kralik et al. 1987b; Fritz 1988; Dallmeyer et al. 1998; Schuster and Nowotny 2015). Low temperature increments of Ar/Ar release patterns of biotite from exotic blocks within the Hallstatt mélange were interpreted to date a Jurassic/Early Cretaceous event at 140–145 Ma (Schorn et al. 2013). A metabasalt olistolite within the Hallstatt mélange yielded a zircon fission-track age around the Jurassic/Cretaceous boundary (143 Ma; Hejl and Grundmann 1989). Kinematic data which correlate with this event suggest westward transport of early formed nappes (Ratschbacher 1986; Fritz 1988; Fritz et al. 1991; Ratschbacher and Neubauer 1989; Neubauer et al. 2000; Frisch and Gawlick 2003; Habermüller 2005). According to Gawlick and Missoni (2019), the Hallstatt mélanges and the ophiolites are far-travelled fragments of the Neotethys Ocean and its western/northwestern outer shelf, brought into their present position by west-/northwest-ward far-distance thrusting.

5.5.1.2. *Austroalpine nappe kinematics and Eo-Alpine evolution*

During a second phase from ca. 100 to 90 Ma, postdating Paleo-Alpine thin skinned west/northwestward thrusting, the deformation center shifted to the site in the Adriatic plate that was formerly weakened by Permian extension and subsequent high-temperature–low-pressure metamorphism (e.g. Schuster and Stüwe 2008). Eo-Alpine intracontinental subduction formed the high-pressure wedge with dominant top to-the-N displacement (Figure 5.6(c)). Eo-Alpine metamorphic isograds indicate a southward increase in metamorphic grade (e.g. Hoinkes et al. 1999), presumed to be related to southward underthrusting of the LCAA (Figure 5.6(c)). The already existing Jurassic to Cretaceous nappe stack was affected by out of sequence thrusts that divided the uppermost Juvavic Nappe System into the Lower and Upper Juvavic nappes. The Eo-Alpine high-pressure event proceeded at least 50 Ma after initial emplacement of Juvavic nappes through the closure of the Meliata basin. This time gap therefore reflects a two-stage evolution, that is, Meliata subduction and emplacement of Permian/Mesozoic cover sediments followed by subsequent intracontinental suturing (Kurz and Fritz 2003) (Figure 5.6). The related Eo-Alpine high-pressure rocks can be traced along a belt in the southern part of the LCAA, immediately north of the "Southern Limit of Alpine Cretaceous Metamorphism" (SAM) (Hoinkes et al. 1999) (Figures 5.1, 5.2 and 5.6(d)) from the southern part of the Ötztal region (Texel complex), to the Schober and the Kreuzeck Mountains south of the Tauern Window, and to the Saualpe and Koralpe (Figure 5.1). This is oblique to the orientation of thrusts related to W-directed emplacement of thin-skinned nappes. The SAM may therefore be interpreted to represent a relic of an intracontinental plate boundary along a former Permian rift/transform zone (Figures 5.6(b) and (c)). During Late Cretaceous to Paleogene times, the SAM was reactivated, now presenting in large parts a ductile normal fault (Hoke 1990; Kurz and Fritz 2003), being related to the exhumation of Eo-Alpine high-grade metamorphic LCAA units.

5.5.2. *Post Eo-Alpine evolution*

5.5.2.1. *Gosau basins*

In the Late Cretaceous, the major nappe structures within the Austroalpine unit were completed. Subsequent to nappe stacking the Austroalpine nappe assemblage has been affected by crustal extension (Ratschbacher et al. 1989; Neubauer et al. 1995; Froitzheim et al. 1997; Willingshofer 2000). This triggered the formation of several sedimentary basins, the Gosau basins, on top of the Austroalpine nappe stack (e.g. Faupl et al. 1987). It is worth mentioning that the deepest portions of the Austroalpine nappes, namely the Lower Austroalpine, were still affected by thrusting related to underplating by the Penninic oceanic lithosphere (Dallmayer et al. 1998; Liu et al. 2001; Ratschbacher et al. 2004).

Stretching and Gosau basin formation started during the Late Cretaceous (Butt 1981; Faupl et al. 1987; Faupl and Wagreich 1992a, 1992b, 2000; Ortner 1992, 1994a, 1994b; Wagreich 1993, 1995; Wagreich and Faupl 1994; Neubauer et al. 1995; Froitzheim et al. 1997; Ebner and Rantitsch 2000) with deposition of clastic sediments resting, in most cases, unconformly on their substratum. The sedimentological evolution is documented up to the Eocene (e.g. Faupl et al. 1987; Wilkens 1989; Wagreich and Faupl 1994; Egger et al. 1996). However, structural and sedimentological differences exist between the Gosau basins on top of the NCA ("Kalkalpine Gosau") and on top of the Central Eastern Alps ("Zentralalpine Gosau") (e.g. Neubauer et al. 1992, 1993, 1995; Wagreich and Faupl 1994). Basically, sedimentation was tectonically controlled.

The sedimentary succession on top of the NCA is subdivided into the Lower Gosau Subgroup (Upper Turonian–Campanian), which is characterized by terrestrial to shallow-marine facies associations, and the Upper Gosau Subgroup (Santonian-Eocene), which comprises deep-water hemipelagic and turbiditic deposits (Wagreich and Faupl 1994; Egger et al. 1996; Faupl and Wagreich 2000). Rapid subsidence started diachronously from the northwest (Santonian) to the southeast (Maastrichtian) after a short phase of deformation and erosion. The rapid deepening was interpreted to be related to subduction-related tectonic erosion (subduction of the Penninic Ocean) in the footwall of the Austroalpine nappe complex (Wagreich 1993, 1995). This also documents the fact that these parts of the NCA nappes have travelled all across the Adriatic microcontinent to approach the Penninic margin shortly before the metamorphic climax of Eo-Alpine wedge tectonics. Subsidence of

Central Alpine Gosau basins is closely related to the exhumation of the surrounding basement complexes (e.g. Neubauer et al. 1995; Kurz and Fritz 2003). Basement domes and the related basins formed within a system of strike-slip and normal faults (Neubauer et al. 1995; Rantitsch et al. 2005). As described by Wilkens (1989), the sedimentary sequence within the Gosau deposits is, in most cases, characterized by a well-developed angular and erosive unconformity at the Cretaceous/Palaeogene boundary (~65 Ma).

Especially the Gosau basins on top of the NCA were affected by shortening during the Paleogene, which is related to N- to NE-oriented contraction. This deformation includes generally N- to NE-vergent folding, and faulting which partitioned into reverse faults of variable sense of displacement, having been linked by NE-striking sinistral transfer faults (Eisbacher et al. 1990; Eisbacher and Brandner 1995, 1996). Paleogene thrusts within the NCA have also affected the Gosau deposits (e.g. Faupl and Wagreich 1992a, 2000). Erosional unconformities within the Gosau Group deposits may be related to these deformational events (e.g. Ortner 1992, 1994a, 1992b; Wagreich 1993, 1995).

5.5.2.2. *Syn-Gosau extensional shearing*

Parts of the Austroalpine nappe edifice were affected by Late Cretaceous northwest to west-directed shearing. Formation ages from mylonitic shear zones at the base of these distinct nappes decrease continuously from the UCAA (ca. 98–95 Ma) to the LCAA (ca. 88–75 Ma) (Dallmeyer et al. 1996, 1998; Müller et al. 1999; Pfingstl et al. 2015). Today, these units are situated to the north or west of the LCAA high-pressure wedge (affected by north-directed shearing as outlined above). These ages were interpreted to date a phase of crustal thickening, coinciding with a phase of widespread crustal extension within the Austroalpine realm (e.g. Ratschbacher et al. 1989; Froitzheim et al. 1994). This was associated with west- and east-directed normal faulting (e.g. Koroknai et al. 1999; Fügenschuh et al. 2000; Sölva et al. 2005; Wiesinger et al. 2006) and the related Gosau basins (Wagreich and Faupl 1994; Neubauer et al. 1995; Willingshofer et al. 1999; Ebner and Rantitsch 2000; Fügenschuh et al. 2000; Froitzheim et al. 2008) (Figures 5.1 and 5.6(d)). These ages therefore need not necessarily represent the timing of nappe assembly, but may indicate the timing of flow partitioning with top-to-the north exhumation in the internal high-pressure wedge and related upper crustal east-west extension in the frontal (external) parts (UCAA) (e.g. Kurz and Fritz 2003). Major tectonic lines, previously interpreted as Early Cretaceous thrusts, are overprinted and sealed by upper greenschist- to amphibolite-facies metamorphism and extensional tectonics. The final Austroalpine architecture is therefore likely determined by exhumation of high-pressure units and dispersal of hanging wall units by extension.

A large number of synkinematic age data clustering around 80 Ma and previously interpreted to date Eo-Alpine nappe stacking may therefore easily be re-interpreted in terms of strike-slip and/or extension. Thus, tectonic reconstructions of the Eastern Alps have to include Late Cretaceous disintegration and exhumation by extension. Recognition of Eo-Alpine structures and metamorphism, however, is hampered by the fact that the spatial distribution of structural elements with Cretaceous age coincide frequently with Miocene extrusion-related structures (for a summary, see Frisch et al. 1998, 2000; Kurz and Fritz 2003). In particular, the Austroalpine nappes in the easternmost part of the Eastern Alps (Semmering and Wechsel Nappe Systems) (Figure 5.1) are strongly obliterated by faulting during Miocene times. However, from geochronological and tectono-metamorphic arguments, there is strong evidence that extrusion tectonics also played a major role in Alpine architecture during the latest Cretaceous.

5.6. Cenozoic tectonics

The general Oligocene to Miocene tectonic evolution of the Eastern Alps is characterized by the collision of the southern European margin and the Adriatic plate, resulting in crustal thickening and Penninic and Sub-Penninic nappe assembly, and the formation of east-west oriented strike-slip fault zones that are physically connected to north-south oriented extensional detachments. This was ascribed to the northward oblique indentation of the European continent by a semi-rigid crustal block (the so-called Adriatic indenter), as classically portrayed by Ratschbacher et al. (1991a, 1991b) and defined as lateral extrusion.

The tectonic and metamorphic evolution of the Penninic and Sub-Penninic Nappe Systems within the Tauern Wondow has been recently re-evaluated and described in detail by Schmid et al. (2013); Scharf et al. (2013a, 2013b, 2016); Bertrand et al. (2015, 2017); Favaro et al. (2015, 2017); Schneider et al. (2015); and Rosenberg et al. (2018).

5.6.1. *Penninic and Sub-Penninic nappe tectonics*

5.6.1.1. *Engadine Window*

The Engadine Window (Figure 5.1) exposes a stack of Penninic nappes, overlain and framed by Austroalpine nappes.

The Ramosch Zone represents the transition between a continental unit (Briançonnais) and an oceanic (Valais) unit (Florineth and Froitzheim 1994). It comprises a serpentinized peridotite associated with ophicarbonates and serpentinite breccias and metagabbros. Pillow basalts of Valais origin are underlying the serpentinite along an Alpine, top-north directed thrust (Froitzheim et al. 1996). Further, a huge mass of Valais-derived Bündnerschiefer forms the deepest unit of the Engadine Window. It comprises up to 10 km of calc-schists interbedded with shales and quartzites (Hitz 1995), grading upward into flysch deposits of Late Cretaceous to Eocene age.

The lower parts of the Bündnerschiefer (Mundin unit) and associated metabasites show a Neo-Alpine blueschist facies metamorphic overprint (e.g. Bousquet et al. 1998, 2002).

The Middlepenninic Tasna nappe comprises a continuous sedimentary sequence from the Permian-Triassic to the Late Cretaceous (Waibel and Frisch 1989), with a preserved transition from the continental crust of the Briançonnais to the oceanic crust of the Valais ocean (Florineth and Froitzheim 1994). The Jurassic and Cretaceous sequences are mainly composed of turbidites with associated debris flows and pelagic limestones.

The conditions of Neo-Alpine metamorphic overprint are of lower greenschist facies (Schuster et al. 2004).

The Arosa Zone as the uppermost Penninic nappe is a highly tectonized ophiolite-bearing unit (Ring et al. 1990) with an ophiolite sequence mostly of serpentinites, gabbros and basalts (Höck and Koller 1987) of presumed Piemont-Ligurian origin. It is covered by a sequence of radiolarian cherts, pelagic limestones, black shales of Hauterivian-Aptian age (Weissert and Bernoulli 1985) and flysch. The Arosa Zone, also exposed at the western margin of the Eastern Alps, continues southward into the Platta nappe with similar chemical and paleogeographic features (Frisch et al. 1994). It is likely to correlate with the Matrei Zone in the Tauern Window (Frisch et al. 1987)

The metamorphism of the ophiolites is twofold; an older HT oceanic, probably spreading-related metamorphism can be separated from a younger HP metamorphism (Schuster et al. 2004).

5.6.1.2. *Tauern Window*

The Tauern Window of the Eastern Alps (Figures 5.1, 5.2 and 5.4) exposes exhumed parts of Europe-derived crust. The southern European margin was accreted to the base of an Adria-derived upper plate, represented today by the Austroalpine nappes (e.g. Schmid et al. 2004, 2013). The Tauern Window is characterized by a complex three-dimensional geometry of tectonic units, including important along-strike changes in its structure (Schmid et al. 2013). This geometry, as described by Schmid et al. (2013) by combining surface geology with the results of deep seismic reflection measurements (Gebrande et al. 2002; Lüschen et al. 2006), basically resulted from crustal-scale collisional accretion in the Alps followed by Late Alpine indentation, crustal-scale folding, orogen-parallel extension and lateral extrusion (Ratschbacher et al. 1991a).

While the surface geological structure of the Tauern Window is well understood, the structure of the deeper parts of the lithosphere and the mantle below still remains an open problem in the lithosphere-scale geometry of the Alps-Carpathians-Dinarides system (e.g. Brückl et al. 2008, 2010).

The Tauern Window (Figures 5.1, 5.2 and 5.4) exposes a Cenozoic-age nappe pile consisting of crustal slices derived from the distal continental margin of Europe (Sub-Penninic Units) and a part of the Penninic ocean, accreted to an upper plate that consists of the Austroalpine Nappe System. This general structure of the nappe system within the Tauern Window was previously described by Kurz et al. (1996, 1998) and re-compiled by Schmid et al. (2013). The present-day architecture of the Tauern Window is primarily characterized by a crustal-scale Late Alpine Duplex, the Venediger Duplex (or Venediger Nappe System) that formed during the Oligocene. This duplex structure was severely overprinted by doming and lateral extrusion, which was most probably triggered by the indentation of the Southalpine Units east of the Giudicarie Belt (Figures 5.1 and 5.2). Indentation initiated around 20–23 Ma ago (Frisch et al. 1998, 2000) and was linked to a lithosphere-scale reorganization of the geometry of the mantle slabs beneath the central Eastern Alps (Schmid et al. 2004, 2013).

The European continent consists of a deeply eroded Variscan (Late Devonian to Carboniferous) metamorphic continental crust, rich in plutonic rocks (north of the Alpine front), covered by Carboniferous to Eocene sedimentary sequences. This crust is still in contact with its lithospheric mantle and dips southwards beneath the Alps (e.g. Kummerow et al. 2004). The Sub-Penninic nappes within the Tauern Window (Figures 5.1 and 5.4) represent the distal European margin, forming ductilely deformed basement and cover nappes, which lost contact with their lithospheric mantle. These form the Venediger Nappe System. This interpretation is based on the conclusion that the crustal material of the Venediger Nappe System was not separated from the European margin by an Oceanic basin (e.g. Froitzheim et al. 1996; Kurz et al. 2001). The eclogitic Sub-Penninic basement units (Eclogite Zone, Modereck Nappe System) also contain material derived from the Penninic Ocean and developed in a subduction and accretion channel (Engi et al. 2001; Kurz and Froitzheim 2002).

The Wolfendorn nappe is restricted to the northwestern Tauern Window where it structurally overlies two imbricates of the Venediger Duplex. The Eclogite Zone occupies a tectonic position above the Venediger Nappe System but just below the Modereck Nappe System. Eclogite facies conditions of 1.9–2.2 GPa and 600–630°C (Hoschek 2001) were reached between 45 and 33–32 Ma (Ratschbacher et al. 2004; Glodny et al. 2005, 2008; Kurz et al. 2008; Nagel et al. 2013). Subsequent decompression and reheating to amphibolite facies conditions ("Tauernkristallisation") affected the entire nappe stack including the Venediger Nappe System (Glodny et al. 2008). The onset of subduction is poorly constrained, but available radiometric data suggest between 55 and 45 Ma ago (Ratschbacher et al. 2004; Berger and Bousquet 2008).

The Modereck Nappe System comprises the units in a similar structural position immediately below the Glockner nappe. The Seidlwinkl nappe, an isoclinal fold nappe, is exposed in the central Tauern Window (Frank 1969; Groß et al. 2020, 2021) and makes up most of the Modereck Nappe System. Kurz et al. (2008) report that parts of the Modereck Nappe System located in the Grossglockner area are, together with the adjacent Eclogite Zone, affected by eclogite facies metamorphism.

The Penninic nappes comprise three paleogeographic elements: the Penninic (Piemont-Ligurian) Ocean, the Briançonnais microcontinent and the Valais Ocean. The Piemont-Ligurian Ocean opened in Late Jurassic times (Ratschbacher et al. 2004; Schmid et al. 2004, and references therein). Its initial seafloor formed by exhumation of the sub-continental mantle of the Adriatic microplate (Froitzheim and Manatschal 1996).

The Briançonnais microcontinent was a part of the European distal margin until it was cut off by the opening of the Valais Ocean in Cretaceous times. The Valais Oceanic crust comprises Cretaceous ophiolites overlain by Cretaceous to Eocene calcareous turbiditic metasediments. Toward the east the Valais Ocean merged into the Piemont-Ligurian Ocean, thus forming a single oceanic basin east of the present-day Engadine Window (Figure 5.1) (e.g. Stampfli 1994; Froitzheim et al. 1996; Kurz 2006). This situation therefore makes any subdivision of the Piemont-Ligurian from the Valais basin somewhat artificial in the area east of the Engadine Window (Kurz 2005, 2006).

This is also based on observations that the facies of cover sequences of the Venediger Nappe System are untypical for the Briançonnais units (e.g. Faupl and Wagreich 1992). Lammerer (1986, 1988) and Vesela et al. (2008, 2011) demonstrated that the lithostratigraphy of these cover sequences is very similar to the Helvetic nappes in Switzerland; Frisch (1975) documented similarities between the lithostratigraphy of the Hochstegen Formation (Venediger nappe), and Helvetic units of the Eastern Alps exposed in the Gresten Zone (Figure 5.1). This was brought up by Trümpy (1992) as well. Therefore, the Venediger Nappe System of the Tauern Window was correlated with the units representing the European margin in the Central Alps by Froitzheim et al. (1996), Froitzheim (2001) and also by Kurz et al. (2001), having already been proposed by Trümpy (1988). As a result, the Glockner nappe was assumed to be of Valais origin (Northpenninic) by Froitzheim et al. (1996), Schmid et al. (2004) and previously by Oberhauser (1995). Accordingly, only the Matrei Zone would correspond to the Southpenninic (Piemont-Ligurian) units of the Central and Western Alps. As stated by Kurz (2005b), the Glockner nappe (referred to as the "Upper Schieferhülle Unit" by Schmid et al. 2004), comprises remnants of an oceanic basement in the sense of a partly incomplete ophiolitic sequence. This was interpreted to correspond to the ophiolite nappes derived from Piemont-Ligurian (Southpenninic) in the Western Alps (such as in the Zermatt-Saas ophiolites of Western Switzerland and the Avers Bündnerschiefer of Eastern Switzerland), besides the volumetrically dominating calc-schists referred to as "Bündnerschiefer" or "schistes lustrés" in the Central and Western Alps. However, the "Bündnerschiefer" and "schistes lustrés" are regarded as being diagnostic for the Lower Penninic nappes derived from the Valais (Northpenninic) paleogeographical domain. The Matrei Zone of the Tauern Window (Figure 5.4), containing calc-schists (besides metapelites) as well, is actually defined by the presence of tectonic slivers (or olistolites according to the interpretation of Frisch et al. 1987; Kurz et al. 1998b) of Austroalpine derivation. According to Schmid et al. (2004), this clearly makes the Matrei Zone an analogue to the Upper Penninic (respectively, Piemont-Ligurian) Arosa Zone of Eastern Switzerland (Ring et al. 1988, 1989, 1990; Oberhauser 1995; Manatschal et al. 2003) (Figure 5.1), characterized by material derived from the Austroalpine domain. In contrast, it was demonstrated by Frisch et al. (1987) that the separation of the Matrei Zone and the Glockner nappe within the Tauern Window is very often difficult because of similarities of their lithostratigraphic sequences. Moreover, the transition from the Glockner nappe to the Matrei Zone is very often rather a stratigraphic than a tectonic one ("both... and"; Frisch et al. 1987). Moreover, the Glockner nappe is directly overlain by the Lower Austroalpine nappes in sections where the Matrei Zone is either missing or cannot be separated from the Glockner nappe below. New

U-Pb geochronological data of magmatic zircons yielded a concordant 157 ± 2 Ma crystallization age for the precursor of a coarse-grained metagabbro from the Glockner nappe (Gleißner et al. 2021). This Late Jurassic intrusion age is coeval with ages from Piemont-Ligurian mafic intrusions in the Western and Central Alps. These new zircon ages strongly suggest the formation of the oceanic crust, now preserved within the Glockner nappe, in the Piemont-Ligurian branch of the Penninic Ocean, being inconsistent with previous presumptions of a formation in the Valais domain (e.g. Schmid et al. 2004, 2013).

Since no continental nappes seem to occur between the Glockner nappe and the Matrei Zone, the Briançonnais would be completely missing in the Tauern cross-section, as proposed in the working hypothesis by Froitzheim et al. (1996). This may be explained in two plate tectonic scenarios: (1) The Briançonnais as a paleogeographic domain ended primarily between the present-day Engadine Window and the Tauern Window and the Northpenninic (Valais) and Southpenninic (Piemont-Ligurian) basins merged eastward into a coalesced basin (Stampfli 1993, 1994; Froitzheim et al. 1996; Channel and Kozur 1997; Stampfli and Marchant 1997; Neubauer et al. 2000; Kurz et al. 2001) (Figure 5.7). The Valais, however, did not probably extend very far to the east, as indicated by the Jurassic U-Pb data from the Glockner nappe (see above). The paleogeographic reconstructions displayed by Trümpy (1988, 1992) (Figure 5.7) also show that the Briançonnais ends east of the Engadine Window and is laterally displaced to the east. However, several related crustal slices are not continuous and may only occur as distinct allochthons; thus the main part of the Briançonnais microcontinent withers toward east (Trümpy 1988). (2) The front of the Briançonnais nappes is located south of the Tauern Window, and the Briançonnais was mostly or completely subducted (Froitzheim et al. 1996). Although this postulate cannot be excluded, it appears rather hypothetical since no evidence for high-pressure metamorphism is found in the very thin Briançonnais-derived sliver that outcrops in the Engadine Window nearby (Tasna nappe) (Florineth and Froitzheim 1994; Schmid et al. 2004). According to this, it was mentioned by Trümpy (1992) that equivalents to the Tasna nappe may occur within some gneiss lamellae of the Tauern Window, the main of these forming the tectonic base of the Modereck Nappe System (Kurz et al. 1996, 1998) (Figure 5.4). As these gneiss lamellae are situated below the Glockner nappe, the latter might be assumed to be of Southpenninic (Piemont-Ligurian) origin. Due to the analogies of cover sequences within the Tasna nappe and the Modereck nappe, it may be further assumed that these nappes originated from the same paleogeographic unit. However, the Tasna nappe and the Modereck Nappe System are interpreted to originate from the Briançonnais (Middlepenninic) (see above) and the distal European continental margin (Sub-Penninic), respectively. Furthermore, the contrasting lithostratigraphy of cover sequences within the Venediger nappe and the Modereck nappe (see summary by Kurz et al. 1998; Schmid et al. 2013) may also indicate a contrasting peleogeographic origin of these units. A very likely explanation is that the basement units of the Venediger nappe and the Modereck nappe were separated during the Late Jurassic to Early Jurassic rifting previous to Penninic seafloor spreading, that is, the Modereck Nappe System forms an extensional allochthon which was separated from the basement of the Venediger nappe (Kurz 2006; Groß et al. 2020). Thus, the Modereck nappe may represent a fragment of continental crust, rifted off the European margin, and was separated from that by a narrow stripe of either oceanic or transitional crust. The remnants of the latter may be represented by the Eclogite Zone, situated between the Venediger Nappe System below and the Modereck Nappe System above (Figure 5.7).

Beside the U-Pb zircon ages from the Glockner nappe meta-gabbros (Gleißner et al. 2021), the age of Piemont-Ligurian oceanic crust is roughly constrained in the Eastern Alps by Callovian to Oxfordian fossils (~164–154 Ma) (Faupl and Wagreich 2000), a 182 ± 31 Ma Sm/Nd plagioclase–clinopyroxene isochron from an ophiolite remnant in the Engadine Window (Figure 5.1) (Thöni 1999), and a 217 ± 59 Ma (186 ± 12 Ma) Sm/Nd whole rock isochron from the Reckner ophiolite complex (Figure 5.1) (Meisel et al. 1997; see summary by Ratschbacher et al. 2004). The Reckner ophiolitic complex represents a piece of exhumed subcontinental mantle with occurrences of ophicalcite, stratigraphically overlain by Jurassic radiolarites and younger pelagic sediments (see summary by Schmid et al. 2013, with references), being typical for ophiolitic sequences derived from the Piemont-Ligurian Ocean (Froitzheim and Manatschal 1996).

The onset of Penninic subduction is poorly constrained in the Eastern Alps as well. The Reckner ophiolitic complex is accreted to the Austroalpine nappe stack from the Turonian onwards (about 90 Ma) (Schmid et al. 2013). Accretion along the Adria margin was active from late Early Cretaceous to Early Paleogene (~110 to 60 Ma) (Faupl and Wagreich 2000), but the presence of pelagic Eocene sediments implies that the final ocean closure occurred no earlier than Early Eocene times (~55 Ma) (Ratschbacher et al. 2004). Radiometric ages obtained for metamorphism in the Lower Austroalpine Radstädter Tauern between 80 and 50 Ma (Liu et al. 2001) indicate that the northern margin of the Austroalpine nappe pile was part of an active margin until Eocene times. This is constrained by $^{40}Ar/^{39}Ar$ geochronology data of Ratschbacher et al. (2004) indicating Eocene HP metamorphism in the Lower Austroalpine during the Cenozoic.

Recent studies have shown that the Gockner Nappe System within the Tauern Window, as previously defined by Kurz et al. (1996) and Schmid et al. (2013), can be subdivided into a Lower Glockner nappe s.str. having been affected by eclogite facies metamorphism previous to greenschist to amphibolite "Tauernkristallisation" overprint, and a higher Rauris nappe that does not show any high-pressure metamorphic assemblages (Groß et al. 2020, 2021). No direct radiometric data are available for the age of eclogite facies overprint of the Glockner Nappe System. However, it is likely that high-pressure overprint in the Glockner Nappe System occurred contemporaneously with that of the Bündnerschiefer of the Engadine Window having been dated at ~42–40 Ma (Wiederkehr et al. 2008, 2009), and prior to the incorporation of the southern European margin into the subduction channel followed by its HP metamorphic overprint.

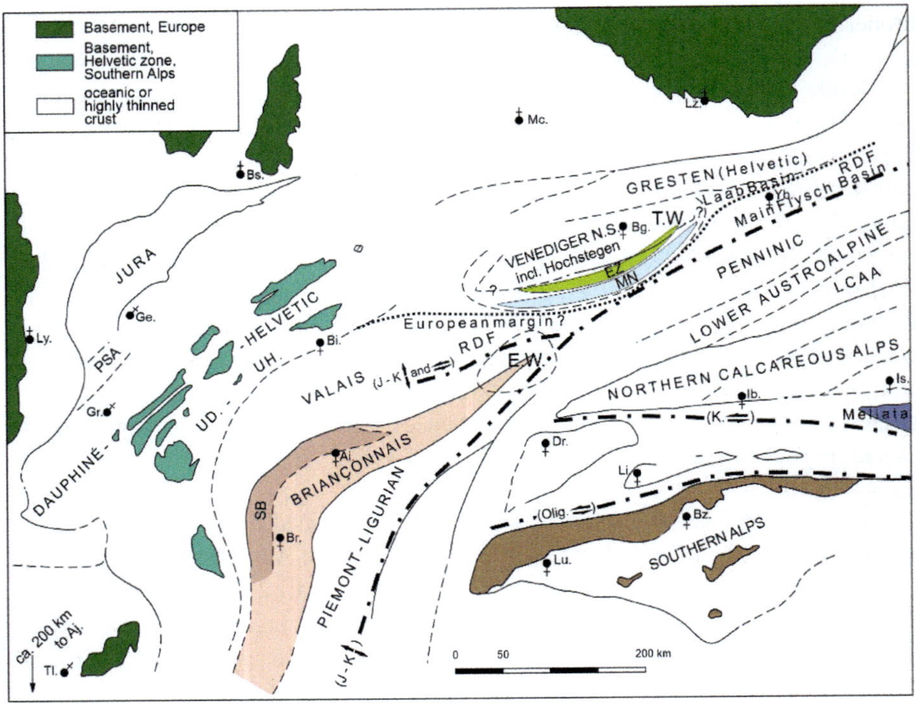

Figure 5.7. *Palinspastic reconstruction of the Alps during the Cretaceous (after Trümpy 1992; Kurz 2006)*

COMMENT ON FIGURE 5.7.– *PSA: Pre-Subalpine; UD-UH: Ultradauphinois, Ultrahelvetic; LCAA: Lower Central Austroalpine; MN: Modereck nappe; EZ: Eclogite Zone; TW: Tauern Window; EW: Engadine Window; RDF: Rhenodanubian Flysch; K.: Cretaceous; Olig.: Oligocene; Ly: Lyon; Bs: Basel; Mc: Munich; Lz: Linz; Aj: Ajaccio; Tl: Toulon; Gr: Grenoble; Ge: Geneve; Br: Briançon (Briançonnais); Ai: Aigle; Bi: Biasca; Bg: Bad Gastein (Zentralgneis); Yb: Ybbsitz;*

Dv: Davos; Li: Lienz; Ib: Innsbruck; Is: Bad Ischl; Lu: Lugano; Bz: Bozen (Southern Alps).

The Penninic oceanic lithosphere preserved in the Tauern Window was, therefore, subducted during the Late Cretaceous to Eocene (Figure 5.8), with parts of it reaching eclogite facies metamorphism (Kurz et al. 2008). Subsequently, the European margin descended into the subduction zone, resulting in eclogite facies metamorphism in the Eclogite Zone at ~42 to 32 Ma (Figure 5.8). Thrust emplacement of the Austroalpine nappes onto the Penninic nappes was followed by thrust emplacement of the Glockner Nappe System onto the Modereck Nappe System (distal European margin) (45–35 Ma), forming a composite ocean-on-continent nappe under high-pressure conditions within the Penninic subduction zone (Groß et al. 2020). This was followed by isoclinal folding of this composite high-pressure nappe and exhumation onto the European margin above a basal thrust (~35–30 Ma) (Kurz 2005a; Groß et al. 2020), and subsequent imbrication of this margin below the basal thrust to form a duplex of Sub-Penninic nappes (Venediger Nappe Duplex). The Eclogite Zone ascended toward the surface within the subduction channel (Kurz et al. 1998; Kurz and Froitzheim 2002; Kurz 2005), while subduction was still active and, hence, heating was prevented within and adjacent to the subduction channel (Kurz et al. 1998; Kurz and Froitzheim 2002). This is evidenced by cooling during decompression and the subsequent blueschist facies metamorphic overprint. Subduction and exhumation of the Modereck Nappe System, representing the southern European margin, occurred approximately at the same time (39 Ma) (Kurz et al. 2008; Groß et al. 2020). Emplacement of the Eclogite Zone onto the Venediger Nappe System, and the subsequent emplacement of the Modereck Nappe System, including the Seidlwinkl nappe sheath fold (Groß et al. 2020) to mid-crustal to lower crustal levels (Figure 5.8), may be dated at 32–31 Ma, as indicated by the $^{40}Ar/^{39}Ar$ phengite ages from eclogite mylonites (Kurz et al. 2008). At this time, the Eclogite Zone and the Modereck nappe show nearly the same pressure conditions of approximately 10–11 kbar.

Timing of eclogite exhumation and nappe assembly coincides with the intrusion of granitoids along the Periadriatic fault south of the Tauern Window (Figure 5.1) (see Chapter 1 in Volume 3 by Berger et al.), having been explained by the breakoff of the Penninic oceanic slab (von Blanckenburg and Davies 1995) and subsequent exhumation of the Eclogite Zone and the adjacent units due to the rebound of the subducted buoyant continental crust of the European margin, or by continuous subduction and contemporaneous slab rollback (Ji et al. 2019). Finally, the detachment of the high-pressure nappe stack was followed by its emplacement onto the European margin along a southward-dipping ramp (Sub-Tauern ramp) (e.g. TRANSALP Working Group 2002; Lüschen et al. 2004) (Figure 5.8). Toward the surface, this ramp connects with the Inntal fault (Ortner et al. 2006) (Figures 5.3 and 5.8(c)). Emplacement onto the European margin was contemporaneous with amphibolite to greenschist facies metamorphism ("Tauernkristallisation") (~30–23 Ma) (for a summary, see Kurz et al. 1996).

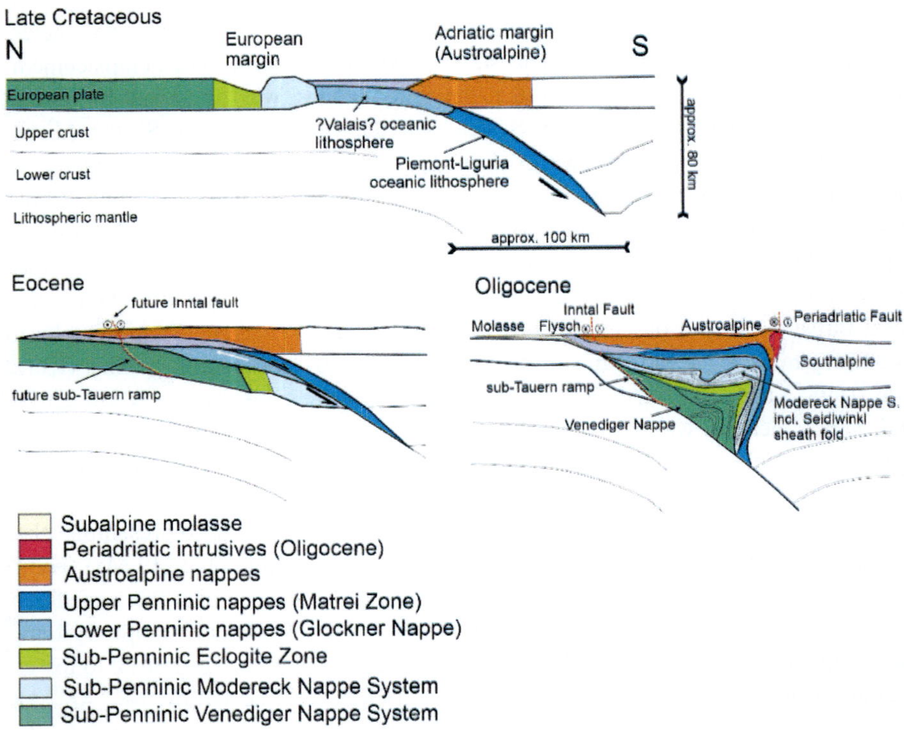

Figure 5.8. *Plate tectonic evolution of the Penninic and Sub-Penninic units of the Eastern Alps during the Late Cretaceous and the Paleogene (modified after Kurz et al. 2008; Handy et al. 2010)*

5.6.2. *Lateral extrusion in the Eastern Alps*

Lateral extrusion of crustal blocks normal to the direction of convergence is a typical process in the evolution of mountain belts, encompassing two very different mechanical processes: (1) tectonic escape in a regime of overall compression (i.e. plane strain horizontal shortening of tectonic wedges driven by forces applied to their boundaries) and (2) extensional collapse in a regime of overall extension (i.e. gravitational spreading away from a potential energy high) (see revised version by Wölfler et al. 2011; Rosenberg 2018; and references therein). In the Eastern Alps, northward oblique indentation of the European continent by a semi-rigid crustal block (the so-called Adriatic indenter), as described by Ratschbacher et al. (1991a, 1991b), was the trigger of such an extrusion process. This was accompanied by more or less orogen-parallel strike-slip faulting, linked by north-south-trending normal faults (Selverstone 2005) along the margins of exhuming crustal domes and along the margins of extensional basins. Refinements of this classical extrusion model were described by Robl and Stüwe (2005) and Rosenberg et al. (2007) by (re-) evaluating the related mechanical processes and were mechanically constrained by Robl et al. (2008). The related fault and basin systems are described in detail below.

5.6.2.1. *Miocene fault zones*

The driving force for lateral extrusion in the Eastern Alps remains enigmatic as there is evidence for both mechanical end members.

Major east-west striking strike-slip fault zones accommodated this extrusion apparently in a regime of tectonic escape, while north-south striking extensional detachments (i.e. Brenner and Katschberg normal fault) appear to indicate extension due to gravitational spreading. In very general terms, the laterally extruding part is an upper crustal block between the Inntal fault, the Salzach-Ennstal-Mariazell-Puchberg fault system (termed SEMP) and Mur-Mürz fault system in the north and the Periadriatic fault system in the south (Figure 5.3). Distinct fault zones were activated by at least Middle Oligocene times (Glodny et al. 2008) and continued to be active throughout the Neogene (Reinecker and Lenhardt 1999) (Figures 5.9 and 5.10). The Inntal fault borders the extruding wedge to the NW of the Tauern Window and is kinematically linked with the Brenner normal fault (Figure 5.3) (Fügenschuh et al. 1997). The eastern border of the Tauern Window is marked by the ESE dipping Katschberg normal fault (Figure 5.3), which transforms the displacement into the Mölltal fault (Kurz and Neubauer 1996; Frisch et al. 2000a; Wölfler et al. 2008; Favaro et al. 2017). The SEMP is the largest of the fault zones bounding the extruding block to the north. It is about 400 km long trending from the Vienna basin in the east to the northern boundary of the Tauern Window (Figure 5.3), where it truncates into the western part of the window (Rosenberg and Schneider 2008). Detailed kinematic analysis reveals sinistral strike-slip (Decker and Peresson 1996; Wang and Neubauer 1998) and an N-down component of ~1–1.2 km (Keil and Neubauer 2011).

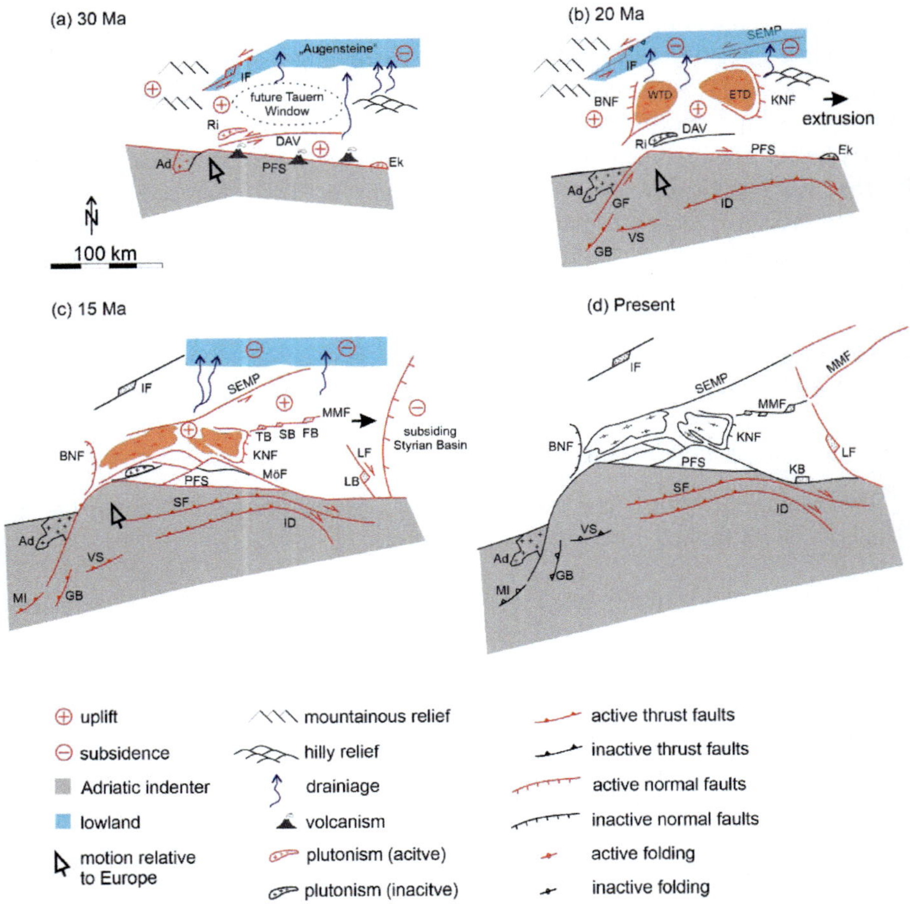

Figure 5.9. *Evolutionary steps of lateral extrusion in the Eastern Alps (compiled from Frisch et al. 1998; Wölfler et al. 2011; Scharf et al. 2013a)*

COMMENT ON FIGURE 5.9.– *(a) The Late Oligocene is characterized by plutonism and volcanism along the Peraidriatic fault system and the formation of the Inntal Basin. The western part of the Eastern Alps was already mountainous, whereas the eastern part formed hilly areas and lowlands with the sedimentation of the "Augenstein" formation. (b) At ~20 Ma, the Brenner and Katschberg normal faults were already active. Indentaion of the Adriatic intender, active folding and rapid exhumation affected the area of the future Tauern Window. (c) Extrusion-related faulting progressively moved to the east. This led to the formation of trantensional pull-apart basins along major fault zones. (d) Present-day situation with active faults at the eastern margin of the Eastern Alps and the Southern Alps. PFS: Periadriatic fault system; DAV: Derfereggen-Antholz-Vals fault; IF: Inntal fault; SEMP: Salzachtal-Ennstal-Mariazell-Puchberg fault system; GF: Giudicaria fault system; GB: Giudicaria Thrust belt; VS: Valsugana thrust; ID: Idrija thrust & fault; KNF: Katschberg normal fault; BNF: Brenner normal fault; MöF: Mölltal fault; PöF: Pöls fault (as northern part of the Pöls-Lavanttal fault system); LF: Lavanttal fault; TB: Tamsweg basin; SB: Seetal basin; FB: Fohnsdorf Basin; LB: Lavanttal Basin; AD: Adamello Pluton; Ri: Rieserferner Pluton; EK: Eisenkappel Pluton; WTD: western Tauern dome; EDT: eastern Tauern dome.*

The southern boundary of the extruding wedge is represented by the Periadriatic fault system, which separates the Southalpine from the Austroalpine units (Figures 5.1 and 5.3). While the Periadriatic fault system generally strikes E–W, it forms a complex system of SW–NE striking strike-slip, normal and reverse faults to the SW of the Tauern Window (Figure 5.3) (e.g. Schmid et al. 1989; Viola et al. 2001; Pomella et al. 2011, 2012; and references therein). During Miocene times, brittle dextral transpressive deformation characterizes the Periadriatic fault system in its eastern part (Polinski and Eisbacher 1992; Fodor et al. 1998; Heberer et al. 2017). The continuation of the Periadriatic fault system to the east is not precisely defined but recent structural studies suggest that it aligns with the Mid-Hungarian fault zone (e.g. Schmid et al. 2008; Ustaszewski et al. 2008). The Southalpine units were affected by intense south-directed reverse faulting (e.g. Castellarin and Cantelli 2000; Lüschen et al. 2004). Shortening in the Southern Alps started between 20 and 17 Ma and is still active in the eastern Southern Alps (e.g. Schmid et al. 1996; Castellarin and Cantelli 2000; Rosenberg and Berger 2009).

Between the Tauern Window and the Periadriatic fault system, a series of major fault zones form an anastamosing network of faults with different senses of displacement. The Mölltal fault and Hochstuhl fault (Figure 5.3) show dextral and the Defereggen-Antholz-Vals fault system (DAV) shows sinistral displacement (Figure 5.3). The Mölltal fault forms a NW trending, subvertical topographic and structural lineament with a main activity between Early to Middle Miocene times (Figures 5.9 and 5.10) (Wölfler et al. 2008; Favaro et al. 2017). The NNW trending Hochstuhl fault (Figures 5.3 and 5.10) offsets the Periadriatic fault system by about 3.5 km and displaces conglomerates deposited during Late Miocene times (Polinski and Eisbacher 1992). The DAV represents a ~80 km long WSW striking fault with a change from sinistral to dextral displacement at ~30 Ma (Müller et al. 2000; Mancktelow et al. 2001).

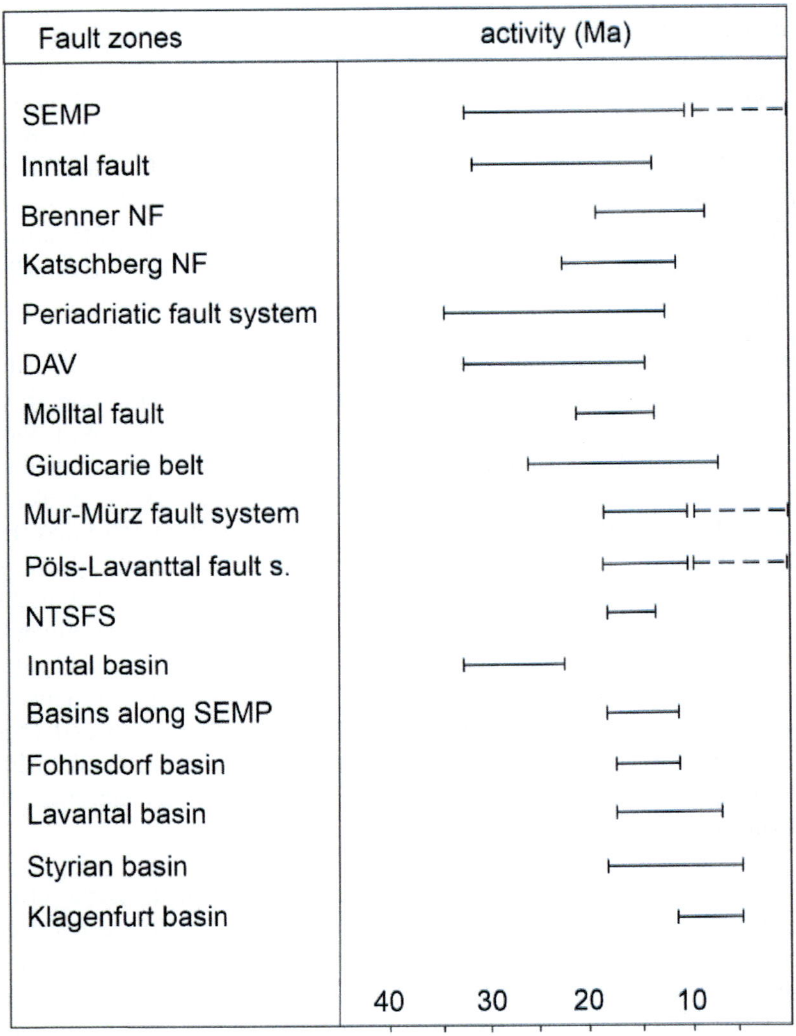

Figure 5.10. *Timing of activity of major fault zones and intramontane basins in the Eastern Alps (modified after Favaro et al. 2017)*

COMMENT ON FIGURE 5.10.– *SEMP: Frank (1987); Ratschbacher et al. (1991a, 1991b); Peresson and Decker (1997a, 1997b); Frisch et al. (1998); Linzer et al. (2002); Urbanek et al. (2002); Glodny et al. (2008); Frost et al. (2009); Schneider et al. (2013). Inntal fault: Ortner and Sachsenhofer (1996); Linzer et al. (2002);*

Ortner (2003); Ortner et al. (2006). Brenner normal fault: Fügenschuh et al. (1997, 2012); Frisch et al. (1998); Glodny et al. (2008); Rosenberg and Garcia (2011); Wolff et al. (2020). Katschberg normal fault: Liu et al. (2001); Dunkl et al. (2003); Scharf et al. (2013a, 2016). Periadriatic fault system: Bögel (1975); Scharbart (1975); Läufer et al. (1997); Frisch et al. (1998); Müller (1998); Viola (2000); Mancktelow et al. (2001); Müller et al. (2001); Stipp et al. (2004); Zwingmann and Mancktelow (2004); Handy et al. (2005). DAV: Borsi et al. (1979); Kleinschrodt (1987); Schulz (1990); Müller et al. (2000; 2001); Mancktelow et al. (2001); Most (2003); Romer and Siegesmund (2003); Handy (2005); Wagner et al. (2006). Mölltal fault: Linzer et al. (2002); Wölfler et al. (2008); Scharf et al. (2013a). Giudicarie belt: Luciane (1989); Elias (1998); Müller et al. (2001); Viola et al. (2001); Pomella et al. (2011; 2012). Mur-Mürz fault system: Metz (1973; 1976); Tollmann (1977); Ratschbacher et al. (1991b); Zeilinger et al. (1999); Reinecker (2000); Sachsenhofer et al. (2000); Strauss et al. (2001); Linzer et al. (2002). Pöls-Lavanttal fault system: Beck-Mannagetta (1952); Exner (1976); Ratschbacher et al. (1991b); Eder and Neubauer (2000); Wölfler et al. (2010); Kurz et al. (2011). Inntal Basin: Ortner & Sachsenhofer (1996). Basins along the SEMP: Dunkl et al. (2005). Fohnsdorf Basin: Strauss et al. (2001); Lavanttal Basin: Reischenbacher et al. (2007). Styrian Basin: Sachsenhofer (1996). Klagenfurt Basin: Nemes et al. (1997).

To the east of the Tauern Window, two major fault zones at about right angles appear to form a conjugate set: the Mur-Mürz fault system and the Pöls-Lavanttal fault system. The Mur-Mürz fault system is characterized by a complex structure with a number of transtensional basins (e.g. Ratschbacher et al. 1991b) (Figure 5.3). The eastern segment is located east of the Fohnsdorf Basin (Figure 5.3) and terminates underneath the Vienna basin. The western segment, west of the Fohnsdorf Basin, originates near the Katschberg normal fault. Based on structural investigations and low-temperature thermochronology, Reinecker (2000) stated that this part of the fault system consists of several fault zones that changed their kinematics from pre-17.2 Ma strike-slip to <16.5 Ma normal faulting. The Pöls-Lavanttal fault system separates the basement complex of the Kor- and Saualpe by dextral offset and is assumed to have been active since Early Miocene times with peaks in activity between 18–16 and 14–12 Ma (Reinecker 2000). Near its southern termination, the Lavanttal fault offsets the Periadriatic fault by about 20 km.

5.6.2.2. *Time constraints on Miocene faulting*

Much of the fault activity is well constrained by fault-related pull-apart basins and their sedimentary fill (Figures 5.3, 5.9 and 5.10). The oldest fault-related sedimentary basin is the Late Eocene to Early Oligocene Inntal Basin that was strongly controlled by tectonic subsidence due to transtension along the Inntal fault (Figures 5.9(a) and (b)) (Ortner and Sachsenhofer 1996). During the Oligocene and Early Miocene, the area to the west of the future Tauern Window and the Inntal fault was already mountainous (Frisch et al. 1998, 2001) (Figures 5.9(a) and (b)). In contrast, the area to the east of the Inntal fault was not exposed before Miocene times (Frisch et al. 1998) but was the site of deposition of Late Oligocene to Early Miocene conglomerates and sandstones, the so-called Augenstein formation (Figure 5.9(a)) (Frisch et al. 2001). The source area of the Augenstein formation consisted predominately of low-grade metamorphic rocks, which were mainly delivered from the south (Figure 5.9(a)), namely the Niedere Tauern and Gurktal Alps (Figure 5.11) (Frisch et al. 1998) that formed a hilly topography during the Oligocene and Late Miocene (Frisch et al. 1998, 2000b).

Other intramontane basins are located along the SEMP, the Mur-Mürz fault, the Pöls-Lavanttal fault system (Lavanttal Basin) and the Periadriatic fault system (Klagenfurt Basin) (Figures 5.2 and 5.9). In contrast to the Inntal Basin, sedimentation in these basins, being described as strike-slip-related pull-apart basins (Frisch et al. 2000a; Linzer et al. 2002), started at ≤18 Ma (e.g. Sachsenhofer et al. 1997; Strauss et al. 2001; Dunkl et al. 2005) (Figure 5.10). Numerous of the basins mentioned above were studied in detail regarding their structural and sedimentary evolution. All these studies agree that subsequent to subsidence a period of basin inversion took place during Late Miocene to Pliocene times (e.g. Fodor et al. 1999; Zeilinger et al. 1999; Sachsenhofer et al. 2000; Bada et al. 2007). This widespread inversion of extensional structures is interpreted to be related to the termination of subduction of European lithosphere beneath the Carpathians. This blocked the eastward escape in front of the northward indenting Adriatic plate (Horváth and Cloething 1996).

Geochronological data from the extrusion-related fault zones are rare and only available from exhumed mylonitic shear zones. By using Rb/Sr data from muscovites and biotites, Glodny et al. (2008) drew two peaks of deformation ages for fault zones within and around the Tauern Window. The first peak is in the Oligocene between 32 and 30 Ma and the second one in the Early Miocene to Middle Miocene between 21 and 15 Ma. According to $^{40}Ar/^{39}Ar$-dating, Schneider et al. (2013) constrained ductile shearing in the western Tauern Window in the range between 33 and 15 Ma for the Ahorn shear zone, 24 and 12 Ma for the Tuxer shear zone and finally 20 and 7 Ma for the Greiner shear zone (Figure 5.3). Data related to cataclastic shearing are even rarer. However, the few available data constrain a time span similar to that for ductile deformation in the Tauern Window. Kralik et al.

(1987a) reported Rb/Sr and K/Ar ages from cataclastic faults of the Austroalpine units to the south-east of the Tauern Window that range from 32 to 30 and 23 to 22 Ma, respectively. Middle Miocene brittle faulting is also well documented along the northern rim of the Mauls tonalite (Zwingmann and Mancktelow 2004) and from the Giudicarie fault system (Pleuger et al. 2012). In very general terms, all available age data indicate a successive rejuvenation of fault activity toward the east (Figure 5.9). All the superordinate structures within and adjacent to the Tauern Window (DAV, Mölltal fault, Periadriatic fault system, SEMP) where active as mylonitic shear zones under amphibolite to greenschist facies conditions during Late Oligocene times (Glodny et al. 2008) (Figures 5.9(a) and 5.10). Some of these shear zones operated up to Middle Miocene times (Periadriatic fault system, SEMP, Brenner normal fault, shear zones within the western Tauern Window), either continuously or intermittently, with pronounced activity at ~ 21 to 15 Ma (Läufer et al. 1997; Müller et al. 2001; Glodny et al. 2008) (Figures 5.9(b) and (c) and 5.10). Furthermore, fault zones at the eastern border of the Eastern Alps are still active today, namely the eastern part of the SEMP and the Mur-Mürz fault system as well as the Pöls-Lavanttal fault system (e.g. Plan et al. 2010; Baroň et al. 2019) (Figure 5.9(d)). In summary, extrusion-related faulting continuously propagated from the western toward the central eastern part of the Eastern Alps (Figure 5.(9)).

5.6.2.3. *Formation of the Tauern Window*

The formation of the Tauern Window itself is attributed to a combination of tectonic and erosional unroofing in response to northward indentation of the Adriatic Plate into the warm and thick Alpine orogenic crust (e.g. Ratschbacher et al. 1991; Rosenberg et al. 2007, 2018; Scharf et al. 2013; Favaro et al. 2017; Reiter et al. 2017) contemporaneous to the lateral extrusion of the Austroalpine Nappe Systems in the hanging-wall of the Penninic and Sub-Penninic edifice within the Tauern Window (see below). Final doming, accompanied by north-south shortening and orogen-parallel stretching, was accommodated along low-angle normal faults at either end of the Tauern Window, respectively, the W- dipping Brenner and the E- dipping Katschberg normal faults (Figure 5.3) from ~23 Ma onwards (Selverstone 1988, 2005; Behrmann 1989; Genser and Neubauer 1989; Fügenschuh et al. 1997) (see also Frisch et al. 1998, 2000; Scharf et al. 2013; Favaro et al. 2015, 2017; Rosenberg et al. 2018; Wolff et al. 2020). Two basement subdomes at either end of the Tauern Window expose the Venediger Duplex, being subsequently modified by approximately east-west trending folds (e.g. Scharf et al. 2013). Austroalpine blocks north and east of the Tauern Window moved toward the east during lateral extrusion. The normal faults are kinematically linked to conjugate strike-slip shear zones and strike-slip faults (Figure 5.3). The most important of these are the

Salzach-Ennstal-Mariazell-Puchberg Fault System (SEMP) along the northern margin of the Tauern Window (Wang and Neubauer 1998; Linzer et al. 2002; Rosenberg and Schneider 2008), the Periadriatic fault, the Giudicarie fault and the Mölltal fault (Kurz and Neubauer 1996; Linzer et al. 2002) (for a detailed review of several fault systems, we refer to Rosenberg et al. 2018; and references therein).

Northward motion of the Southalpine indenter along the sinistral Giudicarie Belt (Figure 5.3) offsets the Periadriatic fault and triggered rapid exhumation of orogenically thickened crust within the entire Tauern Window (Scharf et al. 2013; Reiter et al. 2017). Kinematic models show that lateral extrusion accommodated along these conjugate strike-slip fault systems requires large amounts of north-south shortening in the western part of the Eastern Alps (Rosenberg et al. 2018), being consistent with upright folding and erosion within the Tauern Window. In particular for the western part of the Tauern Window, this is assumed to have contributed to the largest amounts of exhumation. Kinematic restorations indicated that north-south shortening between the NCA and the Southalpine Indenter is in the range of 75 km in the western part of the Eastern Alps, and up to 30 km in the eastern part. Note that 55 km of shortening was accommodated in the area of the Tauern Window (Reiter et al. 2017; Rosenberg et al. 2018). In the eastern part of the Tauern Window, exhumation involved strike-slip and normal faulting that accommodated about 100 km of orogen-parallel extension, contemporaneous with about 30 km of orogen-perpendicular, north–south shortening (Scharf et al. 2013; Favaro et al. 2017).

In general, cooling and exhumation of the units within the Tauern Window was not a uniform process. According to slightly older cooling ages of the eastern Tauern Subdome, it was proposed that cooling and exhumation started earlier in the eastern Tauern Window (Dunkl et al. 2005; Luth and Willingshofer 2008; Scharf et al. 2013a, 2016; Favaro et al. 2015, 2017). Alternatively, Betrand et al. (2017) suggested coeval exhumation of the western and eastern Tauern subdomes, but with different rates, that is, higher rates in the western subdome.

Exhumation of the Rechnitz Window (Figure 5.1) and the Tauern Window generally occurred within the same time range (e.g. Frisch et al. 2000). Subsequent to Barrow-type metamorphic overprint of Penninic and Sub-Penninic nappes (Bousquet et al. 2008) with a peak between 34 and 28 Ma (Cliff et al. 1985; Selverstone et al. 1992; Reddy et al. 1993; Christensen et al. 1994; Inger and Cliff 1994; Schneider et al. 2015), syncollisional exhumation of both, the Rechnitz and Tauern Window occurred from Middle to Late Miocene times, as inferred from zircon fission track (ZFT) and apatite fission track (AFT) data, ranging from 22 to 13 Ma and 7 to 9 Ma, respectively (Dunkl and Demeny 1997).

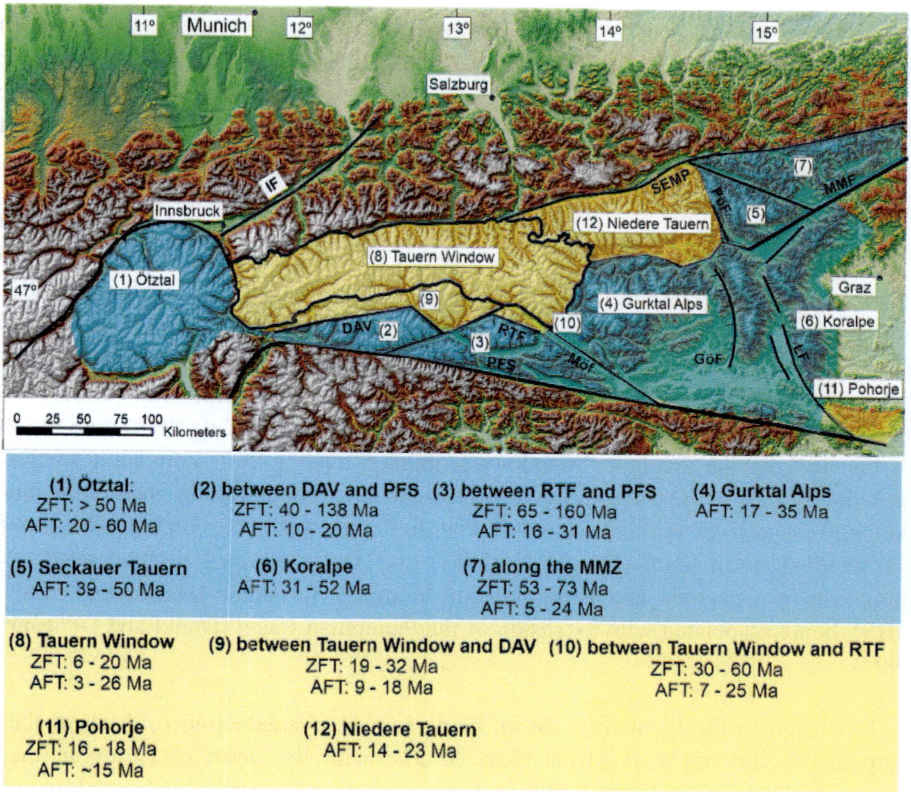

Figure 5.11. *Illustration of thermochronological ages in the Eastern Alps*

COMMENT ON FIGURE 5.11.– *The blue-colored areas are mainly characterized by pre-extrusion zircon and apatite fission track ages. The yellow-colored areas mainly show extrusion-related cooling ages. Data are from (1): Fügenschuh (1995); Fügenschuh et al. (1997); Elias (1998); Viola et al. (2001); Pomella et al. (2011, 2012). (2): Grundmann and Morteani (1985); Coyle (1994); Fügenschuh (1995); Elias (1998); Stöckhert et al. (1999); Mancktelow et al. (2001); Steenken et al. (2002); Most (2003); Di Fiore (2013). (3): Staufenberg (1987); Dunkl et al. (2003); Wölfler et al. (2008, 2015a, 2015b). (4): Hejl (1997); Wölfler et al. (2016); (5): Hejl (1997); Reinecker (2000); (6): Wölfler et al. (2010); Kurz et al. (2011); (7) Van Gelder et al. (2020). (8): Staufenberg (1987); Fügenschuh et al. (1997); Most (2003); Dunkl et al. (2005); Foeken et al. (2007); Wölfler et al. (2008, 2012); Betrrand et al. (2018); Wolff et al. (2020). (9): Grundmann and Morteani (1985); Coyle (1994); Stöckhert et al. (1999); Steenken et al. (2002); Most (2003). (10): Staufenberg (1987); Dunkl et al. (2005); Wölfler et al. (2008). (11) and*

(12): Hejl (1997); Reinecker (2000); Wölfler et al. (2016). PFS: Periadriatic fault system; DAV: Defereggen-Antholz-Vals fault; IF: Inntal fault; SEMP: Salzachtal-Ennstal-Mariazell-Puchberg fault system; DAV: Defereggen-Antholz-Vals fault; RTF: Ragga Teuchl fault; MöF: Mölltal fault; LF: Lavanttal fault (as southern part of the Pöls-Lavanttal fault system); PöF: Pöls fault (as northern part of the Pöls-Lavanttal fault system); MMF: Mur-Mürz fault system; GöF: Görtschitz Valley fault.

Oligocene structures and kinematics within the Rechnitz Window, however, differ from those within the Tauern Window. The Rechnitz Window is located in the northern part of the Styrian Basin (Figure 5.1), merging into the Pannonian Basin. This domain was exclusively affected by extension throughout the Miocene. The Moho is situated at ~30 km depth underneath the Rechnitz Window, in contrast to approximately 55 km underneath the Tauern Window (Spada et al. 2013). The crust in the area of the Rechnitz Window is thinner by one-third compared to the present-day one in the Tauern Window (Rosenberg et al. 2018). In contrast to the doming-related folds in the Tauern Window, fold axes within the Rechnitz Window are trending north-south (Cao et al. 2013) and are interpreted to be related to unroofing by east-west extension and core complex formation (Rosenberg et al. 2018), being associated with extension of the Pannonian Basin (Dunkl and Demeny 1997).

Extension in the Pannonian Basin, being related to subduction rollback in the Carpathians, did not start earlier than 20 Ma, with the onset of rifting in the Pannonian Basin (Horváth et al. 2006). Extension related to strike-slip and oblique-slip faulting in the Eastern Alps started at 18 Ma in the Styrian Basin, and at about 17 Ma in various intramontane basins, such as the Fohnsdorf Basin and at 16 Ma in the Vienna Basin (see review by Wölfler et al. 2011). Subduction roll-back in the Carpathians is therefore assumed to have no affect on the Eastern Alps before about 17 Ma. Thermochronological data from the Tauern Window indicate that rapid cooling started earlier, that is, before 20 Ma (Fügenschuh et al. 1997; Wölfler et al. 2011; see compilation of data by Luth and Willingshofer 2008). Stratigraphic and radiometric ages constrain exhumation of the (Sub-) Penninic nappe complex to have started around 23–21 Ma in the eastern part of the Tauern Window, being related to unroofing along the Katschberg shear zone (Scharf et al. 2013). Subduction roll-back is therefore assumed to have temporarily enhanced rather than triggered exhumation and orogen-parallel kinematics in the central Eastern Alps (Wölfler et al. 2011; Scharf et al. 2013; Schmid et al. 2013). Pre-20 Ma exhumation of the (Sub-) Penninic nappe complex within the Tauern Window was therefore mainly triggered by north–south shortening in front of the Southalpine indenter and related upright folding and erosion (Rosenberg et al. 2018).

5.6.3. *Constraints on vertical movements – late Alpine cooling*

The timing of vertical movement of fault-bound blocks in the Eastern Alps is documented by thermochronological data from the crustal blocks adjacent to the fault zones. From the thermochronological point of view, the Eastern Alps can be subdivided into areas that display pre-extrusion zircon and apatite fission track cooling ages (ZFT and AFT, respectively) and areas that are characterized by cooling ages that are related to Miocene extrusion (Figure 5.11).

Most of the Austroalpine units cooled below 300°C before Cenozoic times (Frank et al. 1987b; Oberhänsli et al. 2004; Handy and Oberhänsli 2004; Luth and Willingshofer 2008; Van Gelder et al. 2020) (blue-colored areas in Figure 5.11). However, the area to the southwest of the Tauern Window and the DAV is an exception with Oligocene to Miocene $^{87}Rb/^{86}Sr$ biotite ages (Borsi et al. 1973, 1978) and a Miocene cooling history as recorded from zircon and apatite fission track data (Grundmann and Morteani 1985; Fügenschuh et al. 1997; Steenken et al. 2002; Most 2003) (Figure 5.11). A major jump in cooling ages only occurs across the DAV with mostly Variscan $^{87}Rb/^{86}Sr$ data (Borsi et al. 1978). According the tectonothermal definition, Frisch et al. (2000a) referred the Austroalpine units to the north of the DAV as a part of the so-called Tauern thermal dome (yellow-colored area between the Tauern Window and the DAV in Figure 5.11). The same applies to the area SW of the Tauern Window and the Ragga-Teuchl fault where ZFT, AFT and apatite (U-Th)/He (AHe) data indicate Miocene cooling contemporaneous to cooling the Penninic and Sub-Penninic units of the Tauern Window (Staufenberg 1987; Wölfler et al. 2008, 2015a, 2015b) (Figure 5.11).

The Austroalpine units to the east of the Tauern Window are characterized by mainly Cretaceous $^{87}Rb/^{86}Sr$, $^{40}K/^{39}Ar$, $^{40}Ar/^{39}Ar$ and ZFT data (e.g. Frank et al. 1987a; Wölfler et al. 2010; Kurz et al. 2011; Rosenberg et al. 2018; Van Gelder et al. 2020). In contrast, AFT and AHe data are heterogeneously distributed, indicating that the Austroalpine units to the east of the Tauern Window are possibly separated into distinct tectonic blocks with different cooling histories throughout the Cenozoic Era. In this context, the Gurktal and Koralpe blocks referred to as “cold spots” by Hejl (1997), cooled through the AFT closure temperature already in the Eocene and Oligocene and have since that time remained in upper crustal sections (Hejl 1997, 1998; Reinecker 2000; Wölfler et al. 2010; Legrain et al. 2014) (Figure 5.11). On the other hand, the Niedere Tauern and the Pohorje block, termed as “hot spots” by Wölfler et al. (2011), cooled through the uppermost crust during the Middle to Late Miocene (Reinecker 2000; Legrain et al. 2014; Wölfler et al. 2016) (Figure 5.11).

To the west of the Tauern Window, in the Ötztal-Stubai area, ZFT data document Cretaceous cooling through the 240°C isotherm (Fügenschuh et al. 1997, 2000; Elias 1998; Viola et al. 2001; Pomella et al. 2011, 2012) (Figure 5.11). AFT data are very heterogeneously distributed, ranging from ~60 Ma in the west to <20 Ma to the east, particularly near the Brenner normal fault (Fügenschuh et al. 1997, 2000; Elias 1998; Viola et al. 2001; Pomella et al. 2012). This trend may be explained by tilting of the hanging-wall Ötztal basement during rotation of the Brenner normal fault (Buck 1988; Rosenberg et al. 2018).

The most recent episode of cooling and exhumation as recorded by low-temperature thermochronology is documented at the southern and southwestern border of the Eastern Alps, namely in the Karawanken mountains and along the Giudicarie fault system. Low-temperature data, in particular AHe ages, indicate a Late Miocene episode of fault-related exhumation leading to the formation of the Karawanken flower structure and the sedimentary infill of the Klagenfurt Basin (Heberer et al. 2017) (Figure 5.3). Middle to Late Miocene AFT ages and AHe cooling ages from the Valsugana and the Val Trompia thrusts, as well as from the Tonale fault and the Giudiacarie fault system (Figure 5.3) in the Southern Alps indicate late orogenic exhumation that is related to the indentation of the Adriatic microplate (Zattin et al. 2006; Heberer et al. 2017).

5.6.4. *Deep structure of the Eastern Alps*

The pioneering studies of Lippitsch et al. (2003) and Kissling et al. (2006) were the first to question the classical concept of a south dipping oceanic subduction of Europe below the Eastern Alps (e.g. Hawkesworth 1975; Lüschen et al. 2004). Instead, these authors argued for a continental subduction of Adria below Europe, which is also observed by recent studies (Zhao et al. 2016; Hua et al. 2017). However, Dando et al. (2011) and Mitterbauer et al. (2011) interpret the imaged slab as a remnant of the European plate, and Brückl et al. (2007, 2010) suggest a fragmentation of the lithosphere beneath the Eastern Alps into three blocks: Europe, Adria and the so-called Pannonian fragment.

The transition from a highly shortened, thickened and exhumed crust in the western part of the Eastern Alps, that is, in the area of the Tauern Window, to a much less shortened and exhumed wedge east of the Katschberg fault, mainly characterized by orogen-parallel displacements, spatially coincides with a change in the deep structure of an inferred European slab revealed by teleseismic tomography (Mitterbauer et al. 2011; Rosenberg et al. 2018; Kästle et al. 2020). This slab does not continue east of 14° geographic longitude and disappears into a low velocity anomaly, where no significant crustal thickening has taken place there during collision. A deeper slab observed in that area is assumed to be derived from Penninic oceanic lithosphere that broke off from the southward dipping European continental margin (Mitterbauer et al. 2011; Qorbani et al. 2015; Rosenberg et al. 2018) (Figure 5.12).

Recent data and high-resolution tomography models (Sun et al. 2019; Zhao et al. 2016) revealed, however, a more complex interaction between the European and the Adriatic slabs that argue for the presence of a slab edge to the east of the Adamello intrusion (Ji et al. 2019). Based on available tomography, this slab edge is considered to have formed by vertical tearing of the European slab in Eocene times, after the onset of subduction of the Adriatic plate beneath the Dinarides.

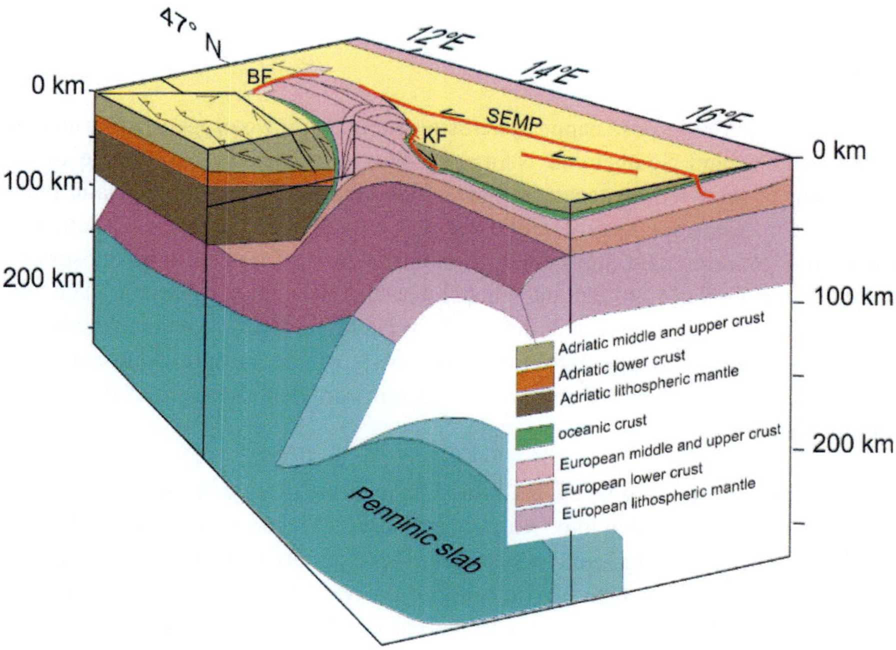

Figure 5.12. *Schematic, evolutionary 3D diagram of the deep structure of the Eastern Alps (modified after Rosenberg et al. 2018), displaying the amplification and exhumation of antiformal stack of Sub-Penninic nappes in the Tauern Window. Referring to Mitterbauer et al. (2011) and Qorbani et al. (2015) the attached European slab does not continue east of 14° longitude, where no significant crustal thickening has taken place there during collision, but a deeper slab is observed*

A recent study by Kästle et al. (2020) provides a detailed review of the above-mentioned models and suggests a new scenario, where the northward dipping slab under the Eastern Alps is caused by imaging a combination of the Adriatic slab and a deeper lying European one. Most of these studies argue for a strong link to the deformation in the Dinarides, where the Adriatic plate is in lower plate position (e.g. Lippitsch et al. 2003; Kissling et al. 2006), so that the Neogene evolution of the Eastern Alps was strongly influenced by the Dinaric slab (e.g. Kissling et al. 2006; Wölfler et al. 2011). Changing dynamic conditions due to collision in the Carpathians during the Late Miocene (Matenco et al. 2010) therefore triggered a deformational response in the Alpine-Carpathian-Pannonian (ALCAPA)-Adria transition. Fortunately, in the near future, the AlpArray experiment (Hetényi et al. 2018) will shed more light into the complex slab geometries beneath the Eastern Alps.

5.7. Conclusion

The Eastern Alps expose nappe systems resulting from two orogenies. Formation of Austroalpine and Southalpine Nappe Systems is related to the closure of an embayment of the Meliata ocean, as part of the Neotethys, into the Adriatic plate from Late Jurassic times onwards, followed by Cretaceous continental collision. The formation of Penninic and Sub-Penninic Nappe Systems is related to the closure of the Penninic (Piemont-Ligurian and Valais) ocean(s) between the Adriatic plate and Europe. The latter also affected the north(-western) Adriatic continental margin (Lower Austroalpine Nappe System). Large parts of the Eastern Alps comprise nappe systems that originate from the southeastern continental margin of the former Adriatic plate.

The Paleo- and Eo-Alpine cycles, which formed the present Austroalpine nappe complex, may be subdivided into two distinct phases: (1) the (ES)E-ward subduction and closure of the Meliata basin, resulting in the assembly of nappe systems in the NCA, including its basement. After closing of the Meliata oceanic basin during the Late Jurassic, the Cretaceous orogeny in the Eastern Alps encompasses the collision between (south)easternmost parts of the Austroalpine continental crust and a continental fragment to the east. (2) Southward underplating of the northern Adriatic continental margin resulted in the imbrication of the Central Austroalpine basement complexes. These units were additionally affected by pronounced metamorphim, increasing from greenschist facies conditions in the northern parts to amphibolite and eclogite facies conditions in the southernmost parts.

Kinematics and its timing suggest that the subduction of the Meliata ocean during Late Jurassic times and the underthrusting of the Lower Central Austroalpine unit during late Early Cretaceous times are not directly related to each other. The closure of the Meliata basin was followed by progressive forward propagation of nappes during continental subduction along the Eo-Alpine high-pressure belt. Accordingly, nappe assembly within the Austroalpine unit would be related to a discontinuous, two-phase evolution.

Continuous underplating was accompanied by extension in the internal parts of the orogen, and the exhumation of high-pressure metamorphic rocks during the Late Cretaceous. Coeval extension in the upper plate resulted in the formation of the Gosau sedimentary basins. Tectonic reconstructions of the Eastern Alps therefore have to include Late Cretaceous disintegration and exhumation by extension. Recognition of Eo-Alpine structures and metamorphism, however, is hampered by the fact that the spatial distribution of structural elements with Cretaceous age coincide frequently with Miocene extrusion-related structures. From geochronological and tectono-metamorphic arguments, there is strong evidence that extrusion tectonics played a major role in Alpine architecture during the latest Cretaceous as well.

The present structure of the Eastern Alps is characterized by a system of fault zones that developed during Late Oligocene to Miocene times, related to orogen-parallel escape of Austroalpine units toward east, a process also termed lateral extrusion. The southward subduction of the European margin beneath the Adriatic Plate (Neo-Alpine cycle) resulted in nappe assembly and high-pressure metamorphism of Sub-Penninic nappes within the Eastern Alps at 33–32 Ma. Although accompanied by massive crustal thickening, this evolution within the lower plate is hardly reflected in the upper plate. Intramontane basin formation and distinct signatures of fault activity indicate orogen-parallel faulting either prior or subsequent to Sub-Penninic nappe assembly. During Oligocene to Middle Miocene times extrusion-related faulting continuously propagated from the western toward the central eastern part of the Eastern Alps. Sub-Penninic high-pressure assemblages of the Lower European Plate were exhumed contemporaneously. This is confined by oblique convergence between the Adriatic and European plates.

5.8. References

Auer, M. and Eisbacher, G.H. (2003). Deep structure and kinematics of the Northern Calcareous Alps (TRANSALP Profile). *Int. J. Earth Sci. (Geol. Rdsch.)*, 92, 210–227.

Bada, G., Horváth, F., Gerner, P., Fejes, I. (1999). Review of the present-day geodynamics of the Pannonian basin: Progress and problems. *Journal of Geodynamics*, 27, 501–527.

Baroň, I., Plan, L., Sokol, L., Grasemann, B., Melichar, R., Mitrovic, I., Stemberk, J. (2019). Present-day kinematic behavior of active faults in the Eastern Alps. *Tectonophysics*, 752, 1–13.

Bauer, C., Rubatto, D., Krenn, K., Proyer, A., Hoinkes, G. (2007). A zircon study from the Rhodope metamorphic complex, N-Greece: Time record of a multistage evolution. *Lithos*, 29, 207–228.

Beck-Mannagetta, P. (1952). Zur Geologie und Paläontologie des Tertiärs des unteren Laventtales. *Jahrb. Geol. Bundesanst.*, 95, 1–102.

Behrmann, J.H. (1988). Crustal-scale extension in a convergent orogen: The Sterzing-Steinach mylonite zone in the Eastern Alps. *Geodin. Acta*, 2, 63–73.

Berger, A. and Bousquet, R. (2008). Subduction-related metamorphism in the Alps: Review of isotopic ages based on petrology and their geodynamic consequences. *Geological Society of London Special Publications*, 298, 117–144.

Bertrand, A., Rosenberg, C.L., Garcia, S. (2015). Fault slip analysis and late exhumation of the Tauern Window, Eastern Alps. *Tectonophysics*, 649, 1–17.

Bertrand, A., Rosenberg, C.L., Fügenschuh, B., Rabaute, A., Herman, F. (2017). Exhumation mechanisms of the Tauern Window (Eastern Alps) inferred from apatite and zircon fission-track thermochronology. *Tectonics*, 36, 207–228. doi:10.1002/2016TC004133.

Bögel, H. (1975). Zur Literatur über die "Periadriatische Naht". *Verh. Geol. B.-A.*, 2–3, 163–199.

Bonev, N. and Stampfli, G. (2008). Petrology, geochemistry and geodynamic implications of Jurassic island arc magmatism as revealed by mafic volcanic rocks in the Mesozoic low-grade sequence, eastern Rhodope, Bulgaria. *Lithos*, 100, 210–233.

Borsi, S., Del Moro, A., Sassi, F.P., Zirpoli, G. (1973). Metamorphic evolution of the Austridic rocks to the south of the Tauern Window (Eastern Alps). Radiometric and geologic data. *Memorie della Societá Geologica Italiana*, 12, 549–571.

Borsi, S., Del Moro, A., Sassi, F., Zanferrari, A., Zirpoli, G. (1978). New geopetrologic and radiometric data on the alpine history of the Austridic continental margin south of the Tauern Window (Eastern Alps). *Mem. Istituto della Regia Univ. di Padova*, 32, 1–17.

Borsi, S., Del Moro, A., Sassi, F.P., Zirpoli, G. (1979). On the age of the Vedrette di Ries (Rieserferner) massif and its geodanamic significance. *Geol. Rundsch.*, 69(1), 41–60.

Bousquet, R., Oberhänsli, R., Goffé, B., Jolivet, L., Vidal, O. (1998). High pressure – Low temperature metamorphism and deformation in the Bündnerschiefer of the Engadine window: Implications for the regional evolution of the eastern Central Alps. *J. Metamorphic Geol.*, 16, 657–674.

Bousquet, R., Goffé, B., Vidal, O., Oberhänsli, R., Patriat, M. (2002). The tectono-metamorphic history of the Valaisan domain from the Western to the Central Alps: New constraints on the evolution of the Alps. *Geol. Soc Amer. Bull.*, 114, 207–225.

Bousquet, R., Oberhänsli, R., Goffé, B., Wiederkehr, K., Koller, F., Schmid, S.M., Schuster, R., Engi, M., Berger, A., Martinotti, G. (2008). Metamorphism of metasedmiments in the scale of an orogen: A key to the Tertiary geodynamic evolution of the Alps. In *Tectonic Aspects of the Alps-Dinarides-Carpathians*, Siegesmund, S., Fügenschuh, B., Froitzheim, N. (eds). *Geol. Soc. Lond. Spec. Publ. No.* Geological Society, London.

Bousquet, R., Oberhänsli, R., Schmid, S., Berger, A., Wiederkehr, M., Robert, C., Möller, A., Rosenberg, C.L., Eilinger, G., Molli, G. et al. (2012a). Metamorphic framework of the Alps-Carte metamorphique des Alpes. *CCGM/CGMW* [Online]. Available at: https://perso.univ-rennes1.fr/romain.bousquet/Alps/Maps/metam.html.

Bousquet, R., Schmid, S.M., Zeilinger, G., Oberhänsli, R., Rosenberg, C.L., Molli, G., Robert, C., Wiederkehr, M., Rossi, P. (2012b). Tectonic framework of the Alps. *CCGM/CGMW* [Online]. Available at: https://ccgm.org/en/catalogue/126-carte-tectonique-des-alpes-9782917310168.html.

Brückl, E., Bleibinhaus, F., Gosar, A., Grad, M., Guterch, A., Hrubcová, P., Keller, G.R., Majdanki, M., Sumanovac, F., Tiira, T. et al. (2008). Crustal structure due to collision and escape tectonics in the Eastern Alps region based on profiles Alp01 and Alp02 from the ALP 2002 seismic experiment. *Journal of Geophysical Research*, 112, B06308.

Brückl, E., Behm, M., Decker, K., Grad, M., Guterch, A., Keller, G.R., Thybo, H. (2010). Crustal structure and active tectonics in the Eastern Alps. *Tectonics*, 29, TC2011.

Buck, W.R. (1988). Flexural rotation of normal faults. *Tectonics*, 7, 959–973.

Butt, A. (1981). Depositional environments of the upper Cretaceous rocks in the northern part of the Eastern Alps. *Cushman Found. Foram. Res. Spec. Publ.*, 20, 1–81.

Cao, S., Neubauer, F., Bernroider, M., Liu, J., Genser, J. (2013). Structures, microfabrics and textures of the Cordilleran-type Rechnitz metamorphic core complex, Eastern Alps. *Tectonophysics*, 608, 1201–1225.

Castellarin, A. and Cantelli, L. (2000). Neo-Alpine evolution of the Southern Eastern Alps. *J. Geodyn.*, 30, 251–274.

Cavazza, W., Roure, F.M., Spakman, W., Stampfli, G.M., Ziegler, P.A. (eds) (2004). *The TRANSMED Atlas: The Mediterranean Region from Crust to Mantle.* Springer, Berlin/Heidelberg/New York.

Channel, J.E.T. and Kozur, H.W. (1997). How many oceans? Meliata, Vardar, and Pindos oceans in Mesozoic Alpine paleogeography. *Geology*, 25, 183–186.

Christensen, J.N., Selverstone, J., Rosenfeld, L., DePaolo, D.J. (1994). Correlation by Rb-Sr geochronology of garnet growth histories from different structural levels within the Tauern Window, Eastern Alps. *Contrib. Mineral. Petrol.*, 118, 1–12.

Cliff, R.A., Droop, G.T.R., Rex, D.C. (1985). Alpine metamorphism in the south-east Tauern Window, Austria II: Heating, cooling and uplift rates. *J. Metamorphic Geol.*, 3, 403–415.

Cliff, R.A., Oberli, F., Meier, M., Droop, G.T.R., Kelly, M. (2015). Syn-metamorphic folding in the Tauern Window, Austria dated by Th–Pb ages from individual allanite porphyroblasts. *J. Metamorph. Geol.*, 33, 427–435.

Cornelius, H.P. and Clar, E. (1934). *Geologische Karte des Großglocknergebietes.* Geologische Bundesanstalt, Vienna.

Dachs, E. and Proyer, A. (2001). Relics of high-pressure metamorphism from the Grossglockner region, Hohe Tauern, Austria. *European Journal of Mineralogy*, 13(1), 67–86.

Dachs, E., Kurz, W., Proyer, A. (2005). Alpine eclogites in the Tauern Window. *Mitteilungen der Österreichischen Mineralogischen Gesellschaft*, 150, 199–226.

Dal Piaz, G.V., Martin, S., Villa, I.M., Gosso, G., Marschalko, R. (1995). Late Jurassic blueschist facies pebbles from the Western Carpathian orogenic wedge and paleostructural implications for western Tethys evolution. *Tectonics*, 14, 874–885.

Dallmeyer, R.D., Neubauer, F., Handler, R., Fritz, H., Müller, W., Pana, D., Putis, M. (1996). Tectonothermal evolution of the internal Alps and Carpathians: Evidence from 40Ar/39Ar mineral and whole rock data. *Eclogae Geol. Helv.*, 89, 203–227.

Dallmeyer, R.D., Handler, R., Neubauer, F., Fritz, H. (1998). Sequence of thrusting within a thick-skinned tectonic wedge: Evidence from 40Ar/39Ar and Rb-Sr ages from the Austroalpine nappe complex of the Eastern Alps. *The Journal of Geology*, 106, 71–86.

Dallmeyer, R.D., Neubauer, F., Fritz, H. (2008). The Meliata suture in the Carpathians: Regional significance and implications for the evolution of high-pressure wedges within collisional orogens. *Geological Society of London Special Publication*, 298, 101–115.

Dando, B., Stuart, G., Houseman, G., Hegedüs, E., Brückl, E., Radovanović, S. (2011). Teleseismic tomography of the mantle in the Carpathian–Pannonian region of central Europe. *Geophys. J. Int.*, 186(1), 11–31.

Davis, H.J. and von Blankenburg, F. (1995). Slab breakoff: A model of lithospheri detachment and its test in the magmatism and deformation of collisional orogenes. *Earth Planet. Sci. Lett.*, 129, 85–102.

Decker, K. (1990). Plate tectonics and pelagic facies: Late Jurassic to Early Cretaceous deep-sea sediments of the Ybbsitz ophiolite unit (Eastern Alps, Austria). *Sedimentary Geology*, 67, 85–99.

Decker, K. and Peresson, H. (1996). Tertiary kinematics in the Alpine-Carpathian-Pannonian system: Links between thrusting, transform faulting and crustal extension. In *Oil and Gas in Alpidic Thrustbelts and Basins of Central and Eastern Europe, EAGE Special Publication*, volume 5, Wessely, G. and Liebl, W. (eds). EAGE Special Publication, Bunnik.

Decker, K., Peresson, H., Faupl, P. (1994). Die miozäne Tektonik der östlichen Kalkalpen: Kinematik, Paläospannung und Deformationsaufteilung während der "lateralen Extrusion" der Zentralalpen. *Jahrb. Geol. Bundesanst.*, 137, 5–18.

Deutsch, A. (1984). Young Alpine dykes south of the Tauern Window (Austria): A K-Ar and Sr isotope study. *Contrib. Mineral. Petrol.*, 45, 45–57.

Dimo-Lahitte, A., Monie, P., Vergely, P. (2001). Metamorphic soles from the Albanian ophiolites, petrology, $^{40}Ar/^{39}Ar$ geochronology, and geodynamic evolution. *Tectonics*, 20, 78–96.

Droop, G.T.R. (1985). Alpine metamorphism in the south-east Tauern Window, Austria: 1. P-T variations in space and time. *J. Met. Geol.*, 3(4) 371–402.

Dunkl, I. and Demeny, A. (1997). Exhumation of the Rechnitz Window at the border of the Eastern Alps and Pannonian Basin during Neogene extension. *Tectonophysics*, 272, 197–211.

Dunkl, I., Frisch, W., Grundmann, G. (2003). Zircon fission track thermochronology of the southeastern part of the Tauern Window and adjacent Austroalpine margin, Eastern Alps. *Eclogae Geol. Helv.*, 96, 209–217.

Dunkl, I., Kuhlemann, J., Reinecker, J., Frisch, W. (2005). Cenozoic relief evolution of the Eastern Alps – Contraints from apatite fission track age-provenance of Neogene intramontane sediments. *Austrian Journal of Earth Sciences (Mitteilungen der Österreichischen Geologischen Gesellschaft)*, 98, 92–105.

Ebner, F. and Rantitsch, G. (2000). Das Gosaubecken von Kainach – Ein Überblick. *Mitt. Ges. Geol. Bergbaustud. Österr.*, 44, 157–172.

Ebner, F., Vozarova, A., Kovacs, S., Krautner, H.-G., Branislav, K., Szederkeny, T., Jamicic, D., Balen, D., Belak, M., Trajanova, M. (2008). Devonian–Carboniferous pre-flysch and flysch environments in the Circum Pannonian region. *Geol. Carpath.*, 59, 159–195.

Eder, N. and Neubauer, F. (2000). On the edge of the extruding wedge: Neogene kinematics and geomorphology along the southern Niedere Tauern, Eastern Alps. *Eclogae Geol. Helv.*, 93, 81–92.

Egger, H., Bichler, M., Homayoun, M., Kirchner, E.C., Surenian, R. (1996). Späteozäne Bentonite aus der Gosau-Gruppe des Untersberg-Vorlandes (Nördliche Kalkalpen, Salzburg). *Jahrb. Geol. Bundesanst.*, 139, 13–21.

Elias, J. (1998). The thermal history of the Ötztal–Stubai complex (Tyrol, Austria/Italy) in the light of the lateral extrusion model. *Tübinger Geowiss. Arb., Reihe A*, 42, 1–172.

Eisbacher, G. and Brandner, R. (1995). Role of high-angle faults during heteroaxial contraction. Inntal thrust sheet, Northern Calcareous Alps, western Austria. *Geol. Paläont. Mitt. Innsbruck*, 20, 389–406.

Eisbacher, G. and Brandner, R. (1996). Superposed fold-thrust structures and high-angle faults, Northwestern Calcareous Alps, Austria. *Eclogae Geol. Helv.*, 89, 553–571.

Eisbacher, G., Linzer, H.G., Meier, L., Polinski, R. (1990). A depth-extropolated structural transect across the Northern Calcareous Alps of western Tirol. *Eclogae Geol. Helv.*, 83, 711–725.

Engi, M., Berger, A., Roselle, G.T. (2001). Role of the tectonic accretion channel in collisional orogeny. *Geology*, 29, 1143–1146.

Exner, C. (1976). Die geologische Position der Magmatite des periadriatischen Lineamentes. *Verh. Geol. B.-A.*, 1976, 3–64.

Faryad, S.W. (1995). Phase petrology and P-T conditions of mafic blueschists from the Meliata Unit, West Carpathians, Slovakia. *Journal of Metamorphic Geology*, 13, 701–714.

Faryad, S.W. (1999). Nature and geotectonic interpretation of the Meliata blueschists. *Geologica Carpathica*, 50, 97–98.

Faryad, S.W. and Henjes-Kunst, F. (1997). Petrological and K-Ar and 40Ar/39Ar age constraints for the tectonothermal evolution of the high-pressure Meliata unit, Western Carpathians (Slovakia). *Tectonophysics*, 280, 141–156.

Faupl, P. and Wagreich, M. (1992a). Cretaceous flysch and pelagic sequences of the Eastern Alps: Correlations, heavy minerals, and paleogeographic implications. *Cretaceous Res.*, 13, 387–403.

Faupl, P. and Wagreich, M. (1992b). Transgressive Gosau (Coniac) auf Branderfleckschichten (Turon) in den Weyrer Bögen (Nördliche Kalkalpen, Oberösterreich). *Jahrb. Geol. Bundesanst.*, 135, 481–491.

Faupl, P. and Wagreich, M. (2000). Late Jurassic to Eocene palaeogeography and geodynamic evolution of the Eastern Alps. *Mitt. Österr. Geol. Ges.*, 92, 79–94.

Faupl, P., Pober, E., Wagreich, M. (1987). Facies development of the Gosau Group of the eastern parts of the Northern Calcareous Alps during the Cretaceous and Paleogene. In *Geodynamics of the Eastern Alps*, Flügel, H.W. and Faupl, P. (eds). Deuticke, Vienna.

Favaro, S., Schuster, R., Handy, M.R., Scharf, A., Pestal, G. (2015). Transition from orogen perpendicular to orogen-parallel exhumation and cooling during crustal indentation – Key constraints from 147 Sm/144 Nd and 87 Rb/87 Sr geochronology (Tauern Window, Alps). *Tectonophysics*, 665, 1–16.

Favaro, S., Handy, M.R., Scharf, A., Schuster, R. (2017). Changing patterns of exhumation and denudation in front of an advancing crustal indenter, Tauern Window (Eastern Alps). *Tectonics*, 36(6), 1053–1071.

Florineth, D. and Froitzheim, N. (1994). Transition from continental to oceanic basement in the Tasna nappe (Engadine window, Graubünden, Switzerland): Evidence for Early Cretaceous opening of the Valais ocean. *Schweizerische Mineralogische und Petrographische Mitteilungen*, 74, 437–448.

Flügel, H.W. and Neubauer, F. (1984). *Erläuterungen zur geologischen Karte der Steiermark.* Geologische Bundesanstalt, Vienna.

Fodor, L., Jelen, B., Marton, E., Skaberne, D., Car, J., Vrabec, M. (1998). Miocene-Pliocene tectonic evolution of the Slovenian Periadriatic fault: Implications for the Alpine-Carpathian extrusion models. *Tectonics*, 17, 690–709.

Fodor, L., Csontos, L., Bada, G., Györfi, I., Benkovics, L. (1999). Tertiary tectonic evolution of the Pannonian basin system and neighbouring orogens: A new synthesis of paleostress data. *Geol. Soc. London Special Pub.*, 156, 295–334.

Fodor, L.I., Gerdes, A., Dunkl, I., Koroknai, B., Pècskay, Z., Trajanova, M., Horwàth, P., Vrabec, M., JeLen, B., Balogh, K. et al. (2008). Miocene emplacement and rapid cooling of the Pohorje pluton at the Alpine-Pannonian-Dinaric junction, Slovenia. *Swiss Journal of Earth Sciences*, 101, 255–271.

Foeken, J.P.T., Persano, C., Stuart, F.M., ter Voorde, M. (2007). Role of topography in isotherm perturbation: Apatite (U-Th)/He and fission track results from the Malta Tunnel, Tauern Window, Austria. *Tectonics*, 26, TC3006.

Frank, W. (1969): Geologie der Glocknergruppe. In *Wissenschaftliche Alpenvereinshefte des Deutschen Alpenvereins*, edition 21.

Frank, W. (1987). Evolution of the Austroalpine elements in the Cretaceous. In *Geodynamics of the Eastern Alps*, Flügel, H.W. and Faupl, P. (eds). Deuticke, Vienna.

Frank, W. and Schlager, W. (2006). Jurassic strike slip versus subduction in the Eastern Alps. *Int. J. Earth Sci.*, 95, 431–450.

Frank, W., Höck, V., Miller, C. (1987a). Metamorphic and tectonic history of the central Tauern Window. In *Geodynamics of the Eastern Alps*, Flügel, H.W. and Faupl, P. (eds). Deuticke, Vienna.

Frank, W., Kralik, S., Scharbert, S., Thöni, M. (1987b). Geochronological data from the Eastern Alps. In *Geodynamics of the Eastern Alps*, Flügel, H.W. and Faupl, P. (eds). Deuticke, Vienna.

Frank, N., Hauzenberger, C., Kurz, W., Dong, Y. (2018). Detrital U/Pb zircon age distribution in Alpine metasedimentary rocks of the Koralpe-Wölz nappe system (Eastern Alps). *Berichte der Geologischen Bundesanstalt*, 128, 25.

Frank, N., Kurz, W., He, D., Schuster, R., Dong, Y., Hauzenberger, C. (2019). Detrital U/Pb zircon age distribution in metasedimentary units of the Eastern Alps. *Geophysical Research Abstracts*, 21, EGU2019-17014-1.

Frey, M., Desmons, J., Neubauer, F. (1999). The new metamorphic map of the Alps: Introduction. *Schweiz. Mineral. Petrogr. Mitt.*, 79, 1–4.

Frisch, W. (1979). Tectonic progradation and plate tectonic evolution of the Alps. *Tectonophysics*, 60, 121–139.

Frisch, W. and Gawlick, H.-J. (2003). The nappe structure of the central Northern Calcareous Alps and its disintegration during Miocene tectonic extrusion – A contribution to understanding the orogenic evolution of the Eastern Alps. *Int. J. Earth Sci. (Geol. Rundsch.)*, 92, 712–727.

Frisch, W., Gommeringer, K., Kelm, U., Popp, F. (1987). The Upper Bündner Schiefer of the Tauern Window – A key to unterstanding Eoalpine Orogenic processes in the Eastern Alps. In *Geodynamics of the Eastern Alps*, Flügel, H.W. and Faupl, P. (eds). Deuticke, Vienna.

Frisch, W., Ring, U., Dürr, S., Borchert, S., Biehler, D. (1994). The Arosa zone and Platta Nappe ophiolites (Eastern Swiss Alps): Geochemical characteristics and their meaning for the evolution of the Penninic ocean. *Jahrb. Geol. Bundesanst.*, 137, 19–33.

Frisch, W., Kuhlemann, J., Dunkl, I., Brügel, A. (1998). Palinspastic reconstruction and topographic evolution of the Eastern Alps during late tertiary tectonic extrusion. *Tectonophysics*, 297, 1–16.

Frisch, W., Dunkl, I., Kuhlemann, J. (2000a). Post-collisional orogen-parallel large-scale extension in the Eastern Alps. *Tectonophysics*, 327, 239–265.

Frisch, W., Székely, B., Kuhlemann, J., Dunkl, I. (2000b). Geomorphological evolution of the Eastern Alps in response to Miocene tectonics. *Zeitschr. Geomorph.*, 44, 103–138.

Frisch, W., Kuhlemann, J., Dunkl, I., Székely, B. (2001). The Dachstein paleosurface and the Augenstein formation in the Northern Calcareous Alps – A mosaic stone in the geomorphological evolution of the Eastern Alps. *International Journal of Earth Sciences*, 90, 500–518.

Fritz, H. (1988). Kinematics and geochronology of early Cretaceous thrusting in the northwestern Paleozoic of Graz (Eastern Alps). *Geodinamica Acta*, 2, 53–62.

Fritz, H., Neubauer, F., Ratschbacher, L. (1991). Compression versus extension in the Paleozoic of Graz (Eastern Alps, Austria). *Zbl. Geol. Paläont. Teil I*, 55–68.

Froitzheim, N. (2001). Origin of the Monte Rosa nappe in the Pennine Alps – A new working hypothesis. *Geological Society of America Bulletin*, 113, 604–614.

Froitzheim, N. and Manatschal, G. (1996). Kinematics of Jurassic rifting, mantle exhumation, and passive margin formation in the Austroalpine and Penninic nappes (eastern Switzerland). *Geol. Soc. Amer. Bull.*, 108, 1120–1133.

Froitzheim, N., Schmid, S.M., Conti, P. (1994). Repeated change from crustal shortening to orogen – Parallel extension in the Austroalpine units of Graubünden. *Eclogae Geol. Helv.*, 87, 559–612.

Froitzheim, N., Schmid, S.M., Frey, M. (1996). Mesozoic paleogeography and the timing of eclogite facies metamorphism in the Alps: A working hypothesis. *Eclogae Geol. Helv.*, 89, 81–110.

Froitzheim, N., Conti, P., van Daalen, M. (1997). Late Cretaceous, synorogenic, low-angle normal faulting along the Schlinig fault (Switzerland, Italy, Austria) and its significance for the tectonics of the Eastern Alps. *Tectonophysics*, 280, 267–293.

Froitzheim, N., Plasienka, D., Schuster, R. (2008). Alpine tectonics of the Alps and Western Carpathians. In *The Geology of Central Europe. Volume 2: Mesozoic and Cenozoic*, McCann, T. (ed.). Geological Society, London.

Frost, E., Dolan, J., Sammis, C., Hacker, B., Cole, J., Ratschbacher, L. (2009). Progressive strain localization in a major strike-slip fault exhumed from midseismogenic depth: Structural observations from the Salzach-Ennstal-Mariazell Puchberg fault system, Austria. *J. Geophys. Res.*, 114, B04406.

Fügenschuh, B., Seward, D., Mancktelow, N. (1997). Exhumation in a convergent orogen: The western Tauern window. *Terra Nova*, 9, 213–217.

Fügenschuh, B., Mancktelow, N.S., Seward, D. (2000). Cretaceous to Neogene cooling and exhumation history of the Oetztal-Stubai basement complex, eastern Alps: A structural and fission track study. *Tectonics*, 19, 905–918.

Fügenschuh, B., Mancktelow, N.S., Schmid, S. (2012). Comment on Rosenberg and Garcia: Estimating displacement along the Brenner fault and orogen-parallel extension in the Eastern Alps. *International Journal of Earth Sciences*, 101, 1451–1455.

Gawlick, H.-J. and Missoni, S. (2019). Middle-Late Jurassic sedimentary mélange formation related to ophioliteobduction in the Alpine-Carpathian-Dinaridic Mountain Range. *Gondwana Research*, 74, 144–172.

Gawlick, H.-J. and Schlagintweit, F. (2006). Berriasian drowning of the Plassen carbonate platform at the type-locality and its bearing on the early Eoalpine orogenic dynamics in the Northern Calcareous Alps (Austria). *Int. J. Earth. Sci.*, 95, 451–462.

Gawlick, H.-J., Frisch, W., Vecsei, A., Steiger, T., Böhm, F. (1999). The change from rifting to thrusting in the northern Calcareous Alps as recorded in Jurassic sediments. *Geol. Rdsch.*, 87, 644–657.

Gawlick, H.-J., Aubrecht, R., Schlagintweit, F., Missoni, S., Plasienka, D. (2015). Ophiolitic detritus in Kimmeridgian resedimented limestones and its provenance from an eroded obducted ophiolitic nappe stack south of the Northern Calcareous Alps (Austria). *Geologica Carpathica*, 66, 473–487.

Gebrande, H., Lüschen, E., Bopp, M., Bleibinhaus, F., Lammerer, B., Oncken, O., Stiller, M., Kummerow, J., Kind, R., Millahn, K. et al. (2002). First deep seismic reflection images of the Eastern Alps reveal giant crustal wedges and transcrustal ramps. *Geophysical Research Letters*, 29(10), 92.1–92.4.

Genser, J. and Neubauer, F. (1989). Low angle normal faults at the eastern margin of the Tauern window (Eastern Alps). *Mitteilungen der Österreichischen Geologischen Gesellschaft*, 81, 233–243.

Gleißner, P., Franz, G., Frei, D. (2021). Contemporaneous opening of the Alpine Tethys in the Eastern and Western Alps: Constraints from a Late Jurassic gabbro intrusion age in the Glockner Nappe, Tauern Window, Austria. *International Journal of Earth Sciences*, 110, 2705–2724. doi: 10.1007/s00531-021-02075-z.

Glodny, J., Ring, U., Kühn A., Gleissner, P., Franz, G. (2005). Crystallization and very rapid exhumation of the youngest Alpine eclogites (Tauern Window, Eastern Alps) from Rb/Sr mineral assemblage analysis. *Contrib. Mineral. Petrol.*, 149, 699–712.

Glodny, J., Ring, U., Kühn, A. (2008). Coeval high-pressure metamorphism, thrusting, strike-slip and extensional shearing in the Tauern Window, Eastern Alps. *Tectonics*, 27, TC4004.

Granado, P., Roca, E., Strauss, P., Pelz, K., Muñoz, J.A. (2018). Structural styles in fold-and-thrust belts involving early salt structures: The Northern Calcareous Alps (Austria). *Geology*, 47, 51–54.

Groß, P., Handy, M.R., John, T., Pestal, G., Pleuger, J. (2020). Crustal-scale sheath folding at HP conditions in an exhumed Alpine subduction zone (Tauern Window, Eastern Alps). *Tectonics*, 39, e2019TC005942.

Groß, P., Pleuger, J., Handy, M., Germer, M., John, T. (2021). Evolving temperature field in a fossil subduction channel during the transition from subduction to collision (Tauern Window, Eastern Alps). *Journal of Metamorphic Geology*, 39, 247–269.

Gruber, P., Faupl, P., Koller, F. (1992). Zur Kenntniss basischer Vulkanitgerölle aus Gosaukonglomeraten der östlichen Kalkalpen – Ein Vergleich mit Vulkaniten aus dem Haselgebirge. *Mitt. Öst. Geol. Ges.*, 84, 77–100.

Grundmann, G. and Morteani, G. (1985). The young uplift and thermal history of the Central Eastern Alps (Austria/Italy), evidence from apatite fission-track ages. *Jahrbuch der Geologischen Bundesanstalt*, 128, 197–216.

Haas, J., Kovàcs, S., Krystyn, L., Lein, R. (1995). Significance of Late Permian-Triassic facies zones in terrane reconstructions in the Alpine-North Pannonian domain. *Tectonophysics*, 242, 19–40.

Habermüller, M. (2005). West-directed thrusting in the Dachstein Nappe: Quantification of the Eo-Alpine deformation around the Echerntal valley (Hallstatt, Austria). Unpublished diploma thesis, University of Vienna.

Handler, R., Dallmeyer, R.D., Neubauer, F. (1997). 40Ar/39Ar ages of detrital white mica from Upper Austroalpine units in the Eastern Alps, Austria: Evidence for Cadomian and contrasting Variscan sources. *Geologische Rundschau*, 86, 69–80.

Handy, M.R. and Oberhänsli, R. (2004). Explanatory notes to the map: Metamorphic structure of the Alps-Age map of the metamorphic structure of the Alps-Tectonic interpretation and outstanding problems. *Mitt. Österr. Min. Ges.*, 149, 201–225.

Handy, M.R., Schmid, S.M, Bousquet, R., Kissling, E., Bernoulli, D. (2010). Reconciling plate-tectonic reconstructions of Alpine Tethys with the geological–geophysical record of spreading and subduction in the Alps. *Earth-Science Reviews*, 102 (3–4), 121–158.

Hauke, M., Froitzheim, N., Nagel, T.J., Miladinova, I., Fassmer, K., Fonseca, R.O.C., Münker, C. (2019). Two high-pressure metamorphic events, Variscan and Alpine, dated by Lu–Hf in an eclogite complex of the Austroalpine nappes (Schobergruppe, Austria). *International Journal of Earth Sciences*, 108(4), 1317–1331.

Hawkesworth, C., Waters, D., Bickle, M. (1975). Plate tectonics in the Eastern Alps. *Earth Planet. Sci. Lett.*, 24(3), 405–441.

Heberer, B., Reverman, R.L., Fellin, M.G., Neubauer, F., Dunkl, I., Zattin, M., Seward, D., Genser, J., Brack, P. (2017). Postcollisional cooling history of the Eastern and Southern Alps and its linkage to Adria indentation. *Int. J. Earth. Sci. (Geol. Rundsch.)*, 106, 1557–1580.

Hejl, E. (1997). “Cold spots” during the Cenozoic evolution of the Eastern Alps: Thermochronological interpretation of apatite fission-track data. *Tectonophysics*, 272, 159–173.

Hejl, E. (1998). Über die känozoische Abkühlung und Denudation der Zentralalpen östlich der Hohen Tauern – Eine Apatit-Spaltspurenanalyse. *Mitt. Österr. Geol. Ges.*, 89, 179–200.

Hejl, E. and Grundmann, G. (1989). Apatit-Spaltspurendaten zur thermischen Geschichte der Nördlichen Kalkalpen, der Flysch- und Molassezone. *Jahrb. Geol. Bundesanst.*, 132, 191–212.

Hetényi, G., Molinari, I., Clinton, J., Bokelmann, G., Bondár, I., Crawford, W.C., Dessa, J.X., Doubre, C., Friederich, W., Fuchs, F. et al. (2018). The AlpArray seismic network: A large-scale European experiment to image the Alpine orogen. *Surv. Geophys.*, 39(5), 1009–1033.

van Hinsbergen, D.J.J., Torsvik, T.H., Schmid, S.M., Maţenco, L.C., Maffione, M., Vissers, R.L.M., Gürer, D., Spakman, W. (2020). Orogenic architecture of the Mediterranean region and kinematic reconstruction of its tectonic evolution since the Triassic. *Gondwana Research*, 81, 79–229. doi:10.1016/j.gr.2019.07.009.

Hitz, L. (1995). The 3D crustal structure of the Alps of eastern Switzerland and western Austria interpreted from network of deep-seismic profiles. *Tectonophysics*, 248, 71–96.

Höck, V. and Koller, F. (1987). The Idalp ophiolite (Lower Engadine Window, Eastern Alps): Petrology and geochemistry. *Ofioliti*, 12, 179–192.

Höck, V. and Miller, C. (1980). Chemistry of mesozoic metabasites in the middle and eastern part of the Hohe Tauern. *Mitteilungen der Österreichischen Geologischen Gesellschaft*, 71(72), 81–80.

Hoinkes, G., Koller, F., Rantitsch, G., Dachs, E., Höck, V., Neubauer, F., Schuster, R. (1999). Alpine metamorphism of the Eastern Alps. *Schweiz. Mineral. Petr. Mitt.*, 79, 155–181.

Hoke, L. (1990). The Altkristallin of the Kreuzeck Mountains, SE Tauern Window, EasternAlps-basement crust in a convergent plateboundary zone. *Jahrb. Geol. Bundesanst.*, 133, 5–87.

Horvath, F. and Cloething, S. (1996). Stress-induced late-stage subsidence anomalies in the Pannonian basin. *Tectonophysics*, 266, 287–300.

Horvath, F., Bada, G., Szafian, P., Tari, G., Adam, A., Cloething, S. (2006). Formation and deformation of the Pannonian Basin: Constraints from observational data, in European lithosphere dynamics. *Mem. Geol. Soc.*, 32, 129–145.

Hua, Y., Zhao, D., Xu, Y. (2017). P-wave anisotropic tomography of the Alps. *J. Geophys Res: Solid Earth*, 122(6), 4509–452.

Inger, S. and Cliff, R.A. (1994). Timing of metamorphism in the Tauern Window, Eastern Alps: RB-Sr ages and fabric formation. *J. Met. Geol.*, 12, 695–707.

Janák, M., Froitzheim, N., Lupták, B., Vrabec, M., Krogh Ravna, E.J. (2004). First evidence for ultrahigh-pressure metamorphism in Pohorje, Slovenia: Tracing deep continental subduction in the Eastern Alps. *Tectonics*, 23, TC5014.

Janák, M., Froitzheim, N., Vrabec, M., Krogh Ravna, E.J., De Hoog, J.C.M. (2006). Ultrahigh-pressure metamorphism and exhumation of garnet periditite in Pohorje, Eastern Alps. *J. Metamorphic Geol.*, 24, 19–31.

Ji, W.Q., Malusà, M.G., Tiepolo, M., Langone, A., Zhao, L., Wu, F.Y. (2019). Synchronous Periadriatic magmatism in the Western and Central Alps in the absence of slab breakoff. *Terra Nova*, 31(2), 120–128.

Kästle, E.D., Rosenberg, C.L., Boschi, L., Bellahsen, N., Meier, F., El-Sharkawy, A. (2020). Slab break-offs in the Alpine subduction zone. *Int. J. Earth. Sci.*, 109, 587–603.

Keil, M. and Neubauer, F. (2011). The Miocene Enns Valley Basin (Austria) and the North Enns Valley Fault. *Austrian J. Earth Sci.*, 104(1), 49–65.

Kissling, E., Schmid, S.M., Lippitsch, R., Ansorge, J., Fügenschuh, B. (2006). Lithosphere structure and tectonic evolution of the Alpine arc: New evidence from high-resolution teleseismic tomography. *Geologica Society of London Memoirs*, 32, 129–145.

Kleinschrodt, R. (1987). Quarzkorngefügeanalyse im Altkristallin südlich des westlichen Tauernfensters (Südtirol/Italien). *Erlanger Geol. Abh.*, 114, 1–82.

Koroknai, B., Neubauer, F., Genser, J., Topa, D. (1999). Metamorphic and tectonic evolution of Austroalpine units at the western margin of the Gurktal nappe complex, Eastern Alps. *Schweiz. Mineral. Petrogr. Mitt.*, 79, 277–295.

Kozur, H. and Mostler, H. (1992). Erster paläontologischer Nachweis von Meliaticum und Süd-Rudabanyaicum in den Nördlichen Kalkalpen (Österreich) und ihre Beziehung zu den Abfolgen der Westkarpaten. *Geologisch-Paläontologische Mitteilungen der Universität Innsbruck*, 18, 87–129.

Kralik, M., Klima, K., Riedmüller, G. (1987a). Dating fault gouges. *Nature*, 327, 315–317.

Kralik, M., Krumm, H., Schramm, J. (1987b). Low grade and very low grade metamorphism in the Northern Calcareous Alps and in the Greywacke zone: Illite-crystallinity data and isotopic ages. In *Geodynamics of the Eastern Alps*, Flügel, H.W. and Faupl, P. (eds). Deuticke, Vienna.

Krenn, K., Bauer, C., Proyer, A., Hoinkes, G. (2010). Tectonometamorphic evolution of the Rhodope orogen. *Tectonics*, 29, TC4001. doi:10.1029/2009TC002513.

Krische, O. and Gawlick, H.-J. (2015). Age and significance of Lower Cretaceous mass flows: Ischl Breccia revisited (Rossfeld Formation, Northern Calcareous Alps, Austria). *Austrian Journal of Earth Sciences*, 108, 128–150.

Krohe, A. (1987). Kinematics of Cretaceous nappe tectonics in the Austroalpine basement of the Koralpe region (eastern Austria). *Tectonophysics*, 136, 171–196.

Kroner, U., Mansky, J.-L., Mazur, S., Aleksandrowski, P., Hann, H.P., Huckride, H., Lacquement, F., Lamarche, J., Ledru, P., Pharaoh, T.C. et al. (2008). Variscan tectonics. In *The Geology of Central Europe, Volume 1: Precambrian and Palaeozoic.*, McCann, T. (ed.). Geological Society, London.

Kummerow, J., Kind, R., Oncken, O., Giese, P., Ryberg, T., Wylegalla, K., Scherbaum, F. (2004). A natural and controlled source seismic profile through the Eastern Alps: TRANSALP. *Earth and Planetary Science Letters*, 225(1–2), 115–129.

Kurz, W. (2005a). Constriction during exhumation: Evidence from eclogite microstructures. *Geology*, 33, 37–40.

Kurz, W. (2005b). Tectonic map and overall architecture of the Alpine orogen: Comment. *Eclogae Geologicae Helvetiae*, 98, 97–98.

Kurz, W. (2006). Penninic paleogeography from the Western toward the Eastern Alps – Still open questions? *International Geology Review*, 48, 996–1022.

Kurz, W. and Fritz, H. (2003). Tectonometamorphic evolution of the Austroalpine Nappe complex in the central Eastern Alps – Consequences for the Eo-Alpine evolution of the Eastern Alps. *Internat. Geology Review*, 45, 1100–1127.

Kurz, W. and Froitzheim, N. (2002). The exhumation of eclogite–facies metamorphic rocks – A review of models confronted with examples from the Alps. *Internat. Geology Review*, 44, 702–743.

Kurz, W. and Neubauer, F. (1996). Deformation partitioning and shear localization during the updoming of the Sonnblick area in the Tauern Window (Eastern Alps, Austria). *J. Struct. Geol.*, 18, 1327–1343.

Kurz, W., Neubauer, F., Genser, J. (1996). Kinematics of Penninic nappes (Glockner Nappe and basement-cover nappes) in the Tauern Window (Eastern Alps, Austria), during subduction and Penninic-Austroalpine collision. *Eclogae Geol. Helv.*, 89, 573–605.

Kurz, W., Neubauer, F., Genser, J., Dachs, E. (1998). Alpine geodynamic evolution of passive and active continental margin sequences in the Tauern Window (Eastern Alps, Austria, Italy): A review. *Geol. Rdsch.*, 87, 225–242.

Kurz, W., Neubauer, F., Genser, J., Unzog, W., Dachs, E. (2001). Tectonic evolution of Penninic Units in the Tauern Window during the Paleogene: Constraints from structural and metamorphic geology. In *Paleogene of the Eastern Alps*, Piller, W.E. and Rasser, M.W. (eds). Österr. Akad. Wiss., Vienna.

Kurz, W., Fritz, H., Tenczer, V., Unzog, W. (2002). Tectonometamorphic evolution of the Koralm complex (Eastern Alps): Constraints from microstructures and textures of the "Plattengneis" – Shear zone. *J. Struct. Geol.*, 24, 1957–1970.

Kurz, W., Wölfler, A., Rabitsch, R., Genser, J. (2011). Polyphase movement on the Lavanttal Fault Zone (Eastern Alps): Reconciling the evidence from different geochronological indicators. *Swiss Journal of Geosciences*, 104, 323–343.

Lammerer, B. (1986). Das Autochthon im westlichen Tauernfenster. *Jahrbuch der Geologischen Bundesanstalt*, 129(1), 51–67.

Lammerer, B. (1988). Thrust-regime and transpression-regime tectonics in the Tauern Window (Eastern Alps). *Geologische Rundschau*, 77, 143–156.

Läufer, A.L., Frisch, W., Steinitz, G., Loeschke, J. (1997). Exhumed fault-bounded Alpine blocks along the Periadriatic lineament: The Eder unit (Carnic Aps, Austria). *Geol. Rdsch.*, 86, 612–626.

Legrain, N., Stüwe, K., Wölfler, A. (2014). Incised relict landscape in the eastern Alps. *Geomorphology*, 221, 124–138.

Lein, R. (1987). Evolution of the northern Calcareous Alps during Triassic times. In *Geodynamics of the Eastern Alps*, Flügel, H.W. and Faupl, P. (eds). Deuticke, Vienna.

Leitner, C. and Finger, F. (2019). Two different basalt provinces (MORB vs. WPB) in the evaporitic Permian Haselgebirge Formatio (Eastern Alps, Austria) and possible tectonic implications. *Mitteilungen der Österreichischen Geologischen Gesellschaft*, 165, 60.

Leitner, C. and Spötl, C. (2017). The Eastern Alps: Multistage development of extremely deformed evaporites. In *Permo-Triassic Salt Provinces of Europe, North Africa and the Atlantic Margins*, Soto, J.I., Flinch, J.F., Tari, G. (eds). Elsevier, Amsterdam.

Linzer, H.G., Ratschbacher, L., Frisch, W. (1995). Transpressional collision structures in the upper crust: The fold-thrust belt of the Northern calcareous Alps. *Tectonophysics*, 242, 41–61.

Linzer, H.-G., Decker, K., Peresson, H., Dell'Mour, R., Frisch, W. (2002). Balancing lateral orogenic float of the Eastern Alps. *Tectonophysics*, 354, 211–237.

Lippitsch, R., Kissling, E., Ansorge, J. (2003). Upper mantle structure beneath the Alpine orogen from high-resolution teleseismic tomography. *J. Geophys. Res.*, 108, 2002JB002016.

Liu, Y., Genser, J., Handler, R., Friedl, G., Neubauer, F. (2001). 40Ar/39Ar muscovite ages from the Penninic-Austroalpine plate boundary, Eastern Alps. *Tectonics*, 20, 526–547.

Luciani, V. (1989). Stratigrafia Sequenziale del Terziaro Nelle Catena del Monte Baldo (Provincie di Verona a Trento). *Sci. Geol. Mem.*, 41, 263–351.

Lüschen, E., Lammerer, B., Gebrande, H., Millahn, K., Nicolich, R., TRANSALP Working Group (2004). Orogenic structure of the Eastern Alps, Europe, from Transalp deep seismic reflection profiling. *Tectonophysics*, 388, 85–102.

Lüschen, E., Borrini, D., Gebrande, H., Lammerer, B., Millahn, K., Neubauer, F., Nicolich, R. (2006). TRANSALP – Deep crustal Vibroseis and explosive seismic profiling in the Eastern Alps. *Tectonophysics*, 414, 9–38.

Luth, S.W. and Willingshofer, E. (2008). Mapping of the post-collisional cooling history of the Eastern Alps. *Swiss J. Geosci.*, 101, 207–223.

Manatschal, G., Müntener, O., Desmurs, L., Bernoulli, D. (2003). An ancient ocean-continent transition in the Alps: The Totalp, Err-Platta and Malenco units in the eastern Central Alps (Graubünden and northern Italy). Excursion of the Swiss Geological Society, September 20 to 24, 2002. *Eclogae Geologicae Helvetiae*, 96, 131–146.

Mancktelow, N.S., Stöckli, D.F., Grollimund, B., Müller, W., Fügenschuh, B., Viola, G., Seward, D., Villa, I.M. (2001). The DAV and periadriatic fault systems in the Eastern Alps south of the Tauern Window. *Int. J. Earth Sci. (Geol. Rdsch.)*, 90, 593–622.

Mandl, G.W. (2000). The Alpine sector of the Tethyan shelf – Examples of Triassic to Jurassic sedimentation and deformation from the Northern Calcareous Alps. *Mitt. Österr. Geol. Ges.*, 92, 61–78.

Mandl, M., Kurz, W., Hauzenberger, C., Fritz, H., Klötzli, U., Schuster, R. (2018). Pre-Alpine evolution of the Seckau Complex (Austroalpine basement/Eastern Alps): Constraints from in-situ LA-ICP-MS U-Pb single bond zircon geochronology, *Lithos*, 296–299, 412–430.

Matenco, L., Krézsek, C., Merten, S., Schmid, S., Cloetingh, S., Andriessen, P. (2010). Characteristics of collisional orogens with low topographic build-up: An example from the Carpathians. *Terra Nova*, 22, 155–165.

Meisel, T., Melcher, F., Tomascak, P., Dingeldey, C., Koller, F. (1997). Re–Os isotopes in orogenic peridotite massifs in the Eastern Alps, Austria. *Chem. Geol.*, 143, 217–229.

Metz, K. (1973). Beiträge zur tektonischen Baugeschichte und Position des Fohnsdorf–Knittelfelder Tertiärbeckens. *Mitt Abt. Geol. Palaeont. Landesmuseum Joanneum*, 33, 4–33.

Metz, K. (1976). Bruchsysteme und Westbewegungen in den östlichen Zentralalpen. *Mitt. Österr. Geol. Ges.*, 69, 27.

Michail, M., Pipera, K., Koroneos, A., Kilias, A., Ntaflos, T. (2016). New perspectives on the origin and emplacement of the Late Jurassic Fanos granite, associated with an intra-oceanic subduction within the Neotethyan Axios-Vardar Ocean. *Int. J. Earth Sci. (Geol. Rundsch.)*, 105, 1965–1983.

Miller, C. and Thöni, M. (1997). Eo-Alpine eclogitization of Permian MORB-type Gabbros in the Koralpe (Eastern Alps, Austria): New geochronological, geochemical and petrological data. *Chem. Geol.*, 137, 283–310.

Mitterbauer, U., Behm, M., Brückl, E., Lippitsch, R., Guterch, A., Keller, G.R., Koslovskaya, E., Rumpfhuber, E.M., Šumanovac, F. (2011). Shape and origin of the East-Alpine slab constrained by the ALPASS teleseismic model. *Tectonophysics*, 510(1), 195–206.

Most, P. (2003). Late Alpine cooling histories of tectonic blocks along the central part of the TRANSALP-traverse (Inntal-Gardertal): Constraints from geochronology. *Tübinger Geowissenschaftliche Arbeiten*, Reihe A 67, 97.

Müller, W. (1998). Isotopic dating of deformation using microsampling techniques: The evolution of the Periadriatic fault system (Alps). PhD Thesis, ETH Zürich.

Müller, W., Dallmeyer, R.D., Neubauer, F., Thöni, M. (1999). Deformation-induced resetting of Rb/Sr and 40Ar/39Ar mineral systems in a low-grade, polymetamorphic terrane (Eastern Alps, Austria). *J. Geol. Soc. Lond.*, 156, 261–278.

Müller, W., Mancktelow, N.S., Meier, M. (2000). Rb-Sr microchrons of synkinematic mica in mylonites: An example from the DAV fault of the Eastern Alps. *Earth Planet. Sci. Lett.*, 180, 385–397.

Müller, W., Prosser, G., Mancktelow, N.S., Villa, I.M., Kelley, S.P., Viola, G., Oberli, F. (2001). Geochronological constraints on the evolution of the Periadriatic Fault System (Alps). *Int. J. Earth Sci. (Geol. Rdsch.)*, 90, 623–653.

Nagel, T. (2006). Structure of Austroalpine and Penninic units in the Tilisuna area (Eastern Rätikon, Austria): Implications for the paleogeographic position of the Allgäu and Lechtal nappes. *Eclogae Geol. Helv.*, 99, 223–235.

Nagel, T., Herwartz, D., Rexroth, S., Münker, C., Froitzheim, N., Kurz, W. (2013). Lu–Hf dating, petrography, and tectonic implications of the youngest Alpine eclogites (Tauern Window, Austria). *Lithos*, 170–171, 179–190.

Neubauer, F. (2002). Evolution of late Neoproterozoic to early Palaeozoic tectonic elements in Central and Southeast European Alpine mountain belts: Review and synthesis. *Tectonophysics*, 352, 87–103.

Neubauer, F. and Vozarova, A. (1990). The Noetsch-Veitsch–North Gemeric zone of Alps and Carpathians: Correlation, paleogeography and geodynamics significance for Variscan orogeny. In *Thirty Years of Geological Cooperation between Austria and Czechoslovakia*, Minarikova, D. and Lobitzer, H. (eds). Ustredni Ustav Geologicky, Prague.

Neubauer, F., Genser, J., Fritz, H., Wallbrecher, E. (1992). Alpine kinematics of the eastern Central Alps. *ALCAPA – Field Guide (IGP/KFU Graz)*, 127–136.

Neubauer, F., Genser, J., Fritz, H., Wallbrecher, E. (1993). Alpine kinematics of the eastern Central Alps. *Field Guide, Structures and Tectonics at Different Lithospheric Levels*, 17–26.

Neubauer, F., Handler, R., Hermann, S., Paulus, G. (1994). Revised lithostratigraphy and structure of the Eastern Greywacke zone (Eastern Alps). *Mitteilungen der Österreichischen Gesellschaft*, 86, 61–74.

Neubauer, F., Dallmeyer, R.D., Dunkl, I., Schirnik, D. (1995). Late Cretaceous exhumation of the metamorphic Gleinalm dome, Eastern Alps: Kinematics, cooling history and sedimentary response in a sinistral wrench corridor. *Tectonophysics*, 242, 79–98.

Neubauer, F., Genser, J., Handler, R. (2000). The Eastern Alps: Result of a two-stage collision process. *Mitt. Österr. geol. Ges.*, 92, 117–134.

Oberhänsli, R. (2004). Metamorphic structure of the Alps. *Mitt. Österr. Miner. Ges.*, 149, 115–226.

Oberhauser, R. (1995). Zur Kenntnis der Tektonik und der Paläogeographie des Ostalpenraumes zur Kreide-, Paläozän- und Eozänzeit. *Jahrbuch der Geologischen Bundesanstalt*, 138, 369–432.

Ortner, H. (1992). Die sedimentäre Entwicklung der Muttekopfgosau (westliche Ostalpen, Tirol). *Zbl. Geol. Paläont. Teil I*, 1991, 2873–2886.

Ortner, H. (1994a). Die Muttekopfgosau (Lechtaler Alpen, Tirol/Österreich). Sedimentologie und Beckenentwicklung. *Geol. Rdsch.*, 83, 197–211.

Ortner, H. (1994b). Die petrographische Entwicklung der Muttekopfgosau (Lechtaler Alpen, Tirol). *Zbl. Geol. Paläont. Teil I*, 1992, 1355–1371.

Ortner, H. (2003). Local and far field stress-analysis of brittle deformation in the western part of the Northern Calcareous Alps, Austria. *Geol. Paläontol. Mitt. Univ. Innsbruck*, 26, 109–131.

Ortner, H. and Sachsenhofer, R.F. (1996). Evolution of the lower Inn Vally Tertiary and constraints on the development of the source area. In *Oil and Gas in Alpidic Thrust Belts and Basins of Central and Eastern Europe*, Liebl, W. and Wessely, G. (eds). EAGE Spec. Publ. No. 5, Bunnik.

Ortner, H., Reiter, F., Brandner, R. (2006). Kinematics of the Inntal shear zone–sub-Tauern ramp fault system and the interpretation of the TRANSALP seismic section, Eastern Alps, Austria. *Tectonophysics*, 414, 241–258.

Pamić, J., Gušić I., Jelaska, V. (1998). Geodynamic evolution of the Central Dinarides. *Tectonophysics*, 297, 251–268.

Peresson, H. and Decker, K. (1997a). Far-field effects of late Miocene subduction in the Eastern Carpathians: E–W compression and inversion of structures in the Alpine-Carpathian–Pannonian region. *Tectonics*, 16, 38–56.

Peresson, H. and Decker, K. (1997b). The tertiary dynamics of the northern Eastern Alps (Austria): Changing paleostress in a collisonal plate boundary. *Tectonophysics*, 272, 125–157.

Pfingstl, S., Kurz, W., Schuster, R., Hauzenberger, C. (2015). Geochronological constraints on the exhumation of the Austroalpine Seckau Nappe (Eastern Alps). *Austrian Journal of Earth Sciences*, 108, 172–185.

Plan, L., Grasemann, B., Spötl, C., Decker, K., Boch, R., Kramers, J. (2010). Neotectonic extrusion of the early Alps: Constraints from U/Th dating of tectonically damaged speleothems. *Geology*, 38, 483–486.

Polinski, R.K. and Eisbacher, H. (1992). Deformation patitioning during polyphase oblique convergence in the Karawanken Mountains, southeastern Alps. *J. Struct. Geol.*, 14, 1203–1213.

Pomella, H., Klötzli, U., Scholger, R., Stipp, M., Fügenschuh, B. (2011). The Northern Giudicarie and the Meran-Mauls fault (Alps, Northern Italy) in the light of new paleomagnetic and geochronological data from boudinaged Eo-/Oligocene tonalites. *Int. J. Earth Sci.*, 100, 1827–1850.

Pomella, H., Stipp, M., Fügenschuh, B. (2012). Thermochronological record of thrusting and strike-slip faulting along the Giudicarie Fault System (Alps, Northern Italy). *Tectonophysics*, 579, 118–130.

Putiš, M., Soták, J., Li, Q.-L., Ondrejka, M., Li, X.-H., Hu, Z., Ling, X., Nemec, O., Németh, Z., Ružička, P. (2019). Origin and age determination of the Neotethys Meliata Basin ophiolite fragments in the Late Jurassic–Early Cretaceous accretionary wedge mélange (Inner Western Carpathians, Slovakia). *Minerals*, 9, 652.

Qorbani, E., Bokelmann, G., Bianchi, I. (2015). Slab detachment under the Eastern Alps seen by seismic anisotropy. *Earth Planet. Sci. Lett.*, 409, 96–108.

Rantitsch, G. and Russegger, B. (2000). Thrust-related very low grade metamorphism within the Gurktal Nappe complex (Eastern Alps). *Jahrb. Geol. Bundesanst.*, 142, 219–225.

Rantitsch, G., Sachsenhofer, R.F., Hasenhüttl, C., Russegger, B., Rainer, T. (2005). Thermal evolution of an extensional detachment as constrained by organic metamorphic data and thermal modeling: Graz Paleozoic Nappe Complex (Eastern Alps). *Tectonophysics*, 411, 57–72.

Ratschbacher, L. (1986). Kinematics of Austro-Alpine cover nappes: Changing translation path due to transpression. *Tectonophysics*, 125(4), 335–356.

Ratschbacher, L., Frisch, W., Neubauer, F., Schmid, S.M., Neugebauer, J. (1989). Extension in compressional orogenic belts: The Eastern Alps. *Geology*, 17, 404–407.

Ratschbacher, L., Merle, O., Davy, P., Cobbold, P. (1991a). Lateral extrusion in the Eastern Alps, Part 1: Boundary conditions and experiments scaled for gravity. *Tectonics*, 10(2), 245–256.

Ratschbacher, L., Frisch, W., Linzer, H.-G., Merle, O. (1991b). Lateral extrusion in the Eastern Alps, Part 2: Structural analysis. *Tectonics*, 10(2), 257–271.

Ratschbacher, L., Dingeldey, C., Miller, C., Hacker, B.R., McWilliams, M.O. (2004). Formation, subduction, and exhumation of Penninic oceanic crust in the Eastern Alps: Time constraints from 40Ar/39Ar geochronology. *Tectonophysics*, 394, 155–170.

von Raumer, J.F. and Neubauer, F. (eds) (1993). *The Pre-Mesozoic Geology in the Alps*. Springer, Heidelberg.

von Raumer, J.F., Stampfli, G., Borel, G., Bussy, F. (2002). Organization of pre-Variscan basement areas at the north-Gondwanan margin. *Int. J. Earth Sci.*, 91, 35–52.

von Raumer, J.F., Bussy, F., Schaltegger, U., Schulz, B., Stampfli, G.M. (2012). Pre-Mesozoic Alpine basements – Their place in the European Paleozoic framework. *Geological Society of America Bulletin*, 125(1–2), 89–108.

Reddy, S.M., Cliff, R.A., East, R. (1993). Thermal history of the Sonnblick Dome south-eastern Tauern Window, Austria: Implications for heterogeneous uplift within the Penninic basement. *Geol. Rundsch.*, 82, 667–675.

Reinecker, J. (2000). Stress and deformation: Miocene to present-day tectonics in the Eastern Alps. *Tüb. Geowiss. Arb., Serie A*, 55, 128.

Reinecker, J. and Lenhardt, W.A. (1999). Present-day stress field and deformation in eastern Austria. *Geol. Rdsch.*, 88, 532–550.

Reiter, F., Freudenthaler, C., Hausmann, H., Ortner, H., Lenhardt, W., Brandner, R. (2018). Active seismotectonic deformation in front of the Dolomites indenter, Eastern Alps. *Tectonics*, 37, 4625–4654.

Robl, J. and Stüwe, K. (2005). Continental collision with finite indenter strength 2: European Eastern Alps. *Tectonics*, 24, TC4014.

Robl, J., Stüwe, K., Hergarten, S., Evans, L. (2008). Extension during continental convergence in the Eastern Alps: The influence of orogen-scale strike-slip faults. *Geology*, 36, 603–606.

Ring, U. (1992). The Alpine geodynamic evolution of Penninic nappes in the eastern Central Alps: Geothermobarometric and kinematic data. *J. Metam. Geol.*, 10, 33–53.

Ring, U., Ratschbacher, L., Frisch, W. (1988). Plate-boundary kinematics in the Alps: Motion in the Arosa suture zone. *Geology*, 16, 696–698.

Ring, U., Ratschbacher, L., Frisch, W., Biehler, D., Kralik, M. (1989). Kinematics of the Alpine plate-margin: Structural styles, strain and motion along the Penninic–Austroalpine boundary in the Swiss–Austrian Alps. *Journal of the Geological Society of London*, 146, 835–849.

Ring, U., Ratschbacher, L., Frisch, W., Dürr, S., Borchert, S. (1990). The internal structure of the Arosa Zone (Swiss–Austrian Alps). *Geol. Rundschau*, 79, 725–739.

Romer, R.L. and Siegesmund, S. (2003). Why allanite may swindle about its true age. *Contrib. Mineral. Petrol.*, 146, 297–307.

Rosenberg, C.L. and Berger, A. (2009). On the causes and modes of exhumation and lateral growth of the Alps. *Tectonics*, 28, TC6001.

Rosenberg, C.L. and Garcia, S. (2012). Estimating displacement along the Brenner fault and orogeny parallel extension in the Eastern Alps. *Int. J. Earth. Sci.*, 100, 1129–1145.

Rosenberg, C.L. and Schneider, S. (2008). The western termination of the SEMP Fault (eastern Alps) and its bearing on the exhumation of the Tauern Window. *Geol. Soc. London, Spec. Publ.*, 298, 197–218.

Rosenberg, C.L., Brun, J.-P., Cagnard, F., Gapais, D. (2007). Oblique indentation in the Eastern Alps: Insights from laboratory experiments. *Tectonics*, 26, TC2003.

Rosenberg, C.L., Schneider, S., Scharf, A., Bertrand, A., Hammerschmidt, K., Rabaute, A., Brun, J.-P. (2018). Relating collisional kinematics to exhumation processes in the Eastern Alps. *Earth Sci. Rev.*, 176, 311–344.

Sachsenhofer, R.F., Lankreijer, A., Cloething, S., Ebner, F. (1997). Subsidence analysis and quantitative basin modelling in the Styrian Basin (Pannonian Basin System, Austria). *Tectonophysics*, 272, 175–196.

Sachsenhofer, R.F., Kogler, A., Polesny, H., Strauss, P., Wagreich, M. (2000). The Neogene Fohnsforf Basin: Basin formation and basin inversion during lateral extrusion in the Eastern Alps (Austria). *Geol. Rdsch.*, 89, 415–430.

Sander, B. (1914). Geologische Studien am Westende der Hohen Tauern. *Denkschriften der Kaiserlichen Akademie der Wissenschaften*, 82, 278–287.

Scharbert, S. (1975). Radiometrische Altersdaten von Intrusivgesteinen im Raum Eisenkappel (Karawanken, Kärnten). *Verh. Geol. Bundesanst.*, 4, 301–304.

Scharf, A., Handy, M.R., Favaro, S., Schmid, S.M., Bertrand, A. (2013a). Modes of orogen-parallel stretching and extensional exhumation in response to microplate indentation and roll-back subduction (Tauern Window, Eastern Alps). *Int. J. Earth Sci.*, 102(6), 1627–1654.

Scharf, A., Handy, M.R., Ziemann, M.A., Schmid, S.M. (2013b). Peak-temperature patterns of polyphase metamorphism resulting from accretion, subduction and collision (eastern Tauern Window, European Alps) – A study with Raman microspectroscopy on carbonaceous material (RSCM). *Journal of Metamorphic Geology*, 31(8), 863–880.

Scharf, A., Handy, M.R., Schmid, S.M., Favaro, S., Sudo, M., Schuster, R., Hammerschmidt, K. (2016). Grain-size effects on the closure temperature of white mica in a crustal-scale extensional shear zone – Implications of in-situ 40Ar/39Ar laser-ablation of white mica for dating shearing and cooling (Tauern Window, Eastern Alps). *Tectonophysics*, 674, 210–226.

Schmid, S.M., Aebli, H.R., Heller, F., Zingg, A. (1989). The role of the Periadriatic Line in the tectonic evolution of the Alps. *Geol. Soc. Spec. Publ.*, 45, 153–171.

Schmid, S.M., Pfiffner, O.A., Froitzheim, N., Schönborn, G., Kissling, E. (1996). Geophysical-geological transect and tectonic evolution of the Swiss–Italian Alps. *Tectonics*, 15, 1036–1064.

Schmid, S.M., Fügenschuh, B., Kissling, E., Schuster, R. (2004). Tectonic map and overall architecture of the Alpine orogen. *Eclogae Geol. Helv.*, 97, 93–117.

Schmid, S.M., Bernoulli, D., Fügenschuh, B., Matenco, L., Schefer, S., Schuster, R., Tischler, M., Ustaszewski, K. (2008). The Alpine-Carpathian–Dinaric orogenic system: Correlation and evolution of tectonic units. *Swiss J. Geosci.*, 101, 139–183.

Schmid, S.M., Scharf, A., Handy, M.R., Rosenberg, C.L. (2013). The Tauern Window (Eastern Alps, Austria): A new tectonic map, with cross-sections and a tectonometamorphic synthesis. *Swiss J. Geosci.*, 106, 1–32.

Schmid, S.M., Fügenschuh, B., Kounov, A., Matenco, L., Nievergelt, P., Oberhänsli, R., Pleuger, J., Schefer, S., Schuster, R., Tomljenovic, B. et al. (2020). Tectonic units of the Alpine collision zone between Eastern Alps and western Turkey. *Gondwana Research*, 78, 308–374. doi:10.1016/j.gr.2019.07.005.

Schmidt, T., Blau, J., Kázmér, M. (1991). Large-scale strike-slip displacement of the Drauzug and the Transdanubian Mountains in early Alpine history: Evidence from permo-mesozoic facies belts. *Tectonophysics*, 200, 213–232.

Schneider, S., Hammerschmidt, K., Rosenberg, C.L. (2013). Dating the longevity of ductile shear zones: Insight from 40Ar/39Ar in situ analyses. *Earth Planet. Sci. Lett.*, 369, 43–58.

Schneider, S., Hammerschmidt, K., Rosenberg, C.L., Gerdes, A., Frei, D., Bertrand, A. (2015). U–Pb ages of apatite in the western Tauern Window (Eastern Alps): Tracing the onset of collision-related exhumation in the European plate. *Earth Planet. Sci. Lett.*, 418, 53–65.

Schorn, A. and Neubauer, F. (2014). The structure of the Hallstatt evaporite body (Northern Calcareous Alps, Austria): A compressive diapir superposed by strike-slip shear? *J. Struct. Geol.*, 60, 70–84.

Schorn, A., Franz Neubauer, F., Genser, J., Bernroider, M. (2013). The Haselgebirge evaporitic mélange in central Northern Calcareous Alps (Austria): Part of the Permian to Lower Triassic rift of the Meliata ocean? *Tectonophysics*, 583, 28–48.

Schulz, B. (1989). Jungalpidische Gefügeentwicklung entlang der Defereggen-Antholz-Vals-Linie (Osttirol, Österreich). *Jahrb. Geol. Bundesanst.*, 132, 775–789.

Schulz, B. (1990). Prograde-retrograde P–T–t-deformation path of Austroalpine micaschists during Variscan continental collision (Eastern Alps). *J. Metam. Geol.*, 8, 629–443.

Schuster, R. and Frank, W. (1999). Metamorphic evolution of the Austroalpine units east of the Tauern Window: Indications for Jurassic strike slip tectonics. *Mitt. Ges. Geol. Bergbaustud. Österr.*, 42, 37–58.

Schuster, R. and Nowotny, A. (2015). Die Einheiten des Ostalpinen Kristallins auf den Kartenblättern GK50 Blatt 103 Kindberg und 135 Birkfeld. *Arbeitstagung der Geol. Bundesanst.*, 2, 10–37.

Schuster, R. and Stüwe, K. (2008). Permian metamorphic event in the Alps. *Geology*, 36, 603–606.

Schuster, R., Koller, F., Hoeck, V., Hoinkes, G., Bousquet, R. (2004). Explanatory notes to the map: Metamorphic structure of the Alps – Metamorphic evolution of the Eastern Alps. *Mitt. Österr. Miner. Ges.*, 149, 175–199.

Schuster, R., Koller, F., Frank, W. (2007). Pebbles of upper-amphibolite facies amphibolites in the Gosau Group of the Eastern Alps: Relics of a metamorphic sole? *Abstract 8th ALPSHOP Davos/Switzerland*, 74.

Schuster., R., Kurz, W., Krenn, K., Fritz, H. (2013). Introduction to the geology of the Alps. *Berichte der Geologischen Bundesanstalt*, 99, 121–135.

Selverstone, J. (1988). Evidence for east–west crustal extension in the Eastern Alps: Implications for the unroofing history of the Tauern Window. *Tectonics*, 7, 87–105.

Selverstone, J. (2005). Are the Alps collapsing? *Annu. Rev. Earth Planet. Sci.*, 33, 113–132.

Selverstone, J., Franz, G., Thomas, S., Getty, S. (1992). Fluid variability in 2 GPa eclogites as an indicator of fluid behavior during subduction. *Contrib. Mineral. Petrol.*, 112, 341–357.

Schweigl, J. and Neubauer, F. (1997). Structural evolution of the central Northern Calcareous Alps: Significance for the Jurassic to Tertiary geodynamics in the Alps. *Eclogae Geol. Helv.*, 90, 303–323

Sölva, H., Thöni, M., Grasemann, B., Linner, M. (2001). Emplacement of Eo-Alpine high-pressure rocks in the Austroalpine Ötztal complex (Texel group, Italy/Austria). *Geodin. Acta*, 14, 345–360.

Sölva, H., Grasemann, B., Thöni, M., Thiede, R., Habler, G. (2005). The Schneeberg Normal Fault Zone: Normal faulting associated with Cretaceous SE-directed extrusion in the Eastern Alps (Italy/Austria). *Tectonophysics*, 401, 143–166.

Spada, M., Bianchi, I., Kissling, E., Piana Agostinetti, N., Wiemer, S. (2013). Combining controlled-source seismology and receiver function information to derive 3-D Moho topography for Italy. *Geophys. J. Int.*, 194, 1050–1068.

Spötl, C., Kralik, M., Kunk, M.J. (1996). Authigenic feldspar as an indicator of paleo-rock/water intercalations in Permian carbonates of the northern Calcareous Alps, Austria. *J. Sed. Res.*, 66(1), 139–146.

Spötl, C., Longstaffe, F.J., Ramseyer, K., Kunk, M.J., Wiesheu, R. (1998). Fluid-rock reactions in an evaporitic mélange, Permian Haselgebirge, Austrian Alps. *Sedimentology*, 45, 1019–1044.

Stampfli, G.A. (1993). Le Briançonnais, terrain exotique dans les Alpes ? *Eclogae Geologicae Helvetiae*, 86, 1–45.

Stampfli, G.M. (1994). Exotic terrains in the Alps: A solution for a single Jurassic ocean. *Schweiz. Mineral. Petrogr. Mitt.*, 74, 449–452.

Stampfli, G.M. (2000). Tethyan oceans. *Spec. Publ. Geol. Soc. London*, 173, 1–23.

Stampfli, G.M. and Borel, G.D. (2002). A plate tectonic model for the Paleozoic and Mesozoic constrained by dynamic plate boundaries and restored synthetic ocean isochrons. *Earth Planet. Sci. Lett.*, 196, 17–33.

Stampfli, G.M. and Marchant, R.H. (1997). Geodynamic evolution of the Tethyan margins of the Western Alps. In *Deep Structure of the Swiss Alps: Results of NRP 20*, Pfiffner, A., Lehner, P., Heitzmann, P., Mueller, St., Steck, A. (eds). Birkhäuser, Basel.

Staufenberg, H. (1987). Apatite fission-track evidence for postmetamorphic uplift and cooling history of the eastern Tauern Window and the surrounding Austroalpine (Central Eastern Alps, Austria). *Jahrb. Geol. B.-A.*, 130, 571–586.

Steenken, A., Siegesmund, S., Heinrichs, T., Fügenschuh, B. (2002). Cooling and exhumation of the Rieserferner Pluton (Eastern Alps, Italy/Austria). *Int. J. Earth Sci.*, 91, 799–817.

Stipp, M., Fügenschuh, B., Gromet, L.P., Stünitz, H., Schmid, S.M. (2004). Contemporaneous plutonism and strike-slip faulting: A case study from the Tonale fault zone north of the Adamello pluton (Italian Alps). *Tectonics*, 23, TC3004.

Strauss, P., Wagreich, M., Decker, K., Sachsenhofer, R.F. (2001). Tectonics and sedimentation in the Fohnsdorf-Seckau Basin (Miocene, Austria): From a pull-apart basin to a half-graben. *Int. J. Earth Sci.*, 90, 549–559.

Stüwe, K. and Schuster, R. (2010). Initiation of subduction in the Alps: Continent or ocean? *Geology*, 38, 175–178.

Sun, W., Zhao, L., Malusà, M.G., Guillot, S., Fu, L.-Y. (2019). 3-D Pn tomography reveals continental subduction at the boundaries of the Adriatic microplate in the absence of a precursor oceanic slab. *Earth and Planetary Science Letters*, 510, 131–141.

Tari, V. and Pamić, J. (1998). Geodynamic evolution of the northern Dinarides and the southern part of the Pannonian Basin. *Tectonophysics*, 297, 269–281.

Thöni, M. (1999). A review of geochronological data from the Eastern Alps. *Schweiz. Mineral. Petrogr. Mitt.*, 79(1), 209–230.

Thöni, M. (2006). Dating eclogite-facies metamorphism in the Eastern Alps – Approaches, results, interpretations: A review. *Mineralogy and Petrology*, 88, 123–148.

Thöni, M. and Jagoutz, E. (1992). Some new aspects of dating eclogites in orogenic belts: Sm-Nd, Rb-Sr and Pb-Pb isotopic results from the Austroalpine Saualpe and Koralpe type-locality (Carinthia/Styria, SE Austria). *Geochim. Cosmochim. Acta*, 56, 347–368.

Thöni, M. and Jagoutz, E. (1993). Isotopic constraints for Eo-Alpine high-P metamorphism in the Austroalpine nappes of the Eastern Alps: Bearing on Alpine orogenesis. *Schweiz. Mineral. Petrogr. Mitt.*, 73, 177–189.

Thöni, M. and Miller, C. (1996). Garnet Sm–Nd data from the Saualpe and Koralpe (Eastern Alps, Austria): Chronological and *P–T* constraints on the thermal and tectonic history. *J. Metamorphic Geol.*, 14, 453–466.

Tollmann, A. (1959). Der Deckenbau der Ostalpen auf Grund der Neuuntersuchung des zentralalpinen Mesozoikums. *Mitt. Ges. Geol. Bergbaustud. Österr.*, 10, 1–62.

Tollmann, A. (1961). Neue Ergebnisse über den Deckenbau der Ostalpen auf Grund fazieller und tektonischer Untersuchungen. *Geol. Rundschau*, 50, 507–513.

Tollmann, A. (1963). *Ostalpensynthese*. Deuticke, Vienna.

Tollmann, A. (1973). *Grundprinzipien der alpinen Deckentektonik*. Deuticke, Vienna.

Tollmann, A. (1976a). *Analyse des klassischen nordalpinen Mesozoikums*. Deuticke, Vienna.

Tollmann, A. (1976b). *Der Bau der Nördlichen Kalkalpen*. Deuticke, Vienna.

Tollmann, A. (1977). *Die Geologie von Österreich (Band I): Die Zentralalpen*. Deuticke, Vienna.

TRANSALP Working Group (2002). First deep seismic reflection images of the Eastern Alps reveal giant crustal wedges and transcrustal ramps. *Geophys. Res. Lett.*, 29, 92.1–92.4.

Tropper, P. and Hoinkes, G. (1996). Geothermobarometry of Al_2SiO_5-bearing metapelites in the western Austroalpine Ötztal basement. *Mineral. Petrol.*, 58, 145–170.

Trümpy, R. (1988). A possible Jurassic–Cretaceous transform system in the Alps and the Carpathians. *Geological Society of America Special Paper*, 218, 93–110.

Trümpy, R. (1992). Ostalpen und Westalpen – Verbindendes und Trennendes. *Jahrbuch der Geologischen Bundesanstalt*, 135, 875–882.

Urbanek, C., Frank, W., Grasemann, B., Decker, K. (2002). Eoalpine versus Tertiary deformation: Dating of heterogeneously partitioned strain (Tauern Window, Austria). *Pangeo Austria: Erdwissenschaften in Österreich*, 28.

Ustaszewski, K., Schmid, S., Fügenschuh, B., Tischler, M., Kissling, E., Spakman, W. (2008). A map-view restoration of the Alpine-Capathian–Dinaridic system for the Early Miocene. *Swiss J. Geosci.*, 101(Supplement 1), 273–S294.

Ustaszewski, K., Schmid, S.M., Lugovic, B., Schuster, R., Schaltegger, U., Bernoulli, D., Hottinger, L., Kounov, A., Fügenschuh, B., Schefer, S. (2009). Late Cretaceous intraoceanic magmatism in the internal Dinarides (northern Bosnia and Herzegovina): Implications for the collision of the Adriatic and European plates. *Lithos*, 108, 106–125.

Van Gelder, I.E., Willingshofer, E., Andriessen, P.A.M., Schuster, R., Sokoutis, D. (2020). Cooling and vertical motion of crustal wedges prior, during and after lateral extrusion in the Eastern Alps: New field kinematic and fission track data from the Mur-Mürz Fault System. *Tectonics*, 39, e2019TC005754.

Veselá, P., Lammerer, B., Wetzel, A., Söllner, F., Gerdes, A. (2008). Post-Variscan to Early Alpine sedimentary basins in the Tauern Window (eastern Alps). *Geological Society, London, Special Publications*, 298, 83–100.

Veselá, P., Söllner, F., Finger, F., Gerdes, A. (2011). Magmato-sedimentary Carboniferous to Jurassic evolution of the western Tauern Window, Eastern Alps (constraints from U-Pb zircon dating and geochemistry). *International Journal of Earth Sciences*, 100, 993–1027.

Viola, G. (2000). Kinematics and timing of the Periadriatic fault system in the Giudicarie region (central Eastern Alps). PhD Thesis, University of Zürich.

Viola, G., Mancktelow, N., Seward, D. (2001). Late Oligocene–Neogene evolution of Europe–Adria collision: New structural and geochronological evidence from the Giudicarie fault system (Italian Eastern Alps). *Tectonics*, 20, 999–1020.

Wagner, R., Rosenberg, C.L., Handy, M.R., Möbus, C., Abertz, M. (2006). Fracture-driven intrusion and upwelling of a mid-crust pluton fed from a transpressive shear zone – The Rieserferner pluton (Eastern Alps). *Geol. Soc. Am. Bull.*, 118(1–2), 219–237.

Wagreich, M. (1993). Subcrustal tectonic erosion in orogenic belts – A model for the Late Cretaceous subsidence of the Northern Calcareous Alps (Austria). *Geology*, 21, 941–944.

Wagreich, M. (1995). Subduction tectonic erosion and Late Cretaceous subsidence along the northern Austroalpine margin (Eastern Alps, Austria). *Tectonophysics*, 242, 63–78.

Wagreich, M. and Faupl, P. (1994). Paleogeography and geodynamic evolution of the Gosau Group of the Northern Calcareous Alps (Late Cretaceous, Eastern Alps, Austria). *Palaeogeogr., Palaeoclimatol., Palaeoecol.*, 110, 235–254.

Waibel, A.F. and Frisch, W. (1989). The lower Engadine Window: Sediment deposition and a calcitretion in relation to the plate-tectonic evolution of the Eastern Alps. *Tectonophysics*, 162, 229–241.

Wang, X. and Neubauer, F. (1998). Orogen-parallel strike-slip faults bordering metamorphic core complexes: The Salzach-Enns fault zone in the Eastern Alps, Austria. *J. Struct. Geol.*, 20, 799–818.

Weissert, H.J. and Bernoulli, D. (1985). A transform margin in the Mesozoic Thethys: Evidence from the Swiss Alps. *Geol. Rundschau*, 74, 665–679.

Wiederkehr, M., Bousquet, R., Schmid, S.M., Berger, A. (2008). From subduction to collision: Thermal overprint of HP/LT meta-sediments in the north-eastern Lepontine Dome (Swiss Alps) and consequences regarding the tectono-metamorphic evolution of the Alpine orogenic wedge. *Swiss Journal of Geosciences*, 101(Supplement 1), S127–S155.

Wiederkehr, M., Sudo, M., Bousquet, R., Berger, A., Schmid, S.M. (2009). Alpine orogenic evolution from subduction to collisional thermal overprint: The 40Ar/39Ar age constraints from the Valaisan Ocean, Central Alps. *Tectonics*, 28, TC6009.

Wiesinger, M., Neubauer, F., Handler, R. (2006). Exhumation of the Saualpe eclogite unit, Eastern Alps: Constraints from $^{40}Ar/^{39}Ar$ ages and structural investigations. *Mineral. Petrol.*, 88, 149–180.

Willingshofer, E. (2000). Extension in collisional orogenic belts: The Late Cretaceous evolution of the Alps and Carpathians. PhD Thesis, University of Amsterdam.

Willingshofer, E. and Cloething, S. (2003). Present-day lithospheric strength of the Eastern Alps and its relationship to neotectonics. *Tectonics*, 22, 1075.

Willingshofer, E., Neubauer, F., Cloething, S. (1999). The significance of Gosau-type basins for the Late Cretaceous tectonic history of the Alpine-Carpathian belt. *Phys. Chem. Earth*, 24, 687–695.

Wilkens, E. (1989). Paläogene Sedimente des Krappfeldes und seiner Umgebung. *Arbeitstagung der Geol. Bundesanst.*, 85–99.

Wolff, R., Hetzel, R., Dunkl I., Anczkiewicz, A.A., Pomella, H. (2020). Fast cooling of normal-fault footwall: Rapid fault slip or thermal relaxation? *Geology*, 48(4), 333–337.

Wölfler, A., Dekant, C., Danišík, M., Kurz, W., Dunkl I., Putiš, M., Frisch, W. (2008). Late stage differential exhumation of crustal blocks in the central Eastern Alps: Evidence from fission track and (U-Th)/He thermochronology. *Terra Nova*, 20, 378–384.

Wölfler, A., Kurz, W., Danisik, M., Rabitsch, R. (2010). Dating of fault zone activity by apatite fission track and apatite (U-Th)/He thermochronometry: A case study from the Lavanttal fault system (Eastern Alps). *Terra Nova*, 22, 274–282.

Wölfler, A., Kurz, W., Fritz, H., Stüwe, K. (2011). Lateral extrusion in the Eastern Alps revisited: Refining the model by thermochronological, sedimentary and seismic data. *Tectonics*, 30, TC4006.

Wölfler, A., Dekant, C., Frisch, W., Danišík, M., Frank, W. (2015a). Cretaceous to Miocene cooling of Austroalpine units southeast of the Tauern Window (Eastern Alps) constrained by multisystem thermochronometry. *Austrian J. Earth Sci.*, 108(1), 18–35.

Wölfler, A., Frisch, W., Fritz, H., Danišík, M., Wölfler, A. (2015b). Ductile to brittle fault zone evolution in Austroalpine units to the southeast of the Tauern Window (Eastern Alps). *Swiss J. Geosci.*, 108, 239–251.

Wölfler, A., Kurz, W., Frith, H., Glotzbach, C., Danišík, M. (2016). Late Miocene increasing exhumation rates in the eastern part of the Eastern Alps – Implications from low-temperature thermochronology. *Terra Nova*, 28, 297–305.

Zattin, M., Cuman, A., Fantoni, R., Martin, S., Scotti, P., Stefani, C. (2006). From Middle Jurassic heating to Neogene cooling: The thermochronological evolution of the southern Alps. *Tectonophysics*, 414, 191–202.

Zeilinger, G., Kuhlemann, J., Reinecker, J., Kazmer, M., Frisch, W. (1999). Das Tamsweger Tertiär im Lungau (Österreich): Fazies und Deformation eines intramontanen Beckens. *N. Jb. Geol. Paläont. Abh.*, 214, 537–569.

Zhao L., Paul, A., Malusà, M.G., Xu, X., Zheng, T., Solarino, S., Guillot, S., Schwartz, S., Dumont, T., Salimbeni. S. et al. (2016) Continuity of the Alpine slab unraveled by high-resolution P wave tomography. *J. Geophys. Res. Solid Earth*, 121(12), 8720–8737.

Zwingmann, H. and Mancktelow, N.S. (2004). Timing of Alpine fault gouges. *Earth Planet. Sci. Lett.*, 223, 415–425.

6

Alpine Corsica: From Tethys Subduction to Mediterranean Extension

Jacques MALAVIEILLE[1] and Giancarlo MOLLI[2]

[1] *Géosciences Montpellier, Université de Montpellier, CNRS, Université des Antilles, Montpellier, France*

[2] *Department of Earth Sciences, University of Pisa, Italy*

6.1. Introduction

The Central Mediterranean geological system has been extensively studied for many years, and there are countless models that have attempted to link field observations with the dynamics of subductions. In particular, the north-eastern Corsica (Alpine Corsica) has recorded the entire Alpine orogenic history, from the subduction of the oceanic Ligurian Tethys to Mediterranean extension, and therefore represents a key domain to unravel the Central Mediterranean tectonic history.

Alpine Corsica is a part of the Alps-Apennines continental collisional system, which evolved during the Tertiary from convergent Adria-Europe boundary conditions to extensional ones, with the opening of the Central Mediterranean back-arc basins. Extension with the final opening of the Gulf of Lion, the anticlockwise rotation of the Corsica-Sardina block and the latest opening of the northern Tyrrhenian Sea would be the consequence of the retreat of the Apenninic-related Adria slab after the formation of the Alpine chain of Corsica.

Geodynamics of the Alps 1,
coordinated by Claudio L. ROSENBERG and Nicolas BELLAHSEN. © ISTE Ltd 2023.

In this contribution, we review the history of Corsican tectonic controversies, and directly pose several questions that lie at their heart. The questions in turn are addressed in terms of detailed field and tectonic data.

6.2. Overall architecture of Corsica

Two different tectonic zones or geological domains are traditionally distinguished in Corsica (Figure 6.1): (1) An "autochthonous" Variscan (or Crystalline) Corsica forms the westernmost side of the island (Durand Delga 1984), and it is represented by pre-Variscan and Variscan basement intruded by areally dominant Late Carboniferous-Permian plutons and subvolcanic bodies (Rossi et al. 2003; Faure et al. 2014). Mesozoic to Tertiary sedimentary covers are limited to (i) Jurassic platform carbonates unconformably covered by Paleocene siliciclastic deposits (Peybernes et al. 2000) in southeastern Corsica (Punta Calcinaia); (ii) Eocene deposits represented by turbiditic sandstones that can be locally observed from north to south at the eastern border of "autochthonous"; (iii) early Miocene sediments around the Bonifacio area (Loyet-Pilot and Ferrandini 2014). Absent or weak alpine deformation characterize this tectonic domain; (2) "Alpine" Corsica, in the northeast of the island, which consists of a stack of continental, transitional (ocean-continent transition or OCT) and oceanic tectonic units that were variably deformed and metamorphosed within a former fossil subduction system (e.g. Mattauer et al. 1981; Jolivet et al. 1990; Caron 1994; Ravna et al. 2010; Vitale Brovarone et al. 2013; Deseta et al. 2014; Magott et al. 2016; Beaudouin et al. 2017).

Four major groups of units build up the "Alpine" Corsica (Figure 6.1): (1) continental-crust units, formed by middly to strongly reworked basement rocks and their Mesozoic to Mid-Eocene sedimentary cover. The latter were originally deposited on the proximal to distal part of the thinned Corsica continental margin (Faure and Malavieille 1981; Durand-Delga 1984; Michard and Martinotti 2002; Rossi et al. 2003; Molli 2008). (2) The "Schistes Lustrés" composite nappe formed by (i) Ligurian Tethys-derived oceanic sequences (mantle ultramafics, gabbros, pillow lavas and associated Jurassic to Cretaceous metasediments); (ii) relicts of the former OCT domain represented by exhumed mantle, ophiolitic metagabbros and metabasalts associated with slices ("extensional allocthons") of continental upper and lower crust and related metasedimentary covers (Caron and Delcey 1979; Durand-Delga 1984; Lahondere 1996; Vitale Brovarone et al. 2011; Meresse et al. 2012); (iii) Triassic-early Jurassic to Cretaceous metasediments detached from the original Corsican continental basement, for example, the Bagliacone Riventosa units and, possibly, the metasediments exposed in the inner core of the Castagniccia antiform (see the lower Castagniccia unit of Vitale Brovarone et al. (2013)).

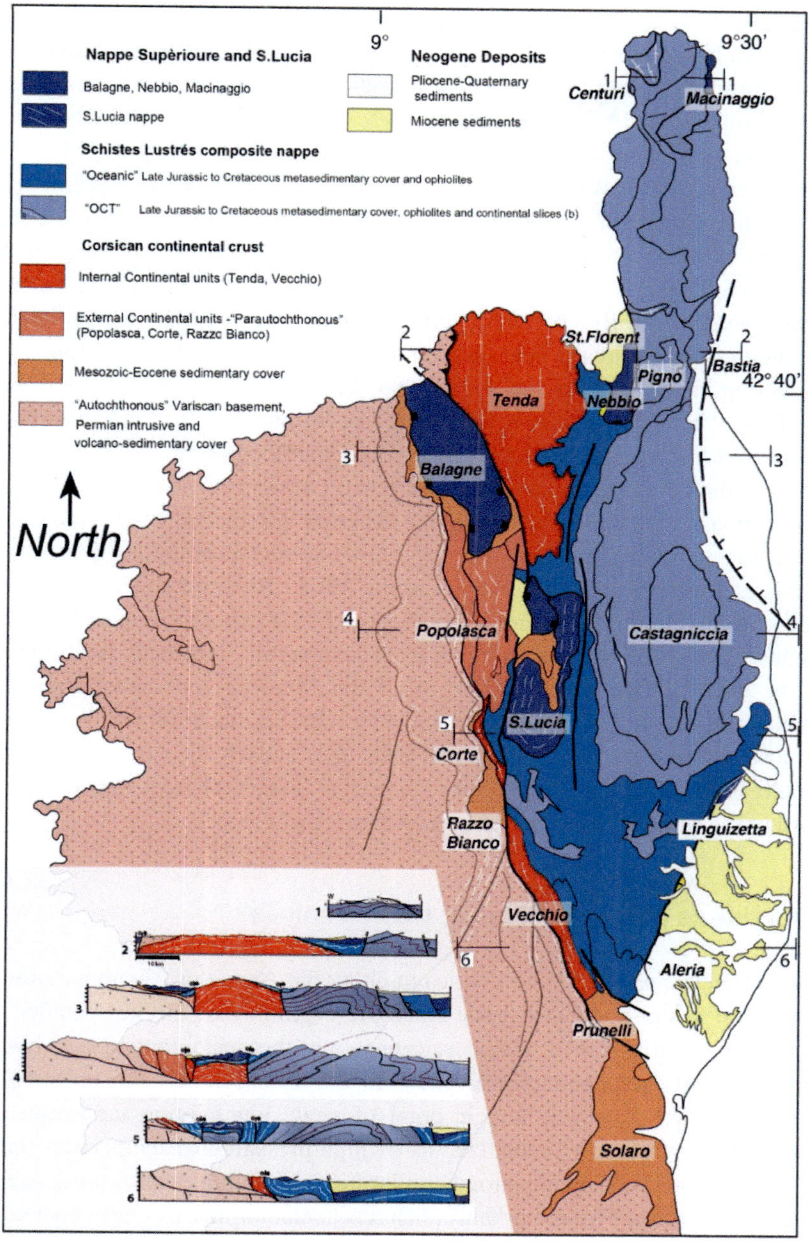

Figure 6.1. *Geologic map and regional cross-sections of northern Corsica showing the main groups of tectonic units distinguished by continental, oceanic or mixed origin (OCT derived and upper plate derived). The geological map is a modified version of Figure 6 in Molli (2008) and Figure 3 in Molli and Malavieille (2011)*

The latter two groups of units show subduction-related peak metamorphism, ranging from low-grade blueschist to eclogite facies. The high pressure assemblages were overprinted at various degrees by retrograde greenschist facies assemblages (Dal Piaz and Zirpoli 1979; Gibbons and Horack 1984; Waters 1990; Fournier et al. 1991; Caron 1994; Daniel et al. 1996; Molli et al. 2006; Martin et al. 2011; Vitale Brovarone et al. 2013; Rossetti et al. 2015). (3) The Upper Nappe System is also called "Nappe Supérieure", which includes the Balagne, Nebbio and Macinaggio units. This system consists of ophiolites, continental derived units (granitoids, low-grade phyllites, Triassic to Jurassic carbonates), Late-Cretaceous siliciclastic and calcareous-marly flysch (Nardi 1968; Durand-Delga 1984), and siliciclastic Paleocene-Eocene sediments deposited on top of the alpine accretionary prism, that is, wedge-top basins (Dallan and Nardi 1984; Puccinelli et al. 2012). The different units of the "Nappe Supérieure" occur at the summit of the nappe stack and experienced very low-grade conditions (prehnite/pumpellyite in basic rocks). The ophiolitic Balagne nappe is found in the external (western) zone directly overriding the continental units ("autochthonous and external continental units" see below). The Nebbio and Macinaggio units occupy a more internal position, resting on top of the Schistes Lustrés composite nappe-system. Vitale Brovarone et al. (2013) also included, in this group, the Santa Lucia Nappe (SLN) exposed west of Corte. The SLN consists of slices of Permian lower crust covered by Cretaceous metasediments (Zibra et al. 2010; Molli et al. 2020). (4) Early Miocene continental and marine sediments of the Francardo, St. Florent and Aleria basins, unconformably sealing the uppermost nappes and their contacts with the underlying units (Dallan and Puccinelli 1995; Ferrandini et al. 1998; Fellin et al. 2005). These sediments provide a record of the Miocene and post-Miocene tectonics of Corsica.

6.3. Corsica as part of the Alps or the Apennines?

The pertinence of the Alpine Corsica to either the Alpine or Apenninic orogenic belt has been a highly debated issue since the beginning of the last century. The interpretation clearly has a major relevance on the tectonic and geodynamic reconstruction of the Mediterranean area. In the 19th century, the problem was principally tackled from a lithologic point of view, highlighting the presence of mainly metamorphic (now recognized as of high-pressure/low temperature facies) rock types in Corsica, showing strong similarities and analogies with those exposed in the Western Alps. The remarkably different metamorphic Corsican rocks from those widely exposed in the Apennines (mainly of sedimentary origin) were considered as the key to interpreting Corsica as the southern extension of the Alpine chain in the Mediterranean (Argand 1924; Staub 1933). Subsequently, the petrographic arguments were coupled with structural elements, such as the westward fold-vergence and sense of transport of the tectonic units, within the pile of nappes. The similarity with the dominant westward vergence in the Alps was quoted as a further support to considering Corsica as a former segment of the Alps – a view consolidated in the aftermath of the plate tectonic revolution (Boccaletti et al. 1971; Alvarez et al. 1974; Mattauer and Proust 1975).

In the mid 1980s, Principi and Treves (1985) presented the alternative interpretation that Corsica actually represents part of the Apennines accretionary wedge formed during Cretaceous to Tertiary west-dipping subduction of the Adria plate and its oceanic portion. In this interpretation, western "autochthonous" Corsica represents the backstop of the Apenninic wedge, while Alpine Corsica is considered the inner portion of Apennines. In detail, the Alpine Corsica should form the deepest part of the Apenninic accretionary system, backthrusted on the "autochthonous" Corsica by corner flow processes (Cowan and Silling 1978) on the rear side of the system (Principi and Treves 1985; Rossetti et al. 2002).

6.4. Formation and exhumation of the high-pressure rocks in Corsica and the vergence of the subduction

The Alpine Corsica records high-pressure/low-temperature (HP-LT) metamorphic fabrics (blueschists and eclogite-facies) related with a fossil subduction, but in which subduction system did these assemblages form? Was it the Alpine east-southeast dipping or the opposite west-northwest dipping Apenninic subduction?

The geologic record of the Corsican-derived continental units and that of the former OCT can help to constrain and answer this key question (Figures 6.2 and 6.3).

The continental crust units are formed by pre-Mesozoic basement rocks and their Mesozoic to Tertiary cover rocks. A widely accepted view (Durand Delga 1984; Jolivet et al. 1998; Rossi et al. 2002; Beaudouin et al. 2017) is that these units are derived from the Corsican continental margin, as are most of the continental slices (see below) of the OCT units found in the Schistes Lustrés composite nappe (Vitale Brovarone et al. 2011; Martin et al. 2011). This view is accepted by all authors, with one unique exception represented by the proposition of Vitale Brovarone and Herwartz (2013) who suggested, as an alternative, the original paleotectonic location of the OCT units as part of the Adria-continental margin. This proposal is ruled out here on the basis of (i) the composition of basement and cover rock types of the

OCT units similar to those of the "autochthonous" Corsica and different from those of the Adria continental margin (e.g. basement and cover of the Tuscan domain Conti et al. 1993; Molli et al. 2020); (ii) the age of their involvement in subduction processes (≥34 Ma for the Corsican OCT; ≤25 Ma for the Tuscan OCT and distal continental margin) (Kligfield et al. 1986; Bianco et al. 2019; Di Vincenzo, Godard, Molli submitted). The Corsican continental units and/or the continental slices of the OCT units show a variable HP-LT metamorphic imprint (Bezert and Caby 1988; Daniel et al. 1996; Molli 2008; Malavieille and Molli 2014; Vitale Brovarone et al. 2014; Beaudoin et al. 2017) with peak conditions ranging from lower grade (high-pressure greenschist to lower blueschist facies) in the geometrically lowermost external continental units, to blueschist and eclogite facies, in the more internal and uppermost OCT units (Figure 6.2; Bézert and Caby 1988; Egal and Caron 1988; Bézert 1990; Caron 1994; Lahondère 1996; Tribuzio and Giacomini 2002; Malasoma et al. 2006; Garfagnoli et al. 2009; Ravna et al. 2010; Martin et al. 2011; Maggi et al. 2012; Molli et al. 2017). Figure 6.2 shows the tectono-metamorphic architecture of Alpine Corsica, with distinction within the Schistes Lustrés composite nappe-system of the oceanic and OCT-derived units based on legacy literature data and the more recent petrological studies by Ravna et al. (2010), Vitale Brovarone et al. (2013, 2014) and Beaudouin et al. (2017).

A pervasive composite mylonitic foliation, with widespread relicts of eclogite fabrics, can be observed in the Eclogite OCT-units (Ec-OCT in Figure 6.2) sandwiched between the low-grade Lawsonite Blueschist units above (Law-Bs OCT of Vitale Brovarone et al. (2013, 2014)) and below (Law-Castagniccia of Vitale Brovarone et al. (2013, 2014)). The stretching lineations and shear sense indicators (Figure 6.2) associated indicate a top-to-north/northwest shearing for the eclogite fabrics (Lahondère 1996; Lahondère and Guerrot 1997), and a dominant top-to-west kinematics for the blueschist main fabric in Bs-OCT units and the retrograde fabric in Ec-OCT (Faure and Malavieille 1981; Mattauer et al. 1981; Malavieille 1983; Harris 1985a, 1985b; Jolivet et al. 1990; Fournier et al. 1991; Daniel et al. 1996; Malavieille et al. 1998; Molli and Malavieille 2011).

The continental Tenda unit is characterized by heterogeneous deformation patterns with preserved low strain domains showing the pristine granitoid magmatic fabric surrounded by mylonitic orthogneisses and/or granitoid mylonites. The dominant fabric is a greenschist facies, exhumation-related foliation preserving widespread relict domains with a blueschist fabric and a dominant top-to-west shearing (Figures 6.4(a) and (b)) (Mattauer et al. 1981; Lahondère et al. 1999; Daniel et al. 1996; Molli et al. 2006; Maggi et al. 2012; Beaudoin et al. 2017).

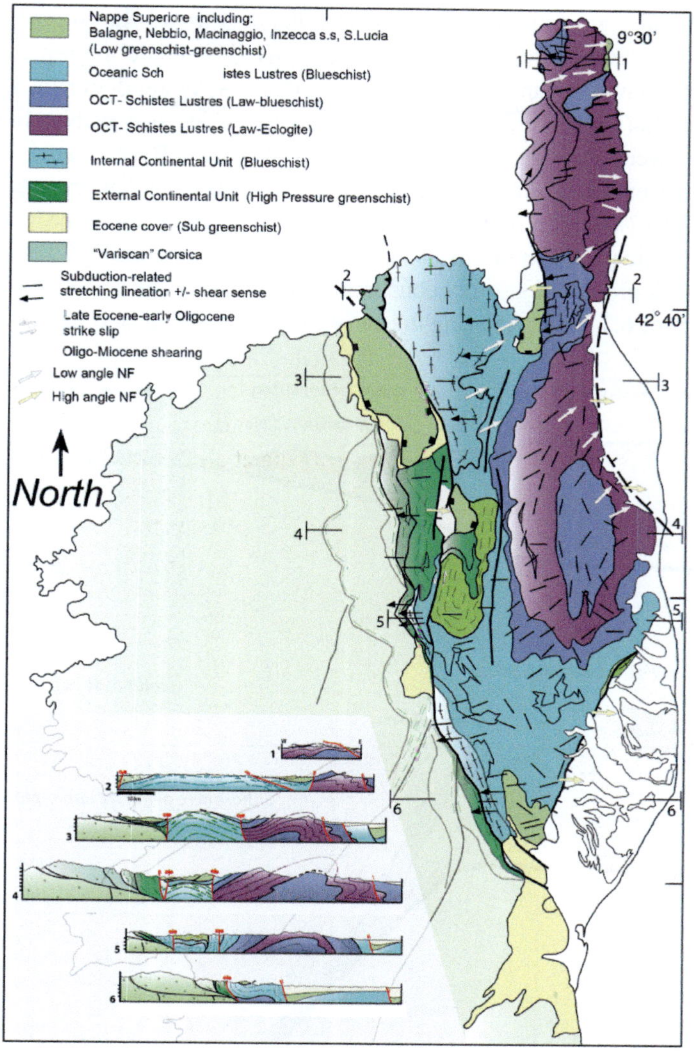

Figure 6.2. *Tectono-metamorphic map and regional cross-sections of northern Corsica showing the main tectonic units characterized by their peak metamorphic assemblages and composition*

COMMENT ON FIGURE 6.2.– *The tectono-metamorphic map is a modified version of Figure 6 in Vitale Brovarone et al. (2013). Structural data are from Malavieille (1983, 1984), Daniel et al. (1996), Molli (2008), Molli and Malavieille (2011), and Beaudouin et al. (2017).*

The external continental units (e.g. Corte, Popolasca and Razzo Bianco units), referred to as ECU, are also characterized by low-grade high-pressure/low-temperature metamorphism associated with structures showing dominant top-to-the-west kinematics (Figures 6.2 and 6.4(c) and (d)). These are observable all along the boundary between Alpine and "autochthonous" Corsica (Bézert and Caby 1988; Egal 1992; Bézert 1990; Malasoma et al. 2006; Molli 2008; Di Vincenzo et al. 2016; Molli et al. 2017). The finite geometries, the internal structures, deformation styles and metamorphic peak conditions of the Corsican continental and OCT units within the nappe stack provide first-order constraints on the role of the Corsican continental margin (Figures 6.3 and 6.4) during the early stages of the Alpine orogenic history.

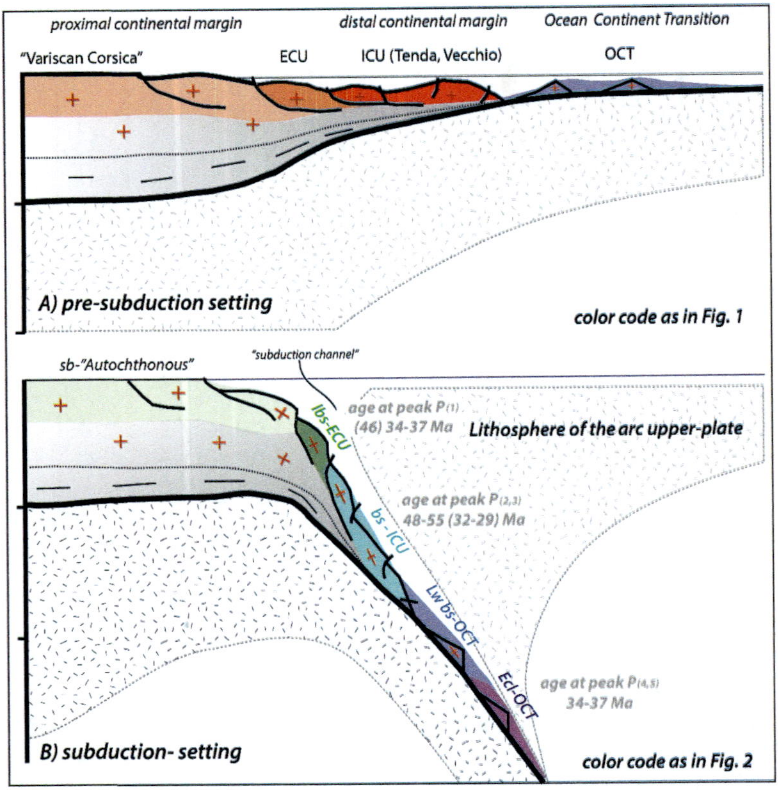

Figure 6.3. *a) Schematic representation of the template in the stretched Corsican continental margin and related OCT transition; b) setting at the stage of the continental subduction with location of different continental units ECU, ICU, OCT-derived. The more recently published geochronological data are inserted and discussed in the text*

As a whole, these structural and metamorphic features suggest a progressive underthrusting of the upper Corsican crust and proximal to distal OCT in a framework of east-dipping continental subduction (Mattauer and Proust 1975; Mattauer et al. 1977; Faure and Malavieille 1980; Mattauer et al. 1981; Gibbons and Horak 1984; Waters 1990; Malavieille et al. 1998; Molli and Tribuzio 2004; Molli et al. 2006; Molli 2008; Molli and Malavieille 2011; Ravna et al. 2010; Handy et al. 2010; Maggi et al. 2012; Agard and Vitale-Brovarone 2013; Beaudouin et al. 2017; van Hinsbergen et al. 2020).

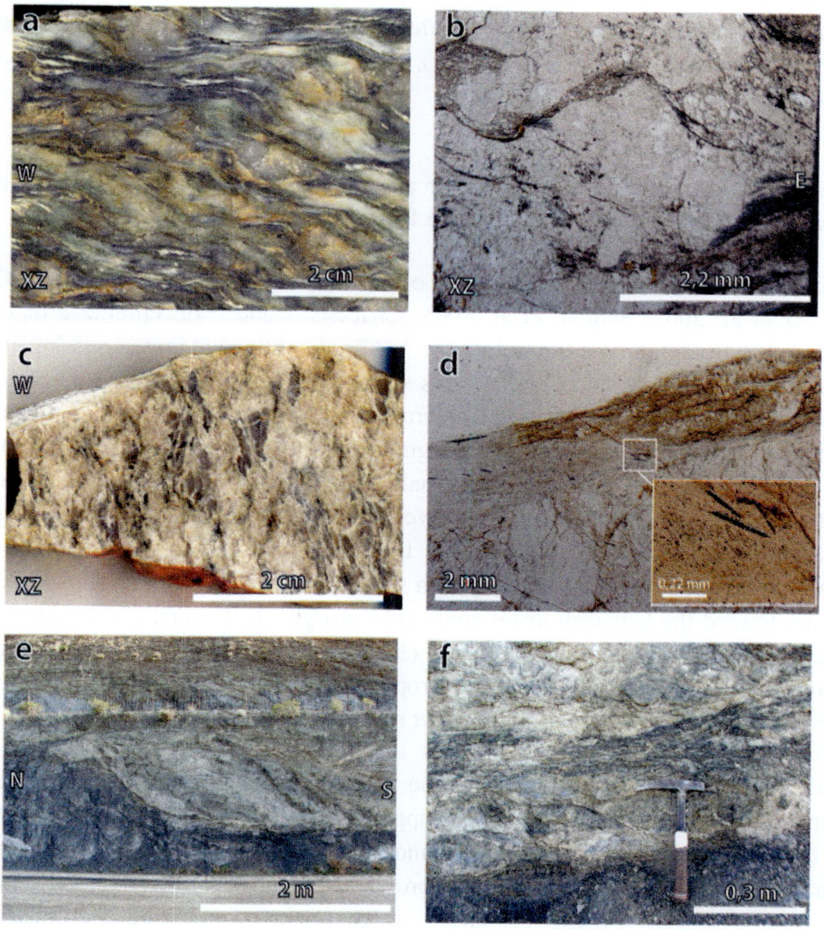

Figure 6.4. *Meso- and microstructures of (a, b) BS Tenda (ICU) and (c, d) the low-grade BS Popolasca (ECU)*

COMMENT ON FIGURE 6.4.– *(a) Blueschist facies mylonitic metagranite showing asymmetric porphyroclasts and shear bands defining a top-to-west shear sense, XZ plane; (b) detail in natural polarized light showing syn-kinematic Na-amphibole in the tails of K-Feldspar asymmetric porphyroclast, XZ plane; (c) protomylonitic metagranite cross-cutted by a millimeter-scale ultramylonite. Structures developed under low-grade blueschist representing typical deformation features of granitoids of the ECU (see Molli et al. (2017)); (d) microscopic view of ultramylonite with detail of ultramylonite and host protomylonite with the sharp shear zone boundary heritage of former host-fracture contact. A detail of Na amphibole, boudinaged within the quartz and albite ultramylonite matrix is reported; (e, f) mesoscopic fabrics of the Central Corsica Fault (CCF), photo year 2006 along the RN193 north of Corte.*

6.5. Was there a micro-continent or an arc between Corsica and Adria and what is the origin of the "Nappe Supérieure"?

Similar to the southwestern portion of the Alpine chain (Ford et al. 2006; Schwartz et al. 2007; Schmid et al. 2005; Chapters 4 and 5 of Volume 2 of this book), an Austroalpine-type (i.e. referrable to Adria) continental unit is missing atop the nappe pile of the Alpine Corsica. This suggests an original paleotectonic frame in which the subduction initiated far from the Austroalpine-Adria continental margin, as was likely the case for other segments of the Alps (e.g. the Central Alps). This early stage of subduction-related paleotectonic setting left a progressively larger portion of oceanic crust and hyperextended continental margin (sensu Manatschal and Bernoulli 1999) between the Adria and the place of inception of Alpine subduction. The subduction system was therefore located in an intraoceanic position, toward the Corsican distal continental margin, in a frame envisaged since the early 1970s by Elter and Pertusati (1973), Dal Piaz (1974), Mattauer and Tapponnier (1978), and more recently reproposed by Malavieille et al. (1998), Molli (2008), Handy et al. (2010) and Manzotti et al. (2014).

In Corsica, the uppermost units of the nappe system (see above), that is, the "Nappe Supérieure", occur as different klippen in external zones (Balagne), as well as in the internal side of the belt (Nebbio and Macinaggio). These units represent the remnants of an original kilometer-wide orogenic Lid preserved by erosion within post nappe synforms (Figures 6.1 and 6.2). The Nappe Supérieure is formed by ophiolitic, as well as continental derived rocks including pre-Mesozoic continental basement slices, Triassic-Jurassic cover, Late Cretaceous sedimentary sequences (Alturaia and Novella sandstones and the Toccone breccias) and Paleocene-Early

Eocene siliciclastic deposits (Annunciata, Palasca and Lozari sandstones; Nardi 1968; Durand-Delga 1984; Dallan and Nardi 1984; Puccinelli et al. 2012).

The debate on the significance, origin and tectonic evolution of the Balagne Nappe, including the Nappe Supérieure, is still open. Some authors, for example, Durand Delga (1984), Principi and Treves (1985), Marroni and Pandolfi (2003), suggest an original location of the Balagne Nappe and the Nappe Supérieure as a whole close to the Corsican continental margin (Balano-Ligure domain of Durand Delga (1984)). This interpretation is mainly supported by the stratigraphic and geochemical signature of the ophiolites and the related sedimentary cover in terms of (Rossi et al. 2002):

a) the presence of continental crust-derived debris at different stratigraphic levels within the ophiolitic sequence and cover series (e.g. the latest Late Jurassic limestone of San Colombano);

b) occurrences of zircon-bearing sandstones within basalts, with zircons identical to those of the neighboring post-Variscan granites of Corsica;

c) continental source area for Late Cretaceous and Eocene siliciclastic deposits indentified in the "autochthonous" Corsica;

d) E-MORB composition of the oceanic basalts formed during the early stage of oceanization in the ocean-continent (Corsica) transition area.

In contrast, other authors (Nardi 1968; Mattauer and Proust 1975; Dallan and Nardi 1984; Malavieille et al. 1998; Molli 2008) interpreted the Balagne Nappe and the Nappe Supérieure of "internal"-derivation, therefore representative of an oceanic portion located far east from the Corsican margin ("ultra-Schistes Lustrés"), toward the "Apenninic" oceanic domain. Molli and Malavieille (2011) envisage the Nappe Supérieure system (Balagne-Nebbio-Macinaggio units) as a relict of an intraoceanic accretionary wedge formed by offscrapping and shallow underplating of an oceanic part of Ligurian Tethys. This accretionary structure developed far from the Corsican continental margin, possibly close to the ribbon Nebbio-Calabria (Alvarez 1991) that represented a microcontinent or continental edge block northernmost part of the larger Alboran-Kabilie, Peloritani, Calabria (AlKaPeCa) domain (Michard et al. 2002; Le Petre et al. 2018).

The ocean-continent transition of the eastern side of Corsican continent is likely presently testified by the eclogitic and blueschist units of the Serra di Pigno, Farinole, San Petrone area (Martin et al. 2011; Meresse et al. 2012; Vitale Brovarone et al. 2011; Vitale Brovarone et al. 2013). The Balagne ophiolite could be representative of the opposite OCT close to the microcontinent located east of the former Corsican margin. The continental basement-derived Late Cretaceous sediments in the supra-ophiolitic cover likely represent the analog of a "forearc basin" partially deposited on the rear portion of a growing accretionary wedge (see below).

The Nappe Supérieure System has attributes consistent with being the relict of the shallowest portion of the former intraoceanic accretionary wedge and of a "forearc basin" partially deposited on top of the growing wedge. The source area of Late Cretaceous siliciclastic deposits could be located in a continental edge block, in the northern prolongation of the AlKaPeCa microcontinent (Michard et al. 2002; Le Petre et al. 2018; Critelli 2018). This microcontinent was left over by rifting processes (Malavieille et al. 1998; Michard et al. 2002; Molli 2008; Handy et al. 2010; Molli and Malavieille 2010; Vitale Brovarone et al. 2013; Schmid et al. 2017), and separated during Late Jurassic from the western and eastern branches of Ligurian Tethys of this transect.

It is of note that, between the OCT units within the Schistes Lustrés, the Centuri unit in the Cap Corse is peculiar for being formed by continental slices of middle-crust with some analogies with those of the lower grade SLN. This suggests an along strike variability of the Corsican OCT with upper or lower crustal signatures within the former extensional allochtons, which were involved in the lower versus upper plate position of the former subduction system.

6.6. Age of subduction-related deformations

The age of the subduction-related metamorphic events, as constrained by geochronological data, is a matter of debate (Brunet et al. 2000; Handy and Oberhänsli 2004; Berger and Bousquet 2008; Vitale Brovarone and Herwartz 2013). Here, we only briefly report and discuss the most recent data (Martin et al. 2011; Maggi et al. 2012; Vitale Brovarone and Herwartz 2013; Rossetti et al. 2015; Di Vincenzo et al. 2016).

Di Vincenzo et al. (2016) dated the sheared granites of the low-grade blueschist facies external continental crust units (ECU) around Corte (central Corsica). The authors document the presence within localized shear zones (see also Molli et al. (2017)) of an earlier celadonite-poor white mica with a minimum age of ~46 Ma. The earlier mica is overgrown by a later celadonite-rich white mica, developed between ~37 and 35 Ma, in turn rimmed by an exhumation-related white mica yielding an age of ca. 33–32 Ma.

Rossetti et al. (2015) dated by Rb-Sr the waning stages of blueschist event in the Tenda continental unit (ICU), that is, in a more distal position of the Corsican continental margin (Figure 6.3), at ca. 32–29 Ma. Previous U/Pb dating of syn-kinematic rutile within a blueschist–facies mylonites by Maggi et al. (2012) yielded an age of 48 ± 18 Ma (MSWD 7.3). The same authors obtained an intercept at 54 ± 8 Ma (MSWD = 48) for coexisting acmite–phengite and coatings of oxides or sulfides; this age was considered the most reliable for the blueschist peak event in the Tenda continental unit (ICU) (Figure 6.3).

Martin et al. (2011) and Vitale Brovarone and Herwartz (2013), based on zircon (U-Pb), garnet (Lu-Hf) and lawsonite (Lu-Hf), suggest a Late Eocene age (37–34 Ma) for the eclogite facies metamorphism in the Schistes Lustrés (Ec-OCT and law-OCT of Vitale Brovarone et al. (2013) and Figure 6.3). The same authors discussed all the older ages (including the Sm–Nd whole rock Grt–Gla–Cpx isochron age of 84 ± 5 Ma of Lahondère and Guerrot (1997)), pointing out their problems due to inheritance (Ar/Ar data in Brunet et al. (2000)) or analytical artifacts.

These recently published geochronological data cannot be easily matched together as shown in Figure 6.3. The prograde and peak ages in ECU may well fit the peak of metamorphism in the Tenda ICU, but in this case the prograde and peak pressure in Ecl-OCT should be older. On the other hand, the peak pressure in Ecl-OCT (as well as in the Law-OCT (Vitale Brovarone and Herwartz 2013)) may be coeval with the peak of metamorphism in ECU, but in this case the BS metamorphism in the Tenda ICU has to be younger than those reported in Maggi et al. (2012). In this respect, it should be considered that the reported ages in Rossetti et al. (2015) were related to post-peak blueschist deformations.

Applying the method proposed by Guyedan et al. (2009), it is possible to infer both subduction velocity and dip angle from the alignment of peak pressure-temperature (PT) data in contemporaneous high-pressure–low-temperature units. Using the data of Vitale Brovarone and Herwartz (2013), Rossetti et al. (2015), Di Vincenzo et al. (2016), and considering the age of HP metamorphism in the ECU and OCT units as coeval, some physical features of the former Corsica subduction zone may be derived. Assuming 0.6 GPa and 300°C for the ECU units (Di Vincenzo et al. 2016), ca. 2 GPa and 500°C in the Ecl-OCT (Ravna et al. 2010; Vitale Brovarone et al. 2014), and that they were coeval at 37–34 Ma, it is possible to estimate a slab dip of ca. 55° and convergence velocity of ca. 1.5 cm/year.

Moreover, assuming a coeval peak pressure and the present finite geometries (see cross-sections in Figures 6.1 and 6.2), as acquired at the time at the end of exhumation in the ECU at 32 Ma (Di Vincenzo et al. 2016; Rossetti et al. 2015), it results in a temperature drop from ca. 500 to 300°C in minimum 2 Ma, which converts to a maximum cooling rate of 100°C Ma^{-1} and an exhumation rate of ca. 3 cm/Ma^{-1} assuming a lithostatic pressure, that is, 70 km of depth at peak for the Ecl-OCT.

Other independent constraints on the ages of deformations in Alpine-Corsica may be derived from stratigraphic data of the deeper and shallower portion of the Corsican wedge.

The inception of oceanic subduction is recorded in the remnants of the former intraoceanic accretionary wedge (Nappe Supérieure system), where Late Cretaceous turbidites were related with a tectonic activity, with reworking and erosion of a source area made up of a continental basement and sedimentary cover part of the edge block that acted as the backstop of the convergent system (Malavieille et al. 1998; Molli and Malavieille 2011).

The internal metamorphic units of the nappe stack (i.e. Schistes Lustrés), including the OCT-derived units, rarely provided any biostratigraphic constraints. Noteworthy exceptions are the findings of Meresse et al. (2012), which indicated a post-Albian deformation and metamorphism in the OCT-derived units within the Schistes Lustrés.

More significant biostratigraphic dates are found in the ECU and in the eastern parts of the "autochthonous" Corsica (Bézert and Caby 1988; Bézert 1990; Waters 1990; Molli et al. 2006; Malasoma et al. 2006; Molli 2008), where remnants of an underfilled foreland basin may be found. This basin shows strong compositional and facies affinity to that of the Western Alps (Michard and Martinotti 2002; Ford et al. 2006), and contains sediments palaeontologically dated by the presence of *Nummulites biarritzensis; Discociclina sp.*, to Early Mid-Eocene (Bezert and Caby 1988).

Therefore, the data from the frontal part of the northern Alpine Corsica suggest that the sedimentation continued until the Bartonian, with deformation post-dating the Early-Mid Eocene (40–37 Ma). This is in agreement with the syn-sedimentary record in the top-wedge satellite basin of Caporalino-Sant'Angelo part of Nappe Supérieure, which suggests an intrawedge tectonic activity during the Bartonian (Puccinelli et al. 2012).

In the central-northern Corsica, sediments of the Venaco Fm. unconformably lie on top of "autochthonous" basement, parautochthonous units (ECU) and Schistes Lustrés. These sediments include a microfauna (in particular, some representatives of *Turborotalia cerroazulensis*), which constrains the end of shortening in the frontal part of the orogen as pre-Priabonian (pre-37 Ma) (Ferrandini et al. 2010). Moreover, in the southeasternmost external domain of Corsica, in similar sediments (Prunelli-Solaro Flysch) showing the same unconformable relationships with the basement of western Corsica and Schistes Lustrés, units are dated at the Lutetian (47–41 Ma) (Egal 1992; Caron 1994).

As a whole, these biostratigraphic available data imply a diachronous involvement of proximal Corsica continental crust in underthrusting and subduction (forming the present ECU): as pre-41 Ma in the south, pre-37 Ma in central Corsica and post-37 Ma in northernmost Corsica. These data attest for a diachronism south to north in the involvement and deformation of continental crust, which in turn was followed by a diachronous end of the subduction (Molli 2008; Molli and Malavieille 2011).

6.7. The role of Apenninic tectonics and the final shaping of Alpine Corsica

Until the Eocene, Corsica was located at the southwesternmost front of the Mediterranean Alps, which was progressively deactivated from south to north in connection to the lateral switch from the "Alpine" (south-southeast vergent) to the opposite north-northwest vergent Apenninic subduction (Michard et al. 2002; Molli 2008; Handy et al. 2010; Molli and Malavieille 2011; Schmid et al. 2017). This change in geodynamics was likely responsible for the transpressional/transtensional reactivation of pre-existing frontal thrusts and the deformation (Figures 6.4(e) and (f)) along the Central Corsica Fault zone (Mattauer and Proust 1975; Waters 1990; Lacombe and Jolivet 2005; Molli 2008; Molli and Malavieille 2011) during the latest Eocene-early Oligocene. The change was possibly in connection with an early stage of Sardinia-Corsica block rotation (Advokaat et al. 2014; van Hinsbergen et al. 2020).

During the Oligocene and early Miocene, the extensional/transtensional tectonics described by Jolivet et al. (1990), Daniel et al. (1996), Jolivet et al. (1998), Rossetti et al. (2015), Beaudoin et al. (2017) and Gueydan et al. (2017) took place when the former "Alpine" wedge became part of the upper plate of the Apenninic subduction. Over this time, Corsica completed the anticlockwise block rotation to its present position (Gattacceca et al. 2007; Maffione et al. 2008). This stage of deformation resulted in a reactivation of previous subduction- and syn-contractional exhumation-related fabrics and structures in greenschist-facies shear-zones, and related low-strain domains (Daniel et al. 1996) developed between 27 and 20 Ma (Rossetti et al. 2015). These structures distributed at the scale of all northern Corsica (Figure 6.2) were later followed by high angles normal to transtensional faults. This tectonics was associated with the development of Miocene basins and their subsequent deformation, resulting in the present day regional morphostructural setting of northern Corsica (Zarki-Jakni et al. 2004; Fellin et al. 2005; Cavazza et al. 2007; Danisk et al. 2012).

6.8. The geodynamic model for the Alpine Corsica: from Tethys subduction to Mediterranean extension

The early stage of evolution of the Corsican belt may be described within a frame of an intraoceanic subduction of the westernmost branch of the Ligurian Tethys (western Liguria Tethys of Handy et al. (2010)) associated with the development of an accretionary wedge, active since Late Cretaceous. This accretionary wedge stage with slicing of oceanic basement and its sedimentary cover was not documented by any geochronological data in the Schistes Lustrés (see previous discussion on geochronological constraints), but only recorded in the structures of the shallowly accreted units of the Nappe Supérieure system (Balagne, Nebbio, Macinaggio). The Mid-Cretaceous to Late Cretaceous siliciclastic continental-derived turbidites and proximal breccias (Alturaia, Novella, Toccone), part of the composite system of Nappe Supérieure, can be related to the forearc and/or rear-arc basins fed by a non-volcanic arc (referred to as Nebbio by Alvarez (1991)). The Corsican non-volcanic arc could represent the northern tip of the Nebbio-Calabria microblock, part of the Meso-Mediterranean or AlKaPeCa domain (Dallan and Nardi 1984; Alvarez 1991; Guerrera et al. 1993; Malavieille et al. 1998; Michard et al. 2002; Molli 2008; Molli and Malavieille 2011; Advokaat et al. 2014; Critelli 2018; Le Petre et al. 2018; van Hinsbergen et al. 2020). In Corsica, relicts of this domain can be found as small slices of basement and cover rocks in the Nebbio nappe (Durand-Delga 1984; Dallan and Nardi 1984; Dallan and Puccinelli 1995), and possibly in the SLN (central Corsica). Moreover, it may be identified in the former source area of the well-developed detritial Cretaceous deposits of the Nappe Supérieure (Critelli 2018). From the Paleocene to the early-Mid Eocene, the Corsican orogenic system (Figure 6.5(a)) entered the stages of incipient to full continental subduction, with the progressive involvement of the OCT domain, the distal continental margin (now represented by the ICU of Tenda and Vecchio units) and the proximal continental margin (ECU units).

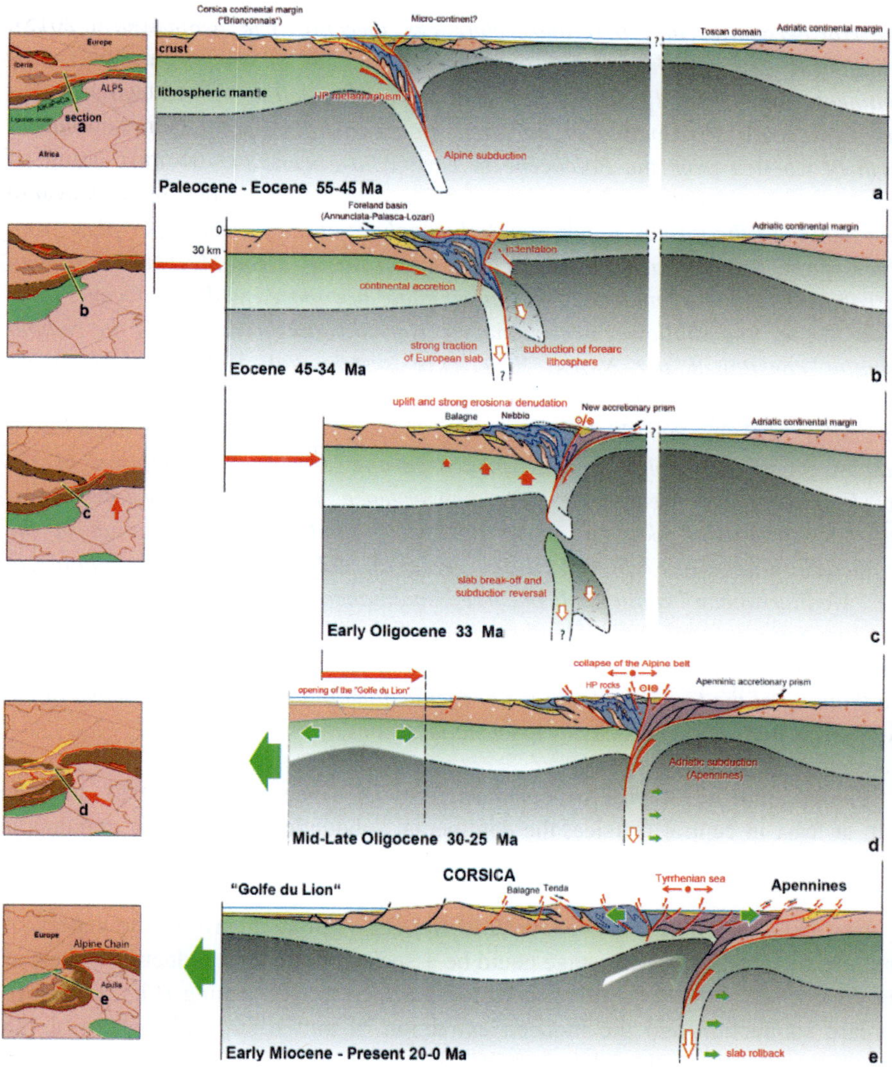

Figure 6.5. *Palaeotectonic sketch maps and evolutionary model for the Corsica/Apennines orogenic system*

COMMENT ON FIGURE 6.5.– *The paleotectonic maps are modified from Molli and Malavieille (2011) and references therein. Others paleotectonic interpretations can be found in the literature (e.g. Dercourt et al. 1986; Dewey et al. 1989; Michard et al. 2002; Faccenna et al. 2004; Lacombe and Jolivet 2005; Rosenbaum and Lister 2005; Argnani 2009; Handy et al. 2010; Malusà et al. 2011; Turco et al. 2012;*

van Hinsbergen et al. 2020 and references therein). The evolutionary model enlightened us: (a) The stages of early involvement of the extended Corsica continental margin and related OCT in a subduction setting, which started since Late Cretaceous. (b) Continental subduction occurred diachronously from south to north of Corsica from the Paleocene to early Mid-Eocene (45–34 Ma). The syn-contractional exhumation of the eclogitic OCT occurred during that period through the removal of the forearc upper-mantle lithosphere of the former upper plate from slices of the Nebbio-Calabria microcontinents AlKaPeCa (or Mesomediterranean microblock) below. Remnants of that upper-plate juxtaposed with the oceanic units of the deeper and shallower accretionary prism, including remnants of the Nebbio domain, constituted the Nappe Supérieure system. (c) Slab breakoff favors subduction reversal, inducing the inception of the Apenninic subduction. The upbending of the former thickened corsican lower plate following slab detachment induces significant uplift and associated strong erosional denudation of the Corsican orogenic wedge. Oblique convergence favors the development of regional-scale strike skip structures (cfr. CCF) and the early rotation of the Corsica-Sardinia block (cfr. van Hinsbergen et al. 2020). (d) Oligocene collapse of the "Mediterranean" alpine belt occurred simultaneously with the development of the Apenninic orogenic wedge and opening of the "Golfe du Lion" in a back-arc setting. (e) The Miocene to present slab retreat of the Apenninic subduction induced the thinning of the formerly thickened orogenic domain up to the development of the Tyrrhenian and southern Tuscany extensional basins.

The prograde metamorphism documented by Di Vincenzo et al. (2016) indicates that, at least in central Corsica, the continental subduction was already ongoing at 46 Ma.

Fast exhumation of eclogitic rocks of the OCT initiated early in the history of continental subduction. This stage could be associated with the subduction of part of the upper-plate AlKaPeCa lithosphere located in a forearc setting (Figure 6.5(b)). Forearc subduction and the induced extraction of part of the lithosphere of the forearc, located above the subducted OCT HP rocks, caused a fast exhumation from around 60–70 km to less than 30 km. This process allowed these rocks to be incorporated in the Alpine orogenic wedge body without experiencing high temperature metamorphism. In Corsica, the mechanism of slab extraction (already suggested in Malavieille et al. (1984) for the Western Alps) can be compared to the Taiwan subduction context (Chemenda et al. 2001; Malavieille et al. 2002) during the initial stage of continental subduction that is still active today (e.g. Malavieille et al. 2019). Indentation of the Nebbio-Calabria backstop (northern tip of AlKaPeCa microcontinent) was coupled with an important backthrusting of forearc sediments on the arc. A possible relict of the indentation system is likely recorded in the SLN in central Corsica. The paleotectonic position of this nappe, its bulk deformation structure and the relationships with the Schistes Lustrés unit have been largely discussed (e.g. Durand-Delga 1984). During Late Mid-Eocene (age range of 37/34 Ma), the system entered in the full stage of continental subduction (Figure 6.5(b)), as recorded by the development of the low-grade blueschist facies metamorphism in the external continental units (now observable in the frontal part of the nappe stack) and by the coeval syn-contractional exhumation of the internal part of the wedge (Tenda and overlying units). The syn-contractional exhumation occurred through basal accretion coupled with erosion, inducing underplating and the development of an antiformal stack (Malavieille et al. 1998; Molli et al. 2006; Malavieille 2010). Early to Mid-Eocene siliciclastic foredeep deposits (Annunciata, Palasca and Lozari sandstones) were frontally accreted and partially underplated and sliced during ongoing contraction along the major foreland propagating basal thrusts (Bézert and Caby 1988; Egal and Caron 1988; Molli et al. 2016). These structures are observable in the Balagne and further south along the boundary between Alpine and "autochthonous" Corsica (Egal and Caron 1988). The Late Eocene subduction of the thicker part of the Corsican continental margin (Figure 6.5(c)) increased buoyancy forces of the lower plate, producing the failure, break-off and detachment of the lithospheric mantle of the subducting European slab (Malavieille et al. 1998; Molli 2008; Beaudouin et al. 2017). Indentation by the oceanic lithosphere of the upper plate induced the subduction reversal marking the inception of the Apenninic subduction. The detachment of the European slab accompanied the subduction reversal. Then, the rebound of the decoupled part of the European subducted plate triggered a strong surface uplift (Figure 6.5(c)) and induced the erosion of the Early–Middle Eocene flysch originally covering a large part of the "autochthonous" (Danisik et al. 2007; Di Vincenzo et al. 2016). The processes described above occurred diachronously from south to north, involving the relicts of the Ligurian Tethys east of the Nebbio-Calabria microblock (eastern Ligurian Tethys). The tectonic regimes related to the new Apenninic subduction (Figures 6.5(c) and (d)) shaped the major tectonic features of the Western Mediterranean, with the back-arc extension of the Liguro-Provencal basin and the opening of the Gulf of Lion (Séranne 1999), the calcalkaline volcanism of Sardinia (Carminati et al. 2012 and reference therein) and, finally, the collapse and disarticulation of the "Mediterranean Alps" (Doglioni et al. 1998; Jolivet et al. 1998; Le Breton et al. 2017 and reference therein). The Miocene to present slab retreat of the Apenninic subduction induced the thinning of the formerly thickened orogenic domain until the development of the Tyrrhenian and southern Tuscany extensional basins (Figure 6.5(e)). Distributed deformation with extensional/transtensional basins developed through partial reactivation of contractional structures at the junction boundary area northern tip of the inverted Apenninic subduction (now corresponding to the offshore domains north/northeast of Corsica and including west Liguria) (Jolivet et al. 1990, 1998;

Daniel et al. 1996; Malavieille et al. 1998; Lacombe and Jolivet 2005; Molli et al. 2006; Beaudouin et al. 2017). A significant anticlockwise block rotation up to 45° occurred concomitantly (Elter and Pertusati 1973; Speranza et al. 2002; Gattacceca et al. 2007; Maffione et al. 2008; Gueydan et al. 2017).

6.9. Acknowledgments

Alberto Vitale Brovarone and an anonymous reviewer are acknowledged for their helpful comments and suggestions. Giorgio Pennacchioni and Joe Clancy White are thanked for the discussions about the contents of this chapter and their suggestions, which improved the readability. Claudio L. Rosenberg and Nicolas Bellahsen are warmly acknowledged for our involvement in this special publication. Chapter accepted in April 2020.

6.10. References

Advokaat, E.L., van Hinsbergen, D., Maffione, M., Langereis, C.G., Vissers, R.L.M., Cherchi, A., Schroeder, R., Madani, H., Columbu, S. (2014). Eocene rotation of Sardinia, and the paleogeography of the western Mediterranean region. *Earth and Planet. Sci. Lett.*, 401, 183–195.

Agard, P. and Vitale Brovarone, A. (2013). Thermal regime of continental subduction: The record from exhumed HP-LT terranes (Oman, N. Caledonia, Corsica). *Tectonophysics*, 602, 206–215.

Alvarez, W. (1991). Tectonic evolution of the Corsica-Apennines-Alps region studied by the method of successive approximations. *Tectonics*, 10, 936–947.

Alvarez, W., Cocozza, T., Wezel, F.C. (1974). Fragmentation of the Alpine orogenic belt by microplate dispersal. *Nature*, 248, 309–314.

Argand, E. (1924). La tectonique de l'Asie. *Comptes Rendus Congrés Géologique International, XIII, Belgique 1922*, 1, 171–372.

Argnani, A. (2009). Plate tectonics and the boundary between Alps and Apennines. *Ital. J. Geosci.*, 128, 317–330.

Beaudoin, A., Augier, R., Jolivet, L., Jourdon, A., Raimbourg, H., Scaillet, S., Cardello, G.L. (2017). Deformation behavior of continental crust during subduction and exhumation: Strain distribution over the Tenda massif (Alpine Corsica, France). *Tectonophysics*, 705, 12–32.

Berger, A. and Bousquet, R. (2008). Subduction-related metamorphism in the Alps: Review of isotopic ages based on petrology and their geodynamic consequences. In *Tectonic Aspects of the Alpine-Dinaride-Carpathian System*, Siegesmund, S., Fugenschuh, B., Froitzheim, N. (eds). Geological Society, London, Spec. Publ., London.

Bézert, P. (1990). Les unités Alpines à la marge du Massif Cristallin Corse : nouvelles données structurales, métamorphiques et contraintes cinématiques. Postgraduate Thesis, Université Montpellier.

Bézert, P. and Caby, R. (1988). Sur l'âge post-bartonien des événements tectono-métamorphiques alpins en bordure orientale de la Corse cristalline. *Bull. Soc. Géol. Fr.*, 8, 965–971.

Boccaletti, M., Elter, P., Guazzone, G. (1971). Plate tectonic models for the development of the western Alps and Northern Apennines. *Nature,* 234, 108–111.

Brunet, C., Monié, P., Jolivet, L., Cadet, J.P. (2000). Migration of compression and extension in the Tyrrhenian Sea, insights from 40Ar/39Ar ages on micas along a transect from Corsica to Tuscany. *Tectonophysics,* 321, 127–155.

Carminati, E., Lustrino, M., Doglioni, C. (2012). Geodynamic evolution of the central and western Mediterranean: Tectonics vs. igneous petrology constraints. *Tectonophysics,* 579, 173–192.

Caron, J.M. (1994). Metamorphism and deformation in Alpine Corsica. *Schweiz Mineral. Petrogr. Mitt.*, 7, 105–114.

Caron, J.M. and Delcey, R. (1979). Lithostratigraphie des schistes lustrés corses : diversité des séries post-ophiolitiques. *Compte Rendu Académie des Sciences Paris*, 208, 1525–1528.

Caron, J.M. and Péquignot, G. (1986). The transition between blueschists and lawsonite-bearing eclogites based on observations from Corsican metabasalts. *Lithos*, 19, 205–218.

Cavazza, W., De Celles, P.G., Fellin, M.G., Paganelli, L. (2007) The Miocene Saint-Florent basin in northern Corsica: Stratigraphy, sedimentology and tectonic implications. *Basin Res.*, 19, 507–527

Chemenda, A., Yang, R.-K., Konstantinovskaya, E., Ivanov, G. (2001). New results from physical modeling of arc-continent collision in Taiwan: Evolutionary model. *Tectonophysics*, 333, 159–178.

Chopin, C., Beyssac, O., Bernard, S., Malavieille, J. (2008). Aragonite garnet intergrowths in eclogite-marble, Alpine Corsica. *Eur. Mineral*, 20(5), 857–865.

Cowan, D. and Silling, R.M. (1978). A dynamic scaled model of accretion at trenches and its implications for the tectonic evolution of subduction complexes. *Jour. Geoph. Res.,* 83, 5389–5396.

Critelli, S. (2018). Provenance of Mesozoic to Cenozoic circum-Mediterranean sandstones in relation to tectonic setting. *Earth Sci. Rev.*, 185, 624–648.

Dal Piaz, G.V. (1974). Le métamorphisme alpin de haute pression et basse température dans l'évolution structurale du bassin ophiolitique alpino-apenninique. Part 1. *Boll. Soc. Geol. It.*, 93, 437–468.

Dal Piaz, G.V. and Zirpoli, G. (1979). Occurrence of eclogites relicts in the ophiolite nappe from Marine d'Albo, Northern Corsica. *Neus Jah. fur Miner.,* 3, 118–122.

Dallan, L. and Nardi, R. (1984). Ipotesi sulla evoluzione dei domini "Liguri" della Corsica nel quadro della paleogeografia e della paleottetonica delle unita` alpine. *Boll. Soc. Geol. It.*, 103, 517–527.

Dallan, L. and Puccinelli, A. (1995). Geologia della regione tra Bastia e St- Florent (Corsica Settentrionale). *Boll. Soc. Geol. It.*, 114, 23–66.

Danelian, T., De Wewer, P., Durand-Delga, M. (2008). Revised radiolarian ages for the sedimentary cover of the Balagne ophiolite (Corsica, France). Implications for the palaeoenvironmental evolution of the Balano-Ligurian margin. *Bull. Soc. géol. Fr.*, 179(3), 289–296.

Daniel, J.M., Jolivet, L., Goffé, B., Poisson, C. (1996). Crustal-scale strain partitioning: Footwall deformation below the Alpine Oligo-Miocene detachment of Corsica. *Jour. Struct. Geol.*, 18, 1841–1859.

Danisik, M., Kuhlemann, J., Dunk, I., Szekely, B., Frisch, W. (2007). Burial and exhumation of Corsica (France) in the light of fission track data. *Tectonics*, 26, TC1001.

Danisik, M., Kuhlemann, J., Dunk, I., Evans, N.J., Szekely, B., Frisch, W. (2012). Survival of ancient landforms in a collisional setting as revealed by combined fission track and (U-Th)/He thermochronometry: A case study from Corsica (France). *Jour. Geol.,* 120, 155–173.

Dercourt, J., Zonenshain, L.P., Ricou, L.-E., Kazmin, V.G., Le Pichon, X., Knipper, A.L., Grandjaquet, C., Sbortshikov, I.M., Geyssant, J., Lepvrier, C. et al. (1986). Geological evolution of the Tethys belt from the Atlantic to the Pamirs since the Lias. *Tectonophysics*, 123, 241–315.

Deseta, N., Andersen, T.B., Ashwal, L.D. (2014). A weakening mechanism for intermediate depth seismicity? Detailed petrographic and microstructural observation from blueschist facies pseudotachylytes, Cap Corse, Corsica. *Tectonophysics,* 610, 138–149.

Dewey, J.F., Helman, M.L., Turco, E., Hutton, D.H.W., Knott, S.D. (1989). Kinematics of the western Mediterranean. In *Alpine Tectonics*, Coward, M.P., Dietrich, D., Park, R.G. (eds). Geol. Soc. Lond. Spec. Publ., London.

Di Vincenzo, G., Grande, A., Prosser, G., Cavazza, W., De Celles, P.G. (2016). 40Ar–39Ar laser dating of ductile shear zones from central Corsica (France): Evidence of Alpine (middle to late Eocene) syn burial shearing in Variscan granitoids. *Lithos*, 262, 369–383.

Doglioni, C., Mongelli, F., Pialli, G. (1998). Boudinage of the Alpine belt in the Apenninic back-arc. *Mem. Soc. Geol. It.*, 52, 457–468.

Durand-Delga, M. (1984). Principaux traits de la Corse Alpine et corrélations avec les Alpes Ligures. *Mem. Soc. Geol. It.*, 28, 285–329.

Egal, E. (1992). Structures and tectonic evolution of the external zone of Alpine Corsica. *J. Struct. Geol.,* 14, 1215–1228.

Egal, E. and Caron, J.M. (1988). Tectonique polyphasée dans l'Eocène autochtone à la bordure ouest de la nappe de la Balagne (Corse). *Bull. Soc. géol. Fr.*, 8, 315–321.

Elter, P. and Pertusati, P.C. (1973). Considerazioni sul limite Alpi-Appennino e sulle sue relazioni con l'arco delle Alpi Occidentali. *Mem. Soc. Geol. It.*, 12, 359–375.

Faccenna, C., Piromallo, C., Crespo-Blanc, L., Jolivet, L., Rossetti, F. (2004). Lateral slab deformation and the origin of Western Mediterranean arcs. *Tectonics*, 23, TC1012.

Faure, M. and Malavieille, J. (1980). Les plis en fourreau du substratum de la Nappe des Schistes lustrés de Corse. Signification cinématique. *C. R. Acad. Sci. Paris*, 290, 1349–1352.

Faure, M. and Malavieille, J. (1981). Etude structurale d'un cisallement ductile : le charriage ophiolitique Corse dans la région de Bastia. *Bull. Soc. Géol. Fr.*, 23, 335–343.

Faure, M., Rossi, P., Gaché, J., Melleton, J., Frei, D., Li, X., Lin, W. (2014). Variscan orogeny in Corsica: New structural and geochronological insights, and its place in the Variscan geodynamic framework. *Int. J. Earth Sci.*, 103, 1553–1551.

Fellin, G., Zattin, M., Picotti, V., Reiner, P., Nicolescu, S. (2005). Relief evolution in northern Corsica (western Mediterranean): Constraints on uplift and erosion on long-term and short-term timescales. *Journ. Geoph. Res.*, 110, F01016.

Ferrandini, M., Ferrandini, J., Loye-Pylot, M.D., Butterlin, J., Cravette., J., Janin, M.C. (1998). Le Miocène du Bassin de Saint-Florent (Corse) : modalités de la transgression du Burdigalien Supérieur et mise en évidence du Serravalien. *Géobios*, 31(1), 125–137.

Ferrandini, J., Ferrandini, M., Rossi, P., Savary-Sismondini, B. (2010). Définition et datation de la Formation de Venaco (Corse) : dépots d'origine gravitaire d'âge Priabonien. *C. R. Gésoscience,* 342, 921–929.

Ford, M., Duchene, S., Vanderhaeghe, O., Gasquet, D. (2006). Coupling of external and internal orogenic processes in the SE Alps during frontal to oblique collision. *Jour. Géol. Soc. Lond.,* 163, 815–826.

Fournier, M., Jolivet, L., Goffé, B., Dubois, R. (1991). The Alpine Corsica metamorphic core complex. *Tectonics,* 10, 1173–1186.

Garfagnoli, F., Menna, F., Pandeli, E., Principi, G. (2009). Alpine metamorphic and tectonic evolution of the Inzecca-Ghisoni area (southern Alpine Corsica, France). *Geol. J.*, 44, 191–210.

Gattacceca, J., Deino, A., Rizzo, R., Jones, D.S., Henry, B., Beaudoin, B., Vadeboin, F. (2007). Miocene rotation of Sardinia: New paleomagnetic and geochronological constraints and geodynamic implications. *Earth Planet Sci Lett.,* 258, 359–377.

Gibbons, W. and Horak, J. (1984). Alpine metamorphism of Hercynian hornblende granodiorite beneath the blueschist facies *schistes lustrés* nappe of NE Corsica. *Journal of Metamorphic Geology*, 2, 95–113.

Gueguen, E., Doglioni, C., Fernandez, M. (1997). On the post-25 Ma geodynamic evolution of the western Mediterranean. *Tectonophysics*, 298, 259–269.

Guerrera, F., Martin-Agarra, A., Perrone, V. (1993). Late Oligocene-Miocene syn-/late-orogenic successions in Western and Central Mediterranean Chains from Betic Cordillera to the Southern Apennines. *Terra Nova,* 6, 525–544.

Gueydan, F., Le Garzic, E., Carry, N. (2009). P/T ratio in high-pressure rocks as a function of dip and velocity of continental subduction. *Litosphere*, 1(5), 282–290.

Gueydan, F., Brun, J.P., Philippon, M., Noury, M. (2017). Sequential extension as a record of Corsica Rotation during Apennines slab roll-back. *Tectonophysics*, 710–711, 149–161.

Handy, M.R. and Oberhänsli, R. (2004). Explanatory notes to the age map of metamorphic structure of the Alps–Tectonic interpretation and outstanding problems. *Mitt. Ost. Miner. Ges.,* 149, 201–226.

Handy, M.R., Schmid, S., Bousquet, R., Kissling, E., Bernoulli, D. (2010). Reconciling plate-tectonic reconstructions of Alpine Tethys with geological-geophysical record of spreading and subductions in the Alps. *Earth Sci. Reviews*, 102, 121–168.

Harris, L.B. (1985a). Progressive and polyphase deformation of the Schistes Lustrés in Cap Corse, Alpine Corsica. *Jour. Struct. Geol.,* 7, 637–650.

Harris, L.B. (1985b). Direction changes in thrusting of the Schistes Lustrés in Alpine Corsica. *Tectonophysics*, 120, 37–56.

van Hinsbergen, D.J.J., Torsvik, T.H., Schmid, S.M., Maţenco, L.C., Maffione, M., Vissers, R.L.M., Gürer, D., Spakman, W. (2020). Orogenic architecture of the Mediterranean region and kinematic reconstruction of its tectonic evolution since the Triassic. *Gondwana Research*, 81, 79–229.

Jolivet, L., Dubois, R., Fournier, M., Michard, A., Jourdan, C. (1990). Ductile extension in Alpine Corsica. *Geology,* 18, 1007–1010.

Jolivet, L., Faccenna, C., Goffé, B., Mattei, M., Rossetti, F., Brunet, C., Storti, F., Funiciello, R., Cadet, J.P., d'Agostino, N. et al. (1998). Mid-crustal shear zones in postorogenic extension: Example from the northern Tyrrhenian Sea. *Jour. Geoph. Res.,* 103, 12123–12160.

Lacombe, O. and Jolivet, L. (2005). Structural relationships between Corsica and the Pyrénées-Provence domain at the time of Pyrennean orogeny. *Tectonics*, 24, TC1003.

Lahondère, D. (1996). Les schistes bleus et les éclogites à lawsonite des unités continentales et océanique de la Corse alpine : nouvelles données pétrologiques et structurales (Corse). *Documents du BRGM*, 240, 1–285.

Lahondère, D. and Guerrot, C. (1997). Datation Nd-Sm du métamorphisme éclogitique en Corse alpine : un argument pour l'existence au Crétacé supérieur, d'une zone de subduction active localisée le long du bloc corso-sarde. *Géol. France*, 3, 3–11.

Lahondère, D., Rossi, P., Lahondère, J.C. (1999). Structuration alpine d'une marge continentale externe : le massif du Tenda (Haute-Corse). Implications géodynamiques au niveau de la transversale Corse-Apennins. *Géol. France*, 4, 27–44.

Le Breton, E., Handy, M., Molli, G., Ustaszewski, K. (2017). Post-20 Ma motion of the Adriatic plate – New constraints from surrounding orogens and implications for crust-mantle decoupling. *Tectonics*, 36, 3135–3154.

Le Petre, R., Frizon de Lamotte, D., Combiere, V., Gimeno-Vives, O., Mohn, G., Escard, R. (2018). The Tell-Rif orogenic system (Morocco, Algeria, Tunisia) and the structural heritage of the southern Tethys margin. *Bull. Soc. Géol. Fr.*, 189, 10.

Loyet-Pilot, M.D. and Ferrandini, M. (2014). Les formation miocènes. *Géochronique,* 132, 41–44.

Maffione, M., Speranza, F., Faccenna, C., Cascella, A., Vignaroli, G., Sagnotti, L. (2008). A synchronous Alpine and Corsica-Sardinia rotation. *Journ. Geophys. Res.*, 113, B03104.

Maggi, M., Rossetti, F., Corfu, F., Theye T., Andersen, T.B., Faccenna, C. (2012). Clinopyroxene–rutile phyllonites from the East Tenda Shear Zone (Alpine Corsica, France): Pressure–temperature–time constraints to the Alpine reworking of Variscan Corsica. *Journ. Geol. Soc.*, 169, 723–732.

Magott, R., Fabbri, O., Fournier, M. (2016). Subduction zone intermediate-depth seismicity: Insights from the structural analysis of Alpine high-pressure ophiolite-hosted pseudotachylyte (Corsica, France). *Journ. Struct. Geol.,* 87, 95–114.

Malasoma, A., Marroni, M., Musumeci, G., Pandolfi, L. (2006). High pressure mineral assemblage in granitic rocks from continental units in Alpine Corsica, France. *Geol. Journal*, 41, 49–59.

Malavieille, J. (1983). Etude tectonique et microtectonique de la nappe de socle de Centuri (zone des Schistes Lustrés de Corse), conséquence pour la géométrie de la chaine alpine. *Bull. Soc. Géol. Fr.*, 25, 195–204.

Malavieille, J. and Molli, G. (2014). Tectonique et géodynamique de la Corse alpine. *Géochronique*, 132, 21–25.

Malavieille, J., Lacassin, R., Mattauer, M. (1984). Signification tectonique des linéations d'allongement des les Alpes occidentales. *Bull. Soc. Géol. Fr.*, 26, 895–906.

Malavieille, J., Chemenda, A., Larroque, C. (1998). Evolutionary model for Alpine Corsica: Mechanism for ophiolite emplacement and exhumation of high-pressure rocks. *Terra Nova*, 10, 317–322.

Malavieille, J., Lallemand, S., Dominguez, S., Deschamps, A., Lu, C.-Y., Liu, C.-S., Schnürle, P., ACT Scientific Crew (2002). Arc-continent collision in Taiwan: New marine observations and tectonic evolution. In *Geology and Geophysics of an Arc-Continent Collision, Taiwan*, Byrne, T.B. and Liu, C.-S. (eds). Geol. Soc. Am. Sp. Paper 358, Boulder, CO.

Malavieille, J., Dominguez, S., Lu, C.-Y., Chen, C.-T., Konstantinovskaya, E. (2019). Deformation partitioning in Mountain Belts, insights from analogue modeling experiments and the Taiwan collisional orogen. *Geological Magazine*, special issue 1 "Subduction zone processes and crustal growth mechanisms at Pacific-Rim continental margins: Modern and ancient analogues", 84–103.

Malusà, M.G., Faccenna, C., Garzanti, E., Polino, R. (2011). Divergence in subduction zones and exhumation of high pressure (Eocene Western Alps). *Earth Planet. Sci. Lett.*, 310, 21–32.

Manatschal, G. and Bernoulli, D. (1999). Architecture and tectonic evolution of non-volcanic margins: Present-day Galicia and ancient Adria. *Tectonics*, 18, 1099–1119.

Manzotti, P., Ballèvre, M., Zucali, M., Robyr, M., Engi, M. (2014). The tectono-metamorphic evolution of the Sesia-Dent Blanche nappes (internal Western Alps): Review and synthesis. *Swiss Journ. Geosc.*, 107, 309–336.

Marroni, M. and Pandolfi, L. (2003). Deformation history of the ophiolite sequence from the Balagne Nappe, northern Corsica: Insights in the tectonic evolution of Alpine Corsica. *Geol. Journal*, 38, 67–83.

Martin, L.A.J., Rubatto, D., Vitale Brovarone, A., Hermann, J. (2011). Late Eocene lawsonite-eclogite facies metasomatism of a granulite sliver associated to ophiolites in Alpine Corsica. *Lithos*, 125, 620–640.

Mattauer, M. and Proust, F. (1975). Sur quelques problèmes géneraux de la chaine alpine en Corse. *Bull. Soc. Géol. Fr.,* 18, 1177–1178.

Mattauer, M. and Proust, F. (1976). La Corse Alpine : un modèle de genèse du métamorphisme haute pression par subduction de croûte continentale sous du matériel océanique. *C. R. Acad. Sci. Paris*, 282, 1249–1252.

Mattauer, M. and Tapponier, P. (1978). Tectonique de plaques et tectonique intracontinentale dans les Alpes Franco-Italiennes. *C. R. Acad. Sci. Paris*, 287, 899–902.

Mattauer, M., Proust, F., Etchecopar, A. (1977). Linéation "*a*" et mécanisme de cisaillement simple liés au chevauchement de la nappe des schistes lustrés en Corse. *Bull. Soc. Géol. Fr.*, 14, 841–847.

Mattauer, M., Faure, M., Malavieille, J. (1981). Transverse lineation and large scale structures related to Alpine obduction in Corsica. *Journ. Struct. Geol.*, 3, 401–409.

Meresse, F., Lagabrielle, Y., Malavieille, J., Ildefonse, B. (2012). A fossil ocean-continent transition of the Mesozoic Tethys preserved in the Schistes Lustrés nappe of northern Corsica. *Tectonophysics*, 579, 4–16.

Michard, A. and Martinotti, G. (2002). The Eocene unconformity of the Briançonnais domain in the French-Italian Alps, revisited (Marguareis massif, Cuneo), a hint for Late Cretaceous-Middle Eocene frontal bulge setting. *Geod. Acta*, 15, 289–301.

Michard, A., Chalouan, A., Feinberg, H., Goffé, B., Montigny, R. (2002). How does the Alpine belt end between Spain and Morocco? *Bull. Soc. Géol. Fr.*, 173, 3–15.

Molli, G. (2008). Northern Apennine-Corsica orogenic system: An updated overview. In *Tectonic Aspects of the Alpine-Dinaride-Carpathian System*, Siegesmund, S., Fügenschuh, B., Froitzheim, N. (eds). Geol. Soc. London Spec. Publ., London.

Molli, G. and Malavieille, J. (2011) Orogenic processes and the Corsica/Apennines geodynamic evolution: Insights from Taiwan. *Inter. Journ. Earth Sci.*, 100, 1207–1224.

Molli, G. and Tribuzio, R. (2004). Shear zones and metamorphic signature of subducted continental crust as tracers of the evolution of the Corsica/Northern Apennine orogenic system. In *Flow Process in Faults and Shear Zones*, Alsop, I., Holdsworth, R.E., McCaffrey, J.W., Hand, M. (eds). Geol. Soc. London Spec. Publ., London.

Molli, G., Tribuzio, R., Marquer, D. (2006). Deformation and metamorphism at the eastern border of the Tenda Massif (NE Corsica): A record of subduction and exhumation of continental crust. *Journ. Struct. Geol.*, 29, 1748–1766.

Molli, G., Menegon, L., Malasoma, A. (2017). Switching deformation mode and mechanisms during subduction of continental crust: A case study from Alpine Corsica. *Solid Earth*, 8, 767–788.

Molli, G., Brogi, A., Caggianelli, A., Capezzuoli, E., Liotta, D., Spina, A., Zibra, I. (2020). Late Paleozoic tectonics in Central Mediterranean: A reappraisal. *Swiss Jour. Geosc.*, 113, 245–274.

Nardi, R. (1968). Le unità alloctone della Corsica e la loro correlazione con le unità delle Alpi e dell'Appennino. *Mem. Soc. Geol. It.*, 7, 323–344.

Peybernes, B., Cugny, P., Fondecave-Vallez, M.J. (2000). Le problème de la limite Crétacé-Tertiaire en Corse-du-Sud. *Bull. Soc. Géol. Fr.*, 171, 451–459.

Piccoli, F., Vitale Brovarone, A., Ague, J.J. (2018). Field and petrological study of metasomatism and high-pressure carbonation from lawsonite eclogite-facies terrains, Alpine Corsica. *Lithos*, 304–307, 16–37.

Principi, G. and Treves, B. (1985). Il sistema corso-appenninico come prisma d'accrezione. Riflessi sul problema generale del limite Alpi-Appennini. *Mem. Soc. Geol. Ital.,* 28, 549–576.

Puccinelli, A., Perilli, N., Cascella, A. (2012). Stratigraphy of the Caporalino-Sant Angelo Unit: A fake Jurassic-Eocene succession of the "Alpine" Corsica. *Riv. Ital. Paleon. Strat.*, 118, 471–491.

Ravna, E.J.K., Andersen, T.B., Jolivet, L., De Capitani, C. (2010). Cold subduction and the formation of lawsonite eclogite – Constraints from prograde evolution of eclogitized pillow lava from Corsica. *J. Metam. Geol.*, 28, 381–395.

Rosenbaum, G. and Lister, G.S. (2005). The western Alps from Jurassic to Oligocene: Spatio-temporal constraints and evolutionary reconstruction. *Earth Sci. Rev.,* 69, 281–306.

Rossetti, F., Faccenna, C., Jolivet, L., Goffé, B., Funicello, R. (2002). Structural signature and exhumation P–T–t paths of the blueschist units exposed in the interior of the Northern Apennine chain, tectonic implication. *Boll. Soc. Geol. It.*, 1, 829–842.

Rossetti, F., Glodny, J., Theye, T., Maggi, M. (2015). Pressure–temperature–deformation–time of the ductile Alpine shearing in Corsica: From orogenic construction to collapse. *Lithos*, 218–219, 99–116.

Rossi, P.H., Cocherie, A., Lahondère, D., Fanning, M. (2002). La marge européenne de la Téthys jurassique en Corse : datation de trondhjémites de Balagne et indices de croûte continentale sous le domaine Balano-Ligure. *C. R. Géoscience*, 334, 313–322.

Rossi, P.H., Durand-Delga, M., Lahondère, J.C., Lahondère, D. (2003). Carte géologique de la France à 1/50.000 : Santo Pietro di Tenda. *BRGM*, Orléans.

Schmid, S., Fügenschuh, B., Kissling, E., Schuster, R. (2005). Tectonic map and overall architecture of the Alpine orogen. *Ecl. Geol. Helv.*, 97, 93–117.

Schmid, S.M., Kissling, E., Diehl, T., van Hinsbergen, D.J.J., Molli, G. (2017). Ivrea mantle wedge, arc of the Western Alps, and kinematic evolution of the Alps–Apennines orogenic system. *Swiss J. Geosc.*, 110(2), 581–612.

Schwartz, S., Lardeaux, J.M., Tricart, P., Guillot, S., Labrin, E. (2007). Diachronous exhumation of HP–LT metamorphic rocks from south-western Alps: Evidence from fission-track analysis. *Terra Nova*, 19, 133–140.

Séranne, M. (1999). The Gulf of Lion continental margin (NW Mediterranean) revisited by IBS: An overview. In *Mediterranean Basins*, Durand, B., Jolivet, L., Horvàth, F., Séranne, M. (eds). Geol. Soc. Lond., Spec. Publ., London.

Speranza, F., Villa, I.M., Sagnotti, L., Florindo, F., Cosentino, D., Cipollari, P., Mattei, M. (2002). Age of the Corsica-Sardinia rotation and Ligure-Provencal basin spreading: New paleomagnetic and Ar/Ar evidence. *Tectonophysics,* 347, 231–251.

Staub, R. (1933). Der Bau der Alpen. *Beitr. Geolog Karte Schweiz*, 46, 1–366.

Tribuzio, R. and Giacomini, F. (2002). Blueschist facies metamorphism of peralkaline rhyolites from the Tenda crystalline massif (northern Corsica): Evidence for involvement in the Alpine subduction event? *J. Metam. Geol.*, 20, 513–526.

Turco, E., Macchiavelli, C., Mazzoli, S., Schettino, A., Pierantoni, P.P. (2012). Kinematic evolution of Alpine Corsica in the framework of Mediterranean mountain belts. *Tectonophysics*, 579, 193–206.

Vitale Brovarone, A. and Herwartz, D. (2013). Timing of HP metamorphism in the Schistes Lustrés of Alpine Corsica: New Lu–Hf garnet and lawsonite ages. *Lithos*, 172–173, 175–191.

Vitale Brovarone, A., Beltrando, M., Malavieille, J., Giuntoli, F., Tondella, E., Groppo, C., Beyssac, O., Compagnoni, R. (2011). Inherited ocean–continent transition zones in deeply subducted terranes: Insights from Alpine Corsica. *Lithos*, 124, 273–290.

Vitale Brovarone, A., Beyssac, O., Malavieille, J., Molli, G., Beltrando, M., Compagnoni, R. (2013). Stacking and metamorphism of continuous segments of subducted lithosphere in a high-pressure wedge: The example of Alpine Corsica (France). *Earth Sci. Rev.*, 116, 35–56.

Vitale Brovarone, A., Picatto, M., Beyssac, O., Lagabrielle, Y., Castelli, D. (2014). The blueschist-eclogite transition in the Alpine chain: PT paths and the role of slow-spreading extensional structures in the evolution of HP-LT mountain belts. *Tectonophysics*, 615–616, 96–121.

Warburton, J. (1986). The ophiolite-bearing Schistes lustrés nappe in Alpine Corsica: A model for emplacement of ophiolites that have suffered HP/LT metamorphism. *Mem. Geol. Soc. Am.,* 164, 313–331.

Waters, C.N. (1990). The Cenozoic tectonic evolution of alpine Corsica. *J. Geol. Soc. Lond.*, 147, 811–824.

Zarki-Janki, B., van der Beek, P., Poupeau, G., Sosson, M., Labrin, E., Rossi, P., Ferrandini, J. (2004). Cenozoic denudation of Corsica in response to Ligurian and Tyrrhenian extension: Results from apatite fission track thermochronology. *Tectonics*, 23.

Zibra, I., Kruhl, J.H., Braga, R. (2010). Late Palaeozoic deformation of post-Variscan lower crust: Shear zone widening due to strain localization during retrograde shearing. *Int. J. Earth Sci.*, 99, 973–991.

List of Authors

Nicolas BELLAHSEN
ISTeP
Sorbonne Université
UMR 7193
CNRS
Paris
France

Harald FRITZ
Institute of Earth Sciences
University of Graz
NAWI Graz Geocenter
Austria

Gerit GRADWOHL
Earth Sciences
University of Graz
Austria

Walter KURZ
Institute of Earth Sciences
University of Graz
NAWI Graz Geocenter
Austria

Moritz LIEBL
Geography and Geology
University of Salzburg
Austria

Jacques MALAVIEILLE
Géosciences Montpellier
Université de Montpellier
CNRS
Université des Antilles
Montpellier
France

Giancarlo MOLLI
Department of Earth Sciences
University of Pisa
Italy

Anne PAUL
Université Grenoble Alpes
Université Savoie Mont Blanc
CNRS, IRD, UGE
ISTerre
Grenoble
France

Jörg ROBL
Geography and Geology
University of Salzburg
Austria

Claudio L. ROSENBERG
ISTeP
Sorbonne Université
UMR 7193
CNRS
Paris
France

Kurt STÜWE
Earth Sciences
University of Graz
Austria

Christian SUE
Université de Franche-Comté
Besançon
and
Université Grenoble Alpes
USMB
CNRS
IRD
IFSTTAR
ISTerre
Grenoble
France

Andreas WÖLFLER
Institute for Geology
Leibniz University Hannover
Germany

Index

S

T

U, V, W

Summary of Volume 2

Chapter 2. Paleogeography and Architecture of the Alpine Tethys Margins
Yves LAGABRIELLE

Chapter 3. One View on the Petrology and Geochemistry of the Mantle and Oceanic Crustal Remnants in the Alps, Reconciling Rifts and Ultra-slow Seafloor Spreading

Othmar MÜNTENER and Anders MCCARTHY

Chapter 4. Oceanic Subduction in the Alps: Lessons from the Rock Record

Philippe AGARD

List of Authors

Index

Summary of Volume 3

Chapter 3. A Review of How Our Understanding on the Relationships between the Alps and the North Alpine Foreland Basin has Evolved

Fritz SCHLUNEGGER and Edi KISSLING

Chapter 4. The Detrital Thermochronological Record of Exhumation of the European Alps since the Early Oligocene

Matthias BERNET, Peter VAN DER BEEK, Pierre G. VALLA and Christoph GLOTZBACH

Chapter 5. The Northern Deformation Front of the European Alps

Hugo ORTNER, Christoph VON HAGKE, Anna SOMMARUGA, Samuel MOCK, Jon MOSAR, Ralph HINSCH and Andreas BEIDINGER

nted and bound by CPI Group (UK) Ltd, Croydon, CR0 4YY

16/09/2026

14963175-0001